D0647063

PRENTICE HALL'S
ENVIRONMENTAL TECHNOLOGY SERIES

Volume 6

Industrial Waste Stream Generation

Prentice Hall's Environmental Technology Series

Volume 1: Introduction to Environmental Technology
Volume 2: Environmental Regulations Overview
Volume 3: Health Effects of Hazardous Materials
Volume 4: Sampling and Analysis
Volume 5: Waste Management Concepts
Volume 6: Industrial Waste Stream Generation

PRENTICE HALL'S
ENVIRONMENTAL TECHNOLOGY SERIES

Volume 6

Industrial Waste Stream Generation

NEAL K. OSTLER, Editor
Salt Lake Community College

Prentice Hall
Upper Saddle River, New Jersey   Columbus, Ohio

Library of Congress Cataloging-in-Publication Data

Industrial waste stream generation / Neal K. Ostler, editor.
p. cm. -- (Prentice Hall's environmental technology series; v. 6)
Includes bibliographical references and index.
ISBN 0-13-238569-4
1. Factory and trade waste--Management. 2. Pollution prevention.
I. Ostler, Neal K. II. Series.
TD897.5.I532 1998
628.4--dc21 97-31973
CIP

Editor: Stephen Helba
Production Editor: Mary Harlan
Design Coordinator: Julia Zonneveld Van Hook
Cover Designer: Brian Deep
Production Manager: Pamela D. Bennett
Marketing Manager: Debbie Yarnell
Editorial/Production Supervision: Custom Editorial Productions, Inc.
Illustrations: Custom Editorial Productions, Inc.

This book was set in Utopia by Custom Editorial Productions, Inc., and was printed and bound by Quebecor Printing/Book Press.

Printed in the United States of America

10 9 8 7 6 5 4 3 2 1

ISBN: 0-13-238569-4

Prentice-Hall International (UK) Limited, *London*
Prentice-Hall of Australia Pty. Limited, *Sydney*
Prentice-Hall of Canada, Inc., *Toronto*
Prentice-Hall Hispanoamericana, S. A., *Mexico*
Prentice-Hall of India Private Limited, *New Delhi*
Prentice-Hall of Japan, Inc., *Tokyo*
Simon & Schuster Asia Pte. Ltd., *Singapore*
Editora Prentice-Hall do Brasil, Ltda., *Rio de Janeiro*

Contents

Preface

Development of This Text

In the early days of the Partnership for Environmental Training and Education (PETE), 1991–1992, it was recognized that identifying opportunities for pollution prevention was a definite contribution the hazardous materials technician or environmental technician could make to industry. It was further recognized that these technicians could make these contributions right out of the starting block with their new employer if properly trained and prepared while completing their studies toward a certificate or an associate's degree in environmental technology.

The creation of PETE was the result of efforts on the part of industry, the academic community, and government agencies (especially the Department of Energy) to establish an appropriate curriculum and provide industry with the ability to create and then to fill technician-level positions throughout the country. As of this writing, more than 350 colleges and universities participate as active members of the national PETE and offer certificate and/or degree programs in fields directly related to PETE-developed curriculum guidelines and professional competencies for the environmental technician.

Pollution Prevention (P_2) and Programs (P_3), although relatively in their infancy in the arena of environmental management, are clearly not new concepts. Investigating one's manufacturing process to identify ways to make it more efficient has always been the task of quality control people. For anyone conducting such research, it is essential to begin with an understanding and a working knowledge of a materials flow diagram and balance sheet.

At the minimum, a technician should have a grasp of material flows, knowledge of where and how to look for P_2 opportunities, and the ability to suitably adapt the language to a variety of different industrial manufacturing processes and their respective waste streams. The outcome I personally desired for my students was for them to be able to interview with potential employers and have a reasonable dialogue with them about the operations.

Eventually the principal players in PETE identified this topic to be recommended for an independent course of study within the Environmental Technology Program. At Salt Lake Community College, the course is identified as ENVT-110: Waste Stream Generation, Reduction, and Treatment, and translates into roughly 50 hours of classroom time. Other colleges participating within the PETE system identify this course as Pollution Prevention.

Case studies were developed for teaching this subject. The instructor of such a course had to gather an assortment of materials on a variety of industries, and then compile and duplicate them for use in the classroom. From this experience, evolved the outline of a text on Industrial Waste Stream Generation.

Particular care has been taken to generalize the content of each chapter so that the material will be as nonperishable as possible. This text is not a procedural document or a how-to manual but is intended to provide the environmental technician with the ability to articulate on materials flows and identify opportunities for pollution prevention that are basic to manufacturing and to industry in general.

Goal of the Text

This text is the sixth volume in Prentice Hall's Environmental Technology Series and is meant primarily to serve as a course textbook for the environmental technician. There are hundreds of different industrial processes with different raw ingredients and different related waste streams, each one of them a study in itself. These numerous processes make this a complex subject for a textbook. Nevertheless, this text has been written with simplified terminology and language and is suitable as an aid to any person involved in a general study of materials flows and will be useful in helping them to identify P_3 opportunities anywhere in industry.

The goals for the text are to introduce students to the concepts of pollution prevention, provide an introductory study of materials flows analysis, and include a relatively wide range of case studies from industry. Each waste stream case study is similarly organized to present (1) general process descriptions, ingredients, and materials balances; (2) hazardous waste stream identification and characterization; (3) regulatory impact; (4) treatment and control technologies, including QA and QC; (5) state-of-the-art pollution prevention technologies; (6) environmental health and safety issues and concerns; and (7) references or additional reading recommendations.

The text includes the following subjects:

- Pollution prevention
- General manufacturing wastes
- Metal finishing wastes
- Construction waste
- Pollution in the printing and publishing industries
- Laboratory chemical and waste management
- Nuclear waste issues
- Pharmaceutical and cosmetic industry wastes

- Chemical wastes
- Pulp and paper industry wastes
- Wood preserving wastes
- Electroplating waste streams
- Hospital wastes
- Source reduction in semiconductor fabrication

In addition to these subjects, the text also provides a substantial number of appendixes to assist the student in developing the skills needed to understand this subject and relate it to industry.

Acknowledgments

The idea for this text, and for the entire series, evolved from the lack of suitable training materials for the environmental technology classroom. To make all of the volumes in this series as authoritative and current as possible, I sought contributing authors to supplement the basic manuscript with chapters pertinent to their individual areas of expertise.

I would like to express my appreciation to each of the contributing authors, all of whom have made significant contributions to this text, and I would entreat the reader to review the About the Authors section for a closer look at each of these professionals.

Since the idea for the Environmental Technology Series was first initiated, I've received continued support from Salt Lake Community College. I wish to thank John Latkiewicz, my division director since the fall of 1993, for his fairness and objectivity as my supervisor, and George Van DeWater, who is involved so closely with PETE and who first hired me.

I also wish to thank the following reviewers for their helpful comments and suggestions: Norman R. Sunderland, Utah State University; Robert R. Treloar, Paradise Valley Community College; Rhonda L. Howard, Coconino Community College; J. K. Roberts, Palo Alto College; and Jeff Bates, Columbus State Community College.

My work in this text could not have been accomplished without the support of my wife, Karen Amundsen, and my children and stepchildren. The series now holds a place of its own on the fireplace mantel, and each family member has a clearer sense of what I'm up to when I'm working late at the office. The family is what it is all about for me.

It has been my pleasure to work with my editor at Prentice Hall, Stephen Helba, the assistants in his office, and the dedicated staff at Custom Editorial Productions. Thank you all for helping to get this sixth volume ready for the presses and into the hands of the students.

Neal K. Ostler

About the Authors

Neal K. Ostler

Neal K. Ostler lectures on environmental technology at Salt Lake Community College and is founder of the Environmental Training Center, where he coordinates and delivers a variety of non-credit workshops and seminars in the environmental health and safety arena. His background and experience includes over 20 years in law enforcement, where he became adept at investigative reporting and emergency response operations. While employed at the Utah State Prison and assigned to maximum security, Ostler designed, developed, and delivered an in-cell Alcohol and Substance Abuse Program (biblio-therapy) for inmates housed in limited-movement facilities that was the first of its kind anywhere in the United States

Ostler first began to develop his haz/mat credentials while employed as a Motor Carrier Investigator with the Utah Department of Commerce. He is certified as "Train-The-Trainer" in a variety of subject areas which include HAZWOPER, Confined Space, DOT HM-181, and Hazard Communications. His credentials include that of Certified Hazardous Materials Manager (CHMM), and he participates in his local chapter of the National Association of Environmental Professionals. In the Fall of 1996, he obtained the credential of Designated Trainer for ISO 14000 Awareness from the Global Environmental Training Foundation. Ostler is a graduate of the University of Utah and has attended over 1,400 hours of related workshops and seminars in addition to completion of the A.S. of Environmental Technology Degree at his own Community College.

Michele Feenstra

Graduating from the Massachusetts Institute of Technology (MIT) with a B.S. degree in materials science and engineering, Michele Feenstra currently provides environmental training to Department of Army personnel at Fort Polk, Louisiana, as an employee of CH2M HILL.

Before moving to Louisiana, Feenstra worked for Texas A&M University as a project manager for the NASA Mid-Continent Technology Transfer Center and as a field engineer for the Texas Manufacturing Assistance Center. Her duties involved transferring federal technologies and providing technical environmental assistance to industry. She also helped in planning for a statewide environmental technical assistance network. As the founder and director of the Kansas State University (KSU) Pollution Prevention Institute, Feenstra provided technical assistance and training on pollution prevention and hazardous waste management to industries throughout Kansas and the Midwest. She built the primary environmental technical assistance program for the state of Kansas. Feenstra also worked for Wichita State University's hazardous materials office, and as a research chemist for Cornell University analyzing phosphate-free detergents for Procter & Gamble.

William N. Christie

William N. Christie holds an M.S. degree in environmental science (1982) from George Washington University in Washington, D.C. His B.S. degree is in biology (1978) from Mary Washington College in Fredericksburg, Virginia. He is also a Certified Environmental Trainer (CET), which he was granted by the National Environmental Training Association (NETA) in 1987, and a Certified Hazardous Materials Manager (CHMM) granted by the Institute of Hazardous Materials Management (IHMM) in 1989.

Christie's professional work experience includes: cancer research, sewage treatment plant operation, environmental scientist and consultant to US EPA, and technical information specialist with US EPA. His twelve years with The Boeing Company have included positions as an environmental instructor with Boeing Aerospace, an environmental administrator with Boeing Aerospace, an emergency response administrator with Corporate Safety, Health, and Environmental Affairs, and an environmental programs administrator for Boeing Commercial Airplane Group (five and a half years). Presently, he is the safety, health, environmental technical consultant for Boeing Commercial Airplane Group, Fabrication Division.

Christie is directly involved with local community colleges and technical colleges by participating in curricula development advisory committees and course instruction for the schools' hazardous materials management programs. He has been actively teaching within The Boeing Company since 1985 and for the community and technical college systems since 1989. In 1994, the National Environmental Training Association recognized him as the "Trainer of the Year" for his years of dedicated service in the field of safety, health, and environmental training. He has had several articles published in professional trade journals and is a contributing author to Prentice Hall's Environmental Technology Series, Volume I—Introduction to Environmental Technology.

Brian F. Goetz

Brian F. Goetz is an environmental consultant and writer who lives in Salt Lake City, Utah. He is also on the staff of Salt Lake Community College, where he developed and implemented a comprehensive seminar on environmental and safety regulations. He has eleven years of experience working with environmental policy and compliance in Ohio, Indiana, and Utah.

Goetz holds a B.S. degree in technology with an emphasis in environmental science from Bowling Green State University, Ohio. Upon graduation, he managed a water-treatment system and state-certified laboratory for the American Water Works Company in Seymour, Indiana, for six years. During that time, he attended Indiana University, obtaining an M.P.A. degree in environmental policy.

Goetz was the coordinator of two recycling programs, one at Bowling Green University and the other in Seymour, Indiana. He also wrote a monthly column, "Talking Trash," for a newspaper in Indiana that focused on environmental and waste-management issues. His other published work includes journal and magazine articles and a chapter for the Environmental Compliance Desk Reference (Thompson Professional Publishing, 1994). He previously contributed a chapter entitled "Citizen Involvement in the Environmental Movement" to Volume 2 of this Environmental Technology Series.

Kong S. Chiu

Kong S. Chiu is an analyst in the US EPA's Office of Water where he works on environmental infrastructure financing programs. Prior to working for the EPA, Chiu served in a selective two year management fellowship with the State of Maryland. As a Maryland Governor's Policy Fellow, he managed comparative risk assessment, pollution prevention, and environmental export projects at Maryland's Departments of Environment and Economic Development. Chiu is a graduate of the University of Chicago and holds an M.S. degree in physics from the University of Massachusetts and a M.P.M. degree in environmental management and finance from the University of Maryland.

Mignon J. Clarke

Mignon J. Clarke has more than nine years of combined technical, analytical, and research support resulting from multidisciplined work environments. She currently serves as an environmental specialist for the Department of the Army, Environmental Management Division, at Fort Benning, Georgia. Her responsibilities include reviewing environmental protection regulatory developments to identify impacts to future installation projects; participating in environmental compliance inspection activities; and reviewing permit applications and reports for compliance with environmental protection regulations.

Clarke has an extensive background in laboratory health and safety, waste minimization, and analytical testing procedures. Her previous professional experience has included such titles as chemical technician, process chemist/compliance coordinator, and environmental, safety, and health professional.

Thomas E. Byrne

Thomas E. Byrne resides in Harriman, Tennessee, and teaches at Roane State. He received his B.S. degree from Tennessee Technological University in biology with a minor in chemistry; M.S. degree from Tennessee Technological University in biology; and Ph.D. from the University of Tennessee at Knoxville in botany and microbiology. He is a registered microbiologist (RM-AAM) with the American Academy of Microbiology, and is a member of the American Society of Microbiology, Environmental Microbiology Section.

Byrne has served as a member of the Executive Committee of the Tennessee Academy of Science and is a life member of the Academy. He has authored many articles for refereed journals, as well as two laboratory manuals. He also serves as a consultant to the Oak Ridge National Laboratory (ORNL) in the area of viral carcinogenesis and radiation assessment and has administered radiation technician training programs for Martin Marietta Energy Systems in Oak Ridge, Tennessee, and Paducah, Kentucky. His most recent consulting duties include environmental radiation assessment at ORNL for the Nuclear Regulatory Commission.

S. Merris Sinha

S. Merris Sinha has over 10 years of broad-based technical, training, and management experience. Sinha, who received a B.S. degree in environmental sciences from Rutgers University in 1983 and an M.S. degree in toxicology from the University of Rochester in 1987, has taught the Hazardous Waste Operations Course, Training Program Development Course, and Annual Review and Refresher Course, in Yokuska, Japan, for the U.S. Navy. She is responsible for training internal staff and corporate clients on regulatory issues and project management and has worked with outside vendors as appropriate to make internal technical training available. Sinha has expertise in assessing, developing, managing, and implementing customized corporate training on environmental, health, and safety (EHS) and team-building issues. She won the "Trainer of the Year" award in 1997 from the National Environmental Training Association.

Some of Sinha's key projects include designing, developing curriculum, and delivering EH&S training for in-house engineers, scientists, geologists, and project managers; developing the "Basic Training Design Skills" Manual for trainers; and delivering "Train-the-Trainer" and "Hazardous Waste Training" in Japan. She also designed and delivered training for industrial clients to meet regulations, internal operating procedures, and quality program. Her publications include such titles as: "Creative EH&S Training: An arsenal of 'fun & effective' tools and techniques" (NETA, 1996); "Environmental training requires innovative techniques" (Eco-Log, 8/93); "How to deliver fun and effective environmental training" (NETA Annual Conference, 1993); and "A simple view of toxicology" (Hazardous Waste Magazine, 12/89).

Dana Acton-Irvin

Dana Acton-Irvin graduated from the Univeristy of Pittsburgh with a B.S. degree in chemistry in 1985. She has twelve years of experience as a consultant. Her primary area of expertise is RCRA hazardous waste management. Irvin instructs and designs environmental health and safety training programs for workers.

Tisha Wilkinson

Tisha Wilkinson is a senior training consultant with TransTech Interactive Training, Inc. in North Vancouver, Canada. She has a B.A. degree (Honors) in geography from Carleton University, Ottawa, where she specialized in environmental planning, and an M.S. degree in environmental studies (MES) from York University, Toronto, where she specialized in environmental training for industries.

Wilkinson uses her educational background, six years experience with the Canadian Federal government, and five years as technical training consultant to design and develop technical training programs for operations and maintenance personnel in pulp and paper, mining, chemical, petrochemical, and other process industries.

Edward Smith

Edward Smith graduated from Colorado College with a B.S. degree in geology in 1983. His twelve years of professional experience includes consulting, regulatory compliance, and remediation in both the public and private sectors of the environmental field. For the past five years, Smith has worked as an environmental specialist for McFarland Cascade. In this position, he helps maintain the environmental compliance of four active wood treating facilities located in the northwestern United States. He also has experience in NPDES, air quality, and land use permitting.

Les D. Lonning

Les D. Lonning received his B.S. degree from Montana State University in 1970 and has over twenty years experience in the wood preserving industry. He is a past president of the American Wood Preserving Association (AWPA) and the Western Wood Preservers Institute (WWPI), and is currently serving on the Board of Directors of the American Wood Preserving Institute. Lonning is also a member of the American National Standards Institute (ANSI) 05.1 Committee on Wood Poles and is the chairman of the Sub-committee on Foreign Wood Species.

Lonning's experience within the industry started in plant production and quality control and expanded into the areas of environmental compliance and remediation by the mid-1980s. He remains active in both the production and environmental areas of the wood preserving industry and is currently the manager of environmental and technical services for McFarland Cascade. He has negotiated numerous environmental Consent Orders with state and federal agencies throughout the northwestern United States and has managed the cleanup activities at both active and inactive wood preserving sites.

Mark Aronson

Mark Aronson is a member of the Biology Faculty at Scott Community College in Bettendorf, Iowa. He has taught environmental science for the past ten years and has been active in the development of environmental curricula for environmental technology students at the local and national level. In 1995, 1996, and 1997, Aronson was selected as a National Science Foundation Fellow for the Advanced Technology Environmental Education Center (ATEEC). He has worked on a number of educational grants from the EPA that focused on the development of 2+2+2 environmental programs in high schools and colleges. He has also conducted scientific research supported by such agencies as the USDA and Department of Interior.

Erwin Gutzwiller

Erwin Gutzwiller is a retired Army officer and medical administrator. He holds a B.S. degree in healthcare administration and is currently completing an M.S. degree in evironmental science and engineering. He has contributed numerous articles to various professional journals in the environmental field, and has participated as a panelist and speaker at conferences and symposiums both nationally and internationally. His work with hazardous materials and hazardous waste ranges from on-scene coordinator for large HAZMAT releases to designing and supervising the cleanup of hazardous waste sites. Gutzwiller has provided professioanl consultation on environmental and waste stream management as well as emergency plan development for numberous industrial activities ranging from heavy industry to the college campus. He develops training courses and seminars for various apects of EPA, DOT, and OSHA requirements. He is a Certified Hazard Control Manager (CHCM) and trained Radiation Safety Officer (RSO).

Kelly Erin O'Brien

Kelly Erin O'Brien holds a B.S.B.A. degree from the University of Phoenix and is a California Registered Environmental Assessor (REA). She has been in the environmental, health, and safety profession since 1980. She has worked at University of Santa Clara, New United Motor Manufacturing, Inc. (NUMMI)—GM and Toyota's joint car manufacturing plant in California—and FMC Corporation. O'Brien has been an independent consultant since 1990 and currently leads ISO 14001 and OSHA Voluntary Protection Program certification efforts for corporate clients. She is also an instructor in the Hazardous Materials Management Programs of the University of California, Santa Cruz Extension and the California Community College system.

1

Pollution Prevention

Neal K. Ostler

Upon completion of this chapter, you will be able to meet the following objectives:

- Discuss the meaning and importance of pollution prevention in general terms.
- Identify the major techniques of pollution prevention.
- Specify a number of ways that the Environmental Protection Agency (EPA) is supporting voluntary compliance with pollution prevention in industry.
- Recognize the EPA's four-step hierarchy of waste management options, which include source reduction, recycling, treatment, and disposal.

INTRODUCTION

This chapter has not only been written for the student of environmental technology, but also as a general introduction and helpful guide to anyone (especially in business and industry) interested in understanding the concepts, policies, and technologies involved in pollution prevention. To achieve this, the chapter has been divided into three parts. Part 1 is a general introduction to the meaning and importance of pollution prevention, offering a discussion of the shift in policy from the legislated "waste minimization" controls at the "end-of-pipe" to reduction of wastes at the source, and explains the potential and real benefits to generators of hazardous waste minimization. Part 2 examines the current regulatory policies and explains various federal and state initiatives to promote voluntary compliance by industry. Part 3 outlines the concepts, benefits, and approaches to the following programs: waste minimization, solid waste reduction, water pollution prevention, and air pollution prevention. Pollution prevention also includes both the practice of the conservation of natural resources, such as energy and water, and the practice of sustainable agriculture. In this chapter, we will confine our discussion to the first four programs.

NOTE: Before starting this chapter, the reader should also review Appendix A, "ISO 14000 Environmental Management System (EMS) and Overcoming Certain Objections," found at the end of this book. This appendix details ISO 14000, a series of standards first published in October 1996 that provide the design for a voluntary environmental management system. This system has been developed and adopted by representatives from more than 50 nations that comprise the International Standards Organization (ISO). This appendix attempts to provide a response to certain questions and objections regarding the value of ISO 14000 in helping to bring about a less-polluted world.

MEANING AND IMPORTANCE OF POLLUTION PREVENTION

Commonly referred to as **P_2,** the concept of **pollution prevention** is relatively new to the environmental community. Prior to P_2, the primary emphasis of federal and state regulatory efforts was to control and clean up the most pressing environmental problems that faced our communities. These efforts resulted in major reductions in pollution; however, regulators also learned from experience that the traditional "end-of-pipe" methods were not only cumbersome and expensive, but also less than fully effective and often resulted in the contaminants simply being transferred from one medium to another. Particulates captured from dirty air often end up in water that has to be treated and finally results in a sludge that must be treated or disposed of as solid and possibly hazardous waste.

If improvements are to be made in the big picture of environmental quality, measures are needed to prevent pollution from occurring in the first place. This is the essence of P_2 and has the potential of affording important economic benefits to the generator of hazardous wastes. Avoiding pollution allows the generator to also avoid costly investments in the management of hazardous waste: The generator will not have to clean it up, nor will the generator have to budget for pollution's collection, transportation, incineration, or storage in a landfill.

Pollution prevention has the marvelous potential of not only protecting the environment, but also strengthening the growth potential of a business or industry in general, through more efficient manufacturing and use of raw materials. Simply put, and in its broadest sense, pollution prevention refers to the reduction in either volume or toxicity of the waste, resulting from an industrial process, prior to either its discharge or disposal.

POLLUTION PREVENTION TECHNIQUES

The most common techniques for pollution prevention are source reduction, which is P_2 in its truest form; recycling activities, which involve finding commercial uses or reuses of a material so that it is not designated as a waste; and treatment to reduce or alter the material's toxicity. This section of the chapter will focus primarily on source reduction and recycling activities. Brief mention will be made of treatment because it is not considered by some as P_2 in its truest sense since wastes are still generated. Some experts include a fourth category of P_2 techniques that they designate as "technology modifications," such as converting to automated or computer-run equipment and/or other equipment modifications. This fourth technique is treated in this chapter as an illustration of practices that can result in source reduction.

Source Reduction

Source reduction is the development of operating practices and procedures that can lead to the elimination of a waste at its source. While such elimination may be the goal of source reduction, it is virtually impossible to entirely eliminate waste. Therefore, the realistic goal of source reduction is interpreted as the implementation of state-of-the-art practices and procedures within a facility's manufacturing, production, and waste-generating operations.

Operating practices and procedures are the real key to achieving the objectives of source reduction and are discovered by a thorough examination of each industrial process or manufacturing operation. (Appendix B provides a step-by-step guide to analyzing the flow of materials in any industrial process or manufacturing operation.) .Particular focus must be placed on the following areas, which will be discussed further in Part 3 of this chapter:

- Product specifications and the basic inputs of raw materials
- The sources of power usage
- Transformations that occur in the process and the impact of quality control parameters
- Production goals and the output of materials

Further research can then be conducted into the discovery and feasibility of implementing such practices and procedures that will simultaneously improve efficiency and also reduce production costs. Generally speaking, good operating practices and procedures require little or no investment, are relatively easy to implement, and often result in significant savings. The practice of improving yields by reducing production losses is not new to an industry where raw materials may account for a large percentage of operating costs.

Examples of good operating practices and procedures might include the following:

- Company policy including the formalization of waste reduction programs that indicate strong support for the program and recognize progress in meeting company goals. Company policy should clearly establish a proactive waste reduction program.
- Personnel practices may be used by management to encourage an objective use of employee participation in pollution prevention. Offering incentives to individuals or departments for their contributions may be an effective means of achieving of the company's pollution prevention goals.
- Material substitution requires serious study and research into the product's specifications. It might be possible to use nonhazardous or less hazardous ingredients that would reduce the use of hazardous materials and thus have a corresponding reduction in the quantity of hazardous waste.

- Inventory practices include proper control of materials as they are purchased, stored, handled, or transferred from one location to another. Purchasing of materials in bulk may result in a cost discount but those savings must be weighed against the initial capital outlay and costs incurred in the materials' storage. Such costs include the construction and maintenance of storage facilities, materials handling personnel, and the potential for reportable releases or spills that not only require costly cleanup procedures but also result in lost revenue due to the spill. Poor inventory control can result in overstocking that, in turn, may lead to an expired shelf life and disposal of nonusable materials.

Simple procedures for handling materials and storage include properly training employees in the operation of transfer equipment, adequate spacing of containers to avoid the potential of punctures and breakage, and hazard communications such as the proper marking and labeling to indicate the contents and associated risks of each container.

- Segregating wastes does not actually reduce the total volume of a solid waste at its source, but, if procedures forbid the mixing of hazardous waste streams with those that are nonhazardous, the overall volume of hazardous waste can be significantly reduced. This measure is often more suitably discussed as a waste minimization technique, resulting in reduced costs for the management and disposal of hazardous wastes. It is further examined later in this chapter.
- Loss prevention includes the development of spill control and containment plans that not only provide for response to accidental releases but also ensure that precautionary measures are taken to ensure that such spills and leaks do not occur in the first place. Such prevention programs include operating protocols to avoid spills caused by human error, regular inspections and maintenance of inventories and equipment to avoid leaks, warning systems in the event of unexpected or accidental releases, and catch basins or dikes to prevent contamination of water systems. Other such programs include development of contingency plans or personnel procedures to be taken in the event of a release, training of personnel in the implementation of those procedures, suitable and adequate materials for the containment and collection of contaminated materials, and policy and procedures for the disposition of wastes. Too much emphasis cannot be placed on the old adage, "An ounce of prevention is worth a pound of cure."
- Optimal use of power does not actually contribute to a reduction in waste at the operations level, but certainly can lead to savings in overall production costs. Large facilities can consider installing of their own power sources, while smaller concerns can seek more favorable rates from public utilities during non-peak or optimal hours of operation. Timers on lights and even a policy of turning them off when they are not in use can result in a savings to the company and also help prevent pollution at the source of the power generation.

Many locations may qualify for permits to burn waste as a fuel source in generating of electrical power. Certainly this option would both reduce the amount of waste needing to be handled and reduce costs associated with power. This option is discussed further under the category of "recycling."

- Scheduling of production can often be an effective technique for waste reduction, especially where various batches are run with different recipes and basic ingredients. Changes in the schedule for production runs can reduce the frequency of tank and equipment cleaning. This can be accomplished by either maximizing the size of the batch or by having a sequence of different batches in an order where cleaning may not be necessary. These techniques can significantly reduce the amounts of solvent used and the wastes resulting from this activity.

- Automation or modification of equipment can lead to immediate and long-term cost savings through their extended lifespan. Conversion to automated equipment may involve an extensive capital outlaw but the greater lifetime may make the investment worthwhile. The benefit that smooth, efficient, programmable equipment offers the environment is the optimal use of power, and the reduced potential of an accidental release or other mishap.

Recycling Programs

It might be possible to **recycle** wastes that cannot be eliminated or reduced through source reduction programs. By definition, a material is recycled if it is used, reused, or reclaimed. Recycling through use or reuse at the production or process level involves either returning the materials to the original process as a substitute for an input material, or as an original input material or ingredient in an entirely different process. **Reclamation,** on the other hand, is the processing of waste for recovery of valuable material, for regeneration, or (if permitted) for burning to generate power or recover energy.

- **Preconsumer recycling** involves using or reusing raw materials, byproducts, products, and waste streams within the original production process before they ever reach consumers. This activity allows the facility to reclaim or reuse certain components and ingredients of its different process waste streams for a beneficial purpose in the same or a different process. Often a consumer use of the end products of this recycling activity can be found, enhancing overall corporate profits as well as reducing losses.
- **Closed-loop recycling** is an additional measure of recycling, that is acceptable under Resource Conservation and Recovery Act (RCRA), whereby the waste is never allowed to escape the original process vessel, or other container in which it was generated. Under RCRA 3010, the waste is not subject to notification requirements until it exits the unit in which it was generated, with the exception of a surface impoundment or vessel that is being used for storage only and remains in the unit more than ninety days and, thus, is no longer operated within the context of a manufacturing process.
- **Post-consumer recycling** involves separating materials that have already served their intended end-use from the solid waste stream and returning them to their producers or centers for recycling. The high cost of managing this type of program makes it not nearly as effective in source reduction as preconsumer recycling; however, it is still beneficial as a source-reduction method in recovering certain discarded materials for use or reuse.

BARRT is an EPA acronym that stands for Best Available Recovery or Recycling Technology. This chapter will not discuss the definition of solid, hazardous, or recycled wastes, as such a discussion involves an in-depth study of RCRA. However, the essential element in the definition of a hazardous waste is that the material no longer has a commercial value, except of course to the Treatment, Storage, and Disposal Facility (TSDFs). What should be understood is that the EPA has taken exhaustive steps in its struggle to define recycling and to establish criteria for such activities, so that many materials with hazardous constituents can remain outside of the classification as hazardous wastes. The EPA wants recycled and recovered materials to meet certain waste constituent standards and that a BARRT will be developed by industry to meet those standards.

The benefits that a facility can realize from recycling include a reduction of raw material costs, elimination or a reduction in waste disposal costs, income from salable materials and wastes, and reclamation of energy.

Waste Treatment

The definition of **treatment** is "any practice, other than recycling, designed to alter the physical, chemical, or biological character or composition of a hazardous substance, pollutant, or contaminant so as to neutralize it or render it nonhazardous throughout a process or activity separate from the production of a product or the providing of a service." The fact is, all wastes cannot be recovered for recycling nor entirely eliminated at the source. As another alternative when source reduction or recycling is not possible, effective treatment methods should be used to either reduce, change, or entirely eliminate the hazardous waste's characteristics.

Because most treatment methods occur after the waste has already been generated, it is not typically considered a pollution prevention technique. However, for the purpose of this text it does constitute the reduction of hazardous waste and prevents such wastes from ending up in a landfill and, thus, offers a savings in related disposal costs. On the other hand, such treatment methods need to be weighed against the cost of additional handling, storage, management, and disposal. Although many do not consider treatment a P_2 concept, ultimately treatment methods result in less or even nonhazardous wastes and, thus, meet the objective of "waste minimization," which is a P_2 goal. So treatment is worthy of consideration when source reduction and recycling methods are not available.

THE BENEFITS OF P_2 TO BUSINESS

In recent years a great deal has been written regarding "green" politics and accounting. In addition to accomplishing P_2 for its own sake, the adoption and practice of P_2 also can result in the following:

- Enhanced public image for businesses both large and small when they have legitimately practiced state-of-the-art pollution prevention techniques.
- Reduced regulatory scrutiny when businesses can demonstrate comprehensive P_2 programs with ongoing assessments, adjustments, upgrades, and other modifications on a regular basis. Regulatory agencies certainly aren't going to ignore businesses with strong P_2 programs, but given their limited inspection and enforcement capabilities, they will likely focus on companies lacking pollution prevention programs. A direct result of sound P_2 practices is that of reduced noncompliance liabilities.
- Economic benefits, from more efficient use of raw materials, power, and equipment, fewer losses from spillage or releases, and reduced costs of management and disposal of solid and hazardous waste.

- Competitive advantages when a product or service can be delivered to the consumer at a lower cost from a company that has achieved a "green" reputation.
- Reduced environmental risks from smaller and fewer varieties of waste streams. Savings also result since fewer employees are needed, as are smaller storage and collection areas, there are fewer wastes to report, there is less chance of an accidental release or spill that the company must clean up, report, or be fined for, and the company needs also a small budget to pay for their disposal. Fewer risks also mean the employer should realize an increase in worker health and safety from fewer accidental exposures. The insurance premiums will also be reduced with fewer risks.

Pollution prevention is not only important as a tool that in environmental management but there is factual evidence that the negligence of a business to "go green" will result in a loss of "green" ($$) to that business.

THE REGULATORY FRAMEWORK

In this section we will discuss the reasons for a shift in regulatory policy toward pollution prevention, the provisions of the Pollution Prevention Act of 1990, the EPA's five-year policy and prevention strategy for P_2, and other pertinent regulations such as RCRA, The Clean Water Act, and trends in state P_2 programs.

Note: Appendix C provides an overview of the current regulatory agenda in the United States.

The Shift in Policy to P_2

Since the creation of the EPA and The Occupational Safety and Health Administration (OSHA) in 1970, substantial progress has been made in improving the quality of the environment. However, by 1994 the following facts still represent the state of our U.S. environment:

- Many Americans still live in areas where the air quality does not meet the standards of the Clean Air Act.
- Roughly one person in four lives within four miles of a hazardous waste site.
- Some sources estimate that as much as 40 percent of the country's rivers and lakes are still not suitable for the habitation of edible fish or for swimming.
- The Comprehensive Environmental Response Compensation and Liability Act (CERCLA) was passed in 1980, yet only a small percentage of the sites on the National Priority have been cleaned up.
- Numerous towns and cities can no longer rely upon their natural sources of drinking water due to pollution (such as aquifers), and must have their water trucked in from neighboring communities.

Clearly it is going to take much more time to clean up an environment that has been spoiled by years of unchecked pollution-generating practices, practices engaged in by most everyone in the economy from the individual consumer and the manufacturer to our government and its legacy of weapon's research and power generation.

Until recently the primary focus of management practices has been "end of pipe" treatment, storage, and disposal of hazardous waste, top of smokestack capture of air pollutants, and pretreatment of industrial waste water. This method of focusing on each of the primary sources of pollutants has segmented the EPA into a "piecemeal" approach to environmental protection that, while providing substantial progress in improving the quality of the environment, has obvious limitations. Further improvement cannot be achieved from methods that merely manage or control pollutants after they have already been generated. In its five-year strategic plan (EPA 200-B-94-0002), the EPA recognizes that it must move beyond strategies that react to today's environmental problems to strategies that anticipate and prevent pollution."

The Pollution Prevention Act (PPA) of 1990

Most companies that generate solid and hazardous wastes are aware of the EPA's shift in policy and a change in focus toward pollution prevention. In 1990, the U.S. Congress passed the Pollution Prevention Act (PPA) to assert this nation's commitment to pollution prevention strategies, waste reduction programs, and recycling activities. The PPA mandates that the EPA develop such strategies and, although viewed largely as a voluntary measure, the EPA is charged with providing grants to the states as an incentive to implement their own P_2 programs.

The PPA defines pollution prevention as "source reduction and other practices that reduce or eliminate the generation of pollution," and identifies such activities as "fundamentally different and more desirable than waste management and pollution control." The law further defines P_2 as any practice that reduces the amount of any hazardous substance, pollutant, or contaminant entering any waste stream or otherwise entering the environment (by any media) prior to recycling, treatment, or disposal, and which further reduces the hazards to public health and the environment associated with the release of such materials. Finally, the law also emphasizes the goal of having a decrease in cross contamination of landfill compartments that have resulted as a consequence of land burial and treatment processes occurring at TSDFs.

The Hierarchy of Waste Management Options

The PPA sets forth the acceptable options of waste management in a descending order of preference as follows:

- **Prevention/source reduction.** Wherever and whenever possible, pollution should be prevented or reduced at the source and never generated or created in the first place.
- **Recycling.** When it is not possible to entirely eliminate the generation of wastes, they should be recycled in a manner that is environmentally safe.
- **Treatment.** To reduce and possibly eliminate the hazardous characteristics of a waste it should be treated only when prevention or recycling opportunities are not available. Treatment is to be in an environmentally sound manner.
- **Disposal.** The incineration or placement of solid and hazardous waste in a landfill or other release to the environment should only be used as a last resort.

Other PPA Measures

Other provisions of the PPA require the EPA to set up an office independent of the agency's "single-medium" programs, to promote a multimedia approach to P_2, and then develop and implement the following partial list:

- Conduct or support P_2 research and establish standard methods of measurement for source reduction.
- Set up an information clearinghouse and develop improved methods for providing public access to this data clearinghouse under federal environmental statutes.
- Investigate and develop federal procurement opportunities to encourage source reduction programs by the states. The PPA authorizes a grant program to provide up to 50 percent in federal matching funds for states that participate in the program.
- Facilitate the adoption of source reduction by private business and other agencies of the federal government.
- The PPA provides that facilities required to report their releases under the Community Right To Know, must now provide additional information on their P_2 efforts for each facility and for each toxic chemical in their inventory.
- The EPA must report to Congress (biennially) on actions needed to implement and promote source reduction strategies, make assessment of the clearinghouse activities, and account for moneys granted under the incentives programs.

The EPA's P_2 Policy and Strategy

Under PPA provisions, the EPA must develop a pollution prevention strategy that reduces pollution at the source. In the years since the PPA's passage two such strategies have been published, a formal statement of agency policy issued by Administrator Carol Browner, and a variety of related programs enacted. A brief discussion of each follows:

The Industrial Toxic Project (ITP)

Published in 1991, the Industrial Toxic Project (ITP) is better known as the "33/50 Initiative." Under this project, the EPA targeted seventeen high-volume "community right to know" chemicals and set a goal to reduce their releases by: 33 percent at the end of 1992 and 50 percent by the end of 1995.

Each of the seventeen chemicals was selected from the Toxic Release Inventory and is regulated under one or more existing environmental statutes. The 33/50 Initiative is intended to complement these other programs and to give companies incentives to identify and implement cost-effective pollution prevention programs created especially for them. All seventeen chemicals are subject to the standards of maximum achievable control technologies under the Clean Air Act Amendments (CAA) of 1990. It is believed that the incentives under the CAA will provide further encouragement for companies to pursue the goals of the 33/50 Initiative.

The EPA expects companies to reach the 33/50 percentage goals by implementing of the P_2 hierarchy outlined previously. The program goals are not enforceable and participation is entirely voluntary. A copy of this initiative can be obtained by contacting the EPCRA hotline at (800) 353-0202. Ask for publication number Pub. EPA/741/R-92/001.

The EPA Policy Statement of 1993

Administrator Carol Browner issued the EPA's Pollution Prevention Policy Statement in June 1993. The purpose of this statement was to offer guidance in how the EPA could achieve the goals of the PPA. Its main points include:

- **Regulation and compliance.** The mainstream activities within the EPA are to reflect the agency's commitment to reduce pollution at the source.
- **Seek new legislation.** The EPA will seek legislation as necessary that encourages the reduction of pollution at its source;
- **Develop new partnerships.** The EPA will work to strengthen the network of national, state, local, and private sector cooperatives that would emphasize multimedia prevention strategies; integrate regulatory, permitting, and inspection programs supported with federal money; present new models for government/private interaction in a common pursuit of new opportunities for reducing waste; and reinforce mutual goals of economic and environmental well-being.
- **Provide incentives for technology transfer.** The EPA will encourage the partnerships mentioned to meet high-priority needs for new P_2 technologies that will increase competitiveness and enhance environmental stewardship.
- **Right to know and public information.** Finally, the EPA will collect and share useful information that helps identify opportunities for source reduction, measure the success of pollution prevention, and recognize the achievers in government, research, education, and industry.

While acknowledging a number of accomplishments, the policy essentially states the intent of the EPA to build P_2 into the framework its mission to protect human health and the environment.

The EPA's Five-Year Strategic Plan

The strategic plan released by the EPA in 1994, states that "pollution prevention should be the strategy of choice in all that the agency does." With the release of this five-year plan, the EPA has publicly recognized and that reducing of waste at its source is far more cost-effective and protective of the environment than end-of-pipe management after waste problems have been created. The plan recognizes that anticipating problems and stopping them before they occur is the preferred strategy by far.

All of the strategies in the plan are essentially consistent with the principles outlined in the Pollution Prevention Policy Statement issued by the EPA the year before and include the following:

1. Progress measures will include that of agency activity, state activity, activity undertaken by the public sector, and environmental improvements. The EPA will reorient its P_2 efforts by focusing on specific activities, such as the development of preventive regulations with qualitative measures of progress based on their environmental value. Measures of progress by the states will include direct outreach to the regulated community to promote P_2 through workshops and facility visits by their amount of P_2-directed grants, and by changes made in their operations, such as permitting and inspections that encourage pollution prevention.

   Progress in the activity of the public sector will be measured by assessing its adoption of pollution prevention approaches. Indications of progress will include voluntary changes in industry and consumer behavior, such as the production and marketing of environmentally preferable products and packaging. The EPA is committed to developing better methods for evaluating the effectiveness of P_2 programs and to a multimedia measurement of the total reduction in the volume and toxicity of pollutants generated across the entire range of its programs.

2. Key activities by the EPA in pursuing its P_2 goals include the following:
   a. **The Common Sense Initiative.** With a focus on selected industry sectors, the EPA will develop multimedia teams to identify and implement environmental management plans that provide greater environmental benefits at a reduced cost of compliance. The EPA will work to develop private partnerships with industry, environmental groups, states, tribes, and local governments to develop these "one-stop shops" for environmental compliance to facilitate the ease and economy of environmental protection. Eventually, a similar approach for all industries would be implemented, once the teams are in place for the pilot industries. (Appendix D provides a more thorough examination of the EPA Common Sense Initiative.)
   b. **Compliance and enforcement efforts.** The EPA will develop auditing policies that encourage consideration of P_2 projects by private parties and the exploration of facility-wide P_2 options as injunctive relief during the settlement process.
   c. **Regulations and permitting.** Reduction at the source will be incorporated into all possible ongoing activities, such as achieving effluent guidelines, implementing Maximum Available Control Technologies (MACT) standards, and in granting and reissuing permits under the Clean Air Act and the Clean Water Act.
   d. **State, tribal, and local partnerships.** The EPA will provide more flexibility in administering federal programs and will support innovation at all levels. The EPA will strengthen its outreach programs for compliance/assistance and other forms of technical assistance by channeling resources into those programs. If state commitments to the federal program interfere with the goals of P_2, the EPA will reassess those commitments.
   e. **Federal partnerships.** Presidential Executive Order 12856 establishes a framework for federal management of its own facilities, requires Toxic Release Inventory Reporting through Community Right To Know, and it mandates goals for the reduction of losses from releases and off-site transfers, and pollution prevention. Order 12856 also requires plans and goals for the elimination or reduction of extremely hazardous materials and toxic substances from federal facilities inventories, and guidance in the acquisition of environmentally friendly preferable products as substitutes.
   f. **Cooperative ventures.** A cooperative agreement has been established between the EPA and the Department of Energy (DOE). Under this agreement, the EPA and DOE will jointly research, design, develop, and implement new programs for the efficient use of energy with a goal of sustainable development of our national resources.
   g. **Environmental justice.** New efforts will be made by the EPA to provide grants that will strengthen the capacity of minority and low-income communities to advance the goals of pollution prevention consistent with emergency planning and Community Right To Know.
   h. **Technology transfer.** The EPA will also support consortia, committees, and other efforts made by higher education in the facilitating joint efforts in P_2

research, design, education, and training projects. An example is the P_2 in Higher Education Committee, supported by the Utah Department of Environmental Quality in the EPA's Region-8.

The EPA's goal is to promote good design and clean technology to reduce pollution overall. To begin with, nearly $5 million in research and technical assistance will be dedicated to P_2 in the metal fabrication industry through the **Environmental Technology Initiative (ETI)** with future funding focused on industries participating in the Common Sense Initiative. The EPA will attempt to move from regulatory compliance and enforcement to the adoption of P_2 and will investigate opportunities to allow industries to coordinate their regulatory compliance programs with their capital development cycles. The EPA will continue to foster the "green chemistry" ethic and encourage industry to avoid generating of toxic wastes.

i. **Legislation.** Rather than shifting waste from one environmental medium to another, as often is the case with a separate and individual focus upon the different acts pertinent to each medium (Safe Air, Safe Water, Safe Drinking Water, RCRA, etc.) the EPA will strive to reorient the fundamental statutory mandates toward pollution prevention. Such legislation will enable the EPA to incorporate its Common Sense Initiative and to establish a one-stop-shop for industry. Better coordination of activities under different statutes can encourage prevention and provide industry with opportunities for source reduction and recycling versus end-of-pipe treatment, storage, and disposal.

3. The Office of Prevention, Pesticides and Toxic Substances' (OPPTS) past role was information gathering and reporting, and then providing that information, technical assistance, and education to empower the public to make informed decisions regarding the risks associated with pesticides and toxic substances. While this role will continue, the five-year future of the OPPTS following the release of the EPA's strategy, includes a new mission: to promote safer designs, wiser use of materials, products, processes, practices and technologies, and disposal methods using pollution prevention as its guide. The OPPTS has reoriented its chemical programs to support an emphasis on source reduction and is the leading advocate for promoting P_2 within the EPA, to states, the general public, and to constituent groups. The OPPTS is also used to instill the green chemistry ethic within industry, with increased emphasis upon wider use of alternative agricultural practices, including Integrated Pest Management (IPM) rather than pesticides.

Other Related Programs

Though not specifically targeted as pollution prevention, there are a variety of other EPA programs that result in accomplishing the goals of P_2. In most areas of the United States, these are voluntary programs such as the following two examples:

- **Green Lights** offers encouragement and incentives for companies to study their lighting needs to see where substantial energy savings may be realized. Savings may be from installing automatic on-off switches in places only occasionally occupied (such as certain areas of warehouses and restrooms); converting to low-energy, high-output lighting (such as switching from incandescent to fluorescent light

fixtures); scheduling operations in sun-lit areas during times of daylight to take advantage of natural lighting; and installing lower-intensity lights or fewer lights in areas where operations do not need brightness, or to provide work-station lights instead of general area lighting.
- **Design for the Environment (DfE)** has established initiatives that promote good environmental design of products and processes. The DfE encourages deployment of risk reduction and P_2 concepts at the earliest design stages.

FEDERAL AND STATE REGULATIONS

The EPA's current objective is to encourage business and industry by giving them incentives; very few pollution prevention measures are mandatory. Some mandatory P_2 requirements do exist, however, and this section will discuss pertinent provisions within the Resource Conservation and Recovery Act (RCRA), the Clean Water Act (CWA), and the Emergency Planning and Community Right-To-Know Act (EPCRA). (We do not discuss either the mandatory or voluntary provisions and programs that have been adopted by separate states.)

RCRA Waste Minimization

The primary purpose of RCRA's regulations originally was to provide guidance in the treatment, storage, and disposal of waste to minimize current and future risk to human health and the environment. With the hazardous and solid waste amendments to RCRA in 1984, Congress declared it to be a national policy that, whenever feasible, the generation of hazardous waste as expeditiously as possible, whenever feasible.

The 1984 amendments established a requirement that the generator of a hazardous waste, which ships its waste off-site for disposal, must "certify" on the shipping manifest that it has programs in place to reduce the volume and/or the toxicity of hazardous waste it generates to the extent that it is economically practical. Permits for hazardous generators waste must also certify a waste minimization "program in place" as a condition for the treatment, storage, or disposal of hazardous waste. The details of what constitutes a "waste minimization" program are discussed at greater length later in this chapter.

Recycled materials are regulated under Subtitle C of RCRA. In order for any substance to be considered a RCRA waste under Subtitle C, it must first be both a solid and a hazardous waste. The recyclable materials provisions of RCRA are particularly complex, and the term itself is used to connote a hazardous waste that is recycled. However, recycling is considered the next preferable method after source reduction in pollution prevention techniques. It is not the purpose of this chapter to provide a working knowledge of RCRA wastes, but many materials are not classified as solid wastes when they can be recycled by one of the following methods:

- Used or reused as ingredients in an industrial process to make a product, provided the materials are not being recycled.
- Used or reused as effective substitutes for commercial products.
- Returned to the original process from which they are generated without first being reclaimed. The material must be returned as a substitute for a raw material feedstock, and the process must use raw materials as principal feedstocks.

Certain wastes are RCRA solid wastes, even if their recycling involves use, reuse, or return to the original process, if the following apply:

- The materials are used in a manner constituting disposal, or used to produce products that are applied to the land unless that is their intended or ordinary manner of use
- The materials are burned for energy recovery, used to produce a fuel, or contained in fuels, unless they themselves are meant to be fuels
- The materials are accumulated speculatively
- The materials are inherently waste-like materials
- The materials are specifically listed in federal or state regulations.

RCRA regulations also require that large-quantity generators and owners of TSDFs complete a biennial report. One provision of that report is self certification and assertion by the facility to have a program in place for waste minimization that compares the company's present efforts to reduce waste with those in previous years. These efforts may be in the form of treatment to reduce the toxicity or other hazardous constituents of the waste, or other recycling efforts. Generators also must certify on hazardous waste shipping papers, known as **manifests,** that they have a waste minimization program in place.

Even if a company claims it is recycling its wastes, it must provide appropriate documentation to demonstrate that the resulting material is not a waste, that there is a known market or disposition for the material, that it has the necessary equipment, that it meets the appropriate terms of the recycling exemption or exclusion, and/or that the material is otherwise exempt from regulation. Whenever involved in the recycling of materials, it is best to maintain careful records to the greatest extent possible.

The Clean Water Act (CWA)

The Clean Water Act (CWA) establishes a system wherein all facilities with point source discharges of contamination are issued a permit that allows them to discharge these contaminants into the water. The system of issuing these permits is called the **National Pollutants Discharge Elimination System** (NPDES).

As part of the CWA's goal, the limitations of these permits to discharge have become much more stringent over time. Decisions regarding the numeric value and type of effluents that are allowed to be discharged are determined by the most stringent value among technology-based effluent guideline limitations, water quality-based limitations, and limitations on a case-by-case basis. Permits also spell out certain state-of-the-art management practices, known as Best Management Practices (BMPs), that may be imposed in NPDES permits to prevent the release of toxic and other pollutants to surface waters. The use of BMPs allows the EPA and the states to issue, renew, and alter permits to require industrial manufacturing and treatment processes to employ pollution control and prevention mechanisms. Examples of activities that can be regulated by mandatory BMPs include the following:

- **Facility runoff** due to rainfall on a plant site. In addition to rinsing away fallout from smokestack emissions, rain can come into contact with harmful materials that are stored in the open and in areas where the materials may be transferred (such as loading docks), and areas where waste may be stored in containers.

Capture and treatment of such stormwater runoff may also be separately regulated under permitting programs that require stormwater pollution prevention plans as a condition for issuance of the permit. The stormwater permit requires identification of all potential areas that might contribute to polluted runoff from rainwater, and a description of the measures and controls that are planned to prevent or minimize water contamination. These controls may include good housekeeping practices, regular inspections, preventive maintenance of equipment, spill prevention and response procedures, testing and groundwater monitoring, training of personnel, and recordkeeping.

- **Material storage areas** at any stage of the production process pose a potential for accidental release of chemicals, hazardous or otherwise. Storage areas may be fully contained or open to the environment, and containers may range from open cubicles such as bins, to boxes and over-packed barrels or drums. BMPs stipulate optimal protection based on the material's risk.
- **Loading/unloading operations** may include pneumatic transfer of dry chemicals from a storage tank to a truck or rail car, or vice versa, pumping of liquids or gases during loading or unloading, conveyor transfer of product, and movement by hand or personnel-operated equipment such as pallet jacks and forklifts. BMPs may require regular cleanup procedures that assure rinse waters are not permitted to be discharged to the public water or sewer system.
- **Sludge and waste storage areas,** depending upon their siting, construction, and operation may potentially leach pollutants into groundwater or surface water via runoff. Such sites include pits, ponds, lagoons, landfills, drum storage, and deep-well injection locations.

Many of the site pollution prevention and control measures may currently be used by industry in the development of their Spill Prevention Control and Countermeasure Plans (SPCC) required by the CWA. A specific discussion of such controls is discussed later in this chapter.

The Clean Air Act (CAA)

The Clean Air Act (CAA) established a permit system setting limits for stationary or identifiable point sources of the discharge of pollutants into the air. The permit requires the holder to monitor emissions, keep records, and to submit periodic reports of the volume of their releases.

Some of the provisions of the CAA include:

- Maximum Available Control Technologies (MACT) standards for 189 air toxins.
- National Ambient Air Quality Standards (NAAQS), which set maximum concentration levels of six criteria-specific air pollutants in the breathing space within a community.
- National Emission Standards for Hazardous Air Pollutants (NESHAPS) is the program of permitting the release of certain toxic pollutants into the air from point sources, with the intent of reducing those discharges through economic incentives.
- Amendments to the CAA passed in 1990 call for the reduction and eventually the complete elimination or phase-out of chlorofluorocarbons (CFCs), which are dangerous ozone-depleting chemicals.

Granting of a CAA permit is always conditional upon the holder's agreement to implement state-of-the-art technologies in controlling and limiting the release of hazardous emissions.

The EPCRA Toxic Chemical Release Inventory

The primary goal of EPCRA Title III is to facilitate public awareness and community planning for chemical emergencies. EPCRA requires companies to report their inventories of hazardous substances (listed and of sufficient quantities) to the local emergency planning committee and to the fire department. EPCRA also requires these companies to report any accidental releases of those chemicals on a particular form (Form R) known as the Toxic Chemical Release Inventory (TRI).

In addition to its primary purpose, the information reported on the TRI can provide appropriate materials or substances for targets of the EPA's 33/50 Industrial Toxics Project and Five-Year Strategic Plans. Seventeen of the most common chemicals were selected from the TRI for initial targets of the 33/50 project.

With passage of the Pollution Prevention Act of 1990, the reporting requirements of Form R include a mandatory source reduction and recycling program with sections added to show both on-site and off-site recycling activities that demonstrate the following:

- Techniques used to identify source reduction opportunities
- Specific chemical source reduction practices in place
- Kinds of substances that are recycled
- Amount of chemicals released, either deliberately in conformance with permits, or accidentally into the environment
- Degree of reduction in hazardous chemicals used as raw ingredients in process and products

The EPA is currently expanding the list of EPCRA substances to include CWA priority pollutants, extremely hazardous substances, RCRA listed hazardous wastes, and criteria and other hazardous air pollutants (HAPS) under the Clean Air Act of 1990. EPCRA's hotline number is (800) 535-0202.

POLLUTION PREVENTION PROGRAMS

Up to this point we have talked about the switch in focus for the management of industrial wastes from end-of-pipe to that of pollution prevention. This switch has occurred for a variety of reasons, among them a dramatic increase in the costs of disposal, public outcry over pollution, competitors who may have obtained an edge in the market from their own P_2 practices, and pressures from EPA- and state-run environmental programs.

In response to these reasons, many companies are incorporating innovative programs of P_2 into their daily operations. These include hazardous waste minimization, source reduction of solid waste, water pollution prevention, and air pollution prevention. Proactive companies have determined that pollution prevention/waste reduction is just plain good business sense. The implementation of such programs will reduce a company's paper workload, future liabilities, and the production of pollution and waste that is costly to manage.

Hazardous Waste Minimization

The highest priority of Congress is to not generate waste in the first place; still, many wastes ultimately cannot be avoided. Thus we will first discuss in this section the necessary components, concepts, and benefits of the minimization of hazardous waste.

Obviously if we do not make hazardous waste in the first place, then we have minimized it. That is the essential concept of reduction of waste at the source. But what is minimization?

It may help to learn what waste minimization is not. Waste treatment, for the purpose of its destruction, its disposal, or to reduce its hazards, does not constitute waste minimization because these efforts occur after the opportunities for waste minimization no longer exits. Waste minimization also does not involve storage and disposal practices necessary to prevent accidental releases of the materials into the environment, nor is it the transfer of hazardous constituents from one environmental medium to another. An example of this is when particles are removed from smokestack emissions by a bag house, and then the contents of the bag are rinsed away and the resulting sludge becomes contaminated water that will need to be further treated. While these end of pipe management practices are important to the big picture of waste management, they are not waste minimization activities. It would also hold true that waste-treatment activities such as compaction, concentration of the wastes, or dilution to reduce their toxicity are not considered waste minimization activities.

So what is waste minimization? The EPA considers the term, as employed by Congress in the passage of RCRA, to include both source reduction (which this chapter treats as a separate, special consideration) and environmentally sound recycling. This does not include recycling activities that closely resemble conventional waste management activities.

Elements of the Program

Waste minimization should be a part of corporate management policy and procedure as well as an integral part of a company's overall strategy to increase both quality and productivity. The company should designate a coordinator for the program and empower that person with responsibility for the program's implementation, monitoring, and evaluation. Employees should be trained on the impact of their individual and collective efforts to minimize waste, including people in every department from research and development to shipping and receiving, marketing and sales. Both qualitative and quantitative goals should be set within a sensible time frame for reducing the volume and toxicity of waste streams. Regardless of their position in the company, individuals should be given incentives for identifying opportunities for waste minimization and a commitment should be made by management to implement all recommendations through assessment and evaluation. A company-wide effort that is fully supported by everyone from the CEO to the bottom level worker will afford the waste minimization program a full chance to prove itself and demonstrate its net worth.

RCRA's permit system requires that each company identify and track the types and amounts of wastes and their constituents, and keep accurate records that include the dates of generation and their accumulation and storage. There are a variety of means by which a company can track its wastes and tabulate the entire costs of management and cleanup. Each company, and each organization within the company, must decide which is the best method to obtain this information, and remember to include each of the following elements to have an accurate representation of the true costs associated with each waste stream:

- Value of materials found in the stream
- Employee exposure and health care
- Liability insurance for site workers/employees
- Compliance with regulatory oversight
- Reduced production capability or potential
- Materials handling and storage
- Recordkeeping and reporting paperwork
- Transportation/shipping
- Treatment, storage, and disposal
- Future corrective or cleanup actions from accidental releases

The accounting system should also allocate the actual costs of the management of wastes to those operations, processes, or activities that are responsible for its generation in the first place rather than charging it to general overhead, and clearly identify the actual generator of solid and hazardous wastes. This will help target appropriate areas for waste minimization and pollution prevention programs.

A waste minimization program assessment involves using information from the accounting system to identify every possible opportunity for reducing wastes and analyzing the cost effectiveness of doing so. Similar to the development of an accounting system, there are numerous methods to performing a waste minimization assessment. Certain reduction programs may require expensive systems design and implementation and not return a savings to the company. Then again, savings is not the only concerns as potential liability for the waste must also be considered.

Anyone directly responsible for waste minimization programs should not try to reinvent the wheel themselves. Many valid and proven techniques and systems have been developed elsewhere that may be easily implemented. Technical information or assistance should be available through trade and professional associations, higher education, other companies, government outreach programs (such as technical assistance programs and clearinghouses for information sharing), and good old library research.

Where possible, separate teams should be assembled for each step in a waste minimization program; from development of an accounting system, to assessments and evaluations. These teams should be empowered to actually implement their recommendations and a periodic review should be made to evaluate both the effectiveness of each team and of the total program. These reviews should be used to identify potential areas of improvement, site-effective recommendations, recognize individuals and areas that are having success, and provide feedback to all company employees.

The EPA has developed numerous documents to facilitate the implementation of waste minimization programs. One such document that may be useful is *Waste Minimization: Environmental Quality with Economic Benefits* (EPA/530-SW-90-044, 1990). This document can be obtained by calling the RCRA hotline at (202) 260-9327

Waste Reduction

The difference between waste minimization and waste reduction can be confusing. Both are regarded as efforts to prevent pollution, but waste reduction includes all actions taken to reduce the amount and/or the toxicity of solid and hazardous waste needing disposal; therefore, waste minimization is a particular approach to waste reduction. However, waste reduction programs also involve other factors such as the prevention of waste, recycling, composting, and both the purchase and manufacture of goods that can be recycled or that result in less waste. Each of these approaches is discussed here.

1. **Waste prevention or source reduction.** Prevention of waste at the source means to generate or produce less waste in the first place. This is true pollution prevention and is the most cost-effective way to reduce waste. Although this approach is discussed previously, the following outline offers a quick review:
   a. Minimize the amount of packaging by purchasing and using reusable or returnable packages.
   b. Adopt measures that involve reusing products and supplies rather than single-use materials. The longer you can keep them out of the waste stream, the better.
   c. Substitute nonhazardous materials for both primary ingredients and/or product constituents.
   d. Purchase equipment and supplies that offer more serviceability and longer life. Often such items will be more costly upfront, but with regular maintenance schedules they offer long-term savings due to fewer replacement expenditures. Product obsolescence needs to be seriously addressed by manufacturing.
   e. Efficient materials handling and inventory practices reduce or eliminate losses due to the expiring material's shelf life.
   f. Eliminate items from the inventory that contribute little or nothing to the process, product, or service.
   g. Changes in product design or marketing strategy can help manufacturers lower costs and reduce wastes. These strategies may involve cutting back on the amount of packaging utilized, making the package reusable, or designing the product to have greater durability and a longer life span.
2. **Recycling.** Today, many businesses have developed comprehensive programs for the collection, sorting, storage, and transportation or pickup of a wide variety of bottles, paper, plastic, cans, cardboard, and other materials. One form of recycling is the establishment of a materials exchange where surplus items can be exchanged, sold, traded, or donated instead of being thrown away.
3. **Composting.** A third approach that can significantly contribute to a company's goal of reducing waste is composting. Grass clippings, yard trimmings, food scraps, waste paper, and other organic matter can be collected and later used as a soil amendment or mulch and, if the quality is high enough, even sold to help offset company expenses. Companies should be careful, however, as there is a science to this; they should also discover what standards their states and counties may have already adopted in this regard.
4. **Purchasing.** The purchase of recycled materials and recyclable materials is an important consideration in any waste reduction effort. Buying products with

recycled components helps to support collection efforts for your recycling program and helps ensure these materials do not end up in a landfill. When you purchase items that have been recycled, you are conserving natural resources and you may be reducing your energy consumption. Specifications should be reviewed to assure that recyclable and recycled materials and products are given fair consideration for purchase.

Any waste reduction program needs strong management support. In addition to a statement of support by management at the outset, the creation of waste reduction teams needs management support throughout the program. Such support should be in the form of the endorsement of program goals, communicating the team's importance to employees and offering incentives for employees and/or group participation. The roles and activities of the waste reduction team parallel those of waste minimization and include setting goals, gathering and analyzing information through a waste assessment, promoting of the program to management and employees through education, monitoring the program, and periodic evaluations and reporting of the program's status to management.

WATER POLLUTION PREVENTION

Over the years BMPs have been developed for many different types of facilities consistent with the provision of the CWA. In recent years, the BMPs have become a twin concept to that of pollution prevention and incorporated into the permitting process to prevent the release of hazardous pollutants into U.S. surface waters. By their very nature, BMPs are P_2 practices.

The NPDES under the CWA probably comes the closest to being a federally mandated P_2 program. For a company or municipality to be granted a permit to pre-treat or treat waste water and then release it to surface waters, it must agree to implement state-of-the-art or BMPs. Historically, the enforcement of these requirements has focused on good housekeeping measures and management techniques intended to prevent leaks, minimize the result of accidental spills, and provide for proper waste disposal. Under authority of the CWA, however, this requirement may now encompass the complete framework of source reduction and waste minimization to include operational changes, production modification, material substitution, conserving of materials and energy, and the use and reuse of water.

A program to implement BMPs should be viewed as a concurrent effort to comply with pollution prevention with particular focus upon the element of water. Therefore, it should follow much the same approach as other P_2 programs, including the selection of teams for planning, gathering data through assessments, selection and implementation of BMPs, indoctrination and training of personnel in the use and value of BMPS, and evaluating of their effectiveness. Activities of these teams will include the following:

- Writing policy statements
- Identifying of on-going and potential release sites
- Writing preparedness, prevention, spill control, countermeasure, and other contingency plans
- Implementing good housekeeping practices
- Training for spill response

- Conducting regular inspections and preventive maintenance
- Developing security plans to prevent accidental or intentional damage to systems that could result in releases
- Extensive recordkeeping and reporting

AIR POLLUTION PREVENTION

The Clean Air Act has many provisions that permit different industries to release only very small quantities of hazardous air pollutants (HAPS). The 1990 Amendments to the CAA requires the EPA to promulgate MACTs for both new and existing HAPS sources.

The discussion of MACTs is almost identical to the discussion of BMPs. To obtain a permit for new construction, a company must have state-of-the-art controls in place to prevent the release of HAPS and must also install them during modification of existing sources. With passage of the TITLE IV Amendments, a business or industry was given additional incentives to reduce its air emissions by the establishment of an emission "allowance market" permitting a company to buy or sell extra allowances on the open market.

Air pollution prevention often requires less planning and more budgeting for capital improvements to purchase air pollution control devices. However, the content of smokestack emissions can also be greatly modified by material substitutions and by implementing of efficient production processes. Sludges remaining from these air pollution devices also demonstrate a media crossover of concern from that of air to that of water pollution and control.

SUMMARY

- Historically, end-of-pipe management or waste treatment has been the primary focus of environmental protection. Due to the large number of existing waste sites, it is apparent that improvements and developments in treatment technologies will continue to be a requirement of the environmental industry. However, as we continue into the next century, the development and implementation of an effective pollution prevention program may just be the single-most important component of a company's successful environmental program.
- In the previous twenty-five years, a "command and control" system of compliance has led to an environmental industry in the United States created largely as the result of statutory protections and policy making by Congress and by the EPA. However, today the most progressive corporations no longer view their environmental responsibilities from merely a regulatory compliance perspective, but more from that of cost efficiency and good public image.
- In recent years many Fortune 500 corporations have been leading the way in pollution prevention and waste minimization efforts, not because the EPA requires it but because they can realize significant gains in corporate profits. Measurable costs savings can be realized by efficient industrial processes and the reduction of wastes both at the source and at the end of pipe. To top this off, at a time when there is increasing environmental awareness and a tremendous growth in a global

environmental ethic, companies now acknowledge that they can realize enormous gains in public goodwill through their pollution prevention efforts.

- Pollution prevention ultimately can result in a combination of economic and competitive advantages that almost defies a business to ignore its benefits and still be left with the issues of regulatory environmental compliance and liability for its wastes.

QUESTIONS FOR REVIEW

1. Explain how pollution prevention is different from "end of pipe" methods of minimizing waste.
2. What is the primary goal of source reduction? Define the term and provide four examples of good operating practices and procedures that will support its primary goal.
3. How are preconsumer and post-consumer recycling programs different, and how does each help accomplish pollution prevention?
4. How does this text define "waste treatment" and what is the controversy regarding this pollution prevention technique?
5. Make a list of pollution prevention's benefits to industry.
6. How does the EPA rank the acceptable options of waste management according to the Pollution Prevention Act of 1990?
7. What are three primary strategies of the EPA to obtain voluntary compliance with pollution prevention by industry? Briefly describe each.
8. In what way do each of the following acts require industry to practice of pollution prevention techniques while compliance with the Pollution Prevention Act is largely voluntary?
   - Resource Conservation and Recovery Act
   - The Clean Water Act
   - The Emergency Planning and Community Right To Know Act, and
   - The Clean Air Act
9. Write a brief paragraph describing the difference between "waste minimization" and "pollution prevention."
10. Prepare a checklist of elements that should be included when identifying and tracking the types and amounts of wastes and their constituents, and in tabulating the costs of management and cleanup of those wastes.

ACTIVITIES

1. Develop a list of businesses in your community (or an imaginary one) where surplus items can be exchanged, sold, traded, or donated instead of being thrown away.
2. Crawl the World Wide Web in search of material exchange sites and develop a list of bookmarks. Prepare a general list and prepare another one for sites that specialize in the more common items of glass, rubber, plastic, aluminum, and building materials.
3. Crawl the World Wide Web and find a success story or an illustration of how a company improved its bottom line by implementing a pollution prevention program. Summarize this case history and prepare to take ten minutes to present it to a live audience.
4. Find a small- to medium-size company in your area that will allow you to conduct an activity to identify its pollution prevention opportunities.

2

General Manufacturing

Michele Feenstra

Upon completion of this chapter, you will be able to meet the following objectives:

- Have a general understanding of the manufacturing process.
- Understand environmentally conscious manufacturing and pollution prevention and why they are practiced.
- Calculate simple material balances.
- Have a general understanding of environmental regulations and the regulatory process.
- Understand the definitions of regulated materials.

INTRODUCTION

Industry leaders today are facing what could be their greatest challenge—to maintain economic growth during a period of increasing environmental concern and regulation. To meet this challenge, many manufacturers are learning to provide services and produce goods in a less wasteful, more efficient manner. In the process, many are discovering that what is good for the environment is also good for business, since more efficient production results in increased profit.

One way to avoid harm to human health and the environment is to stop releasing industrial byproducts and wastes. This goal is known as pollution prevention, or environmentally conscious manufacturing. Manufacturers can redesign products and processes, upgrade their equipment, and improve operating practices.

For example, managers and engineers can rethink their choice of raw materials and modify product packaging, purchase energy-efficient equipment, and implement procedures to minimize spills.

Environmentally conscious manufacturing is the result of many forces, most of which can be tracked straight to the bottom line. A large industrial firm could pay up to $450 million a year to comply with environmental regulations, including required pollution control and cleanup, according to a 1990 survey by Stanford University. As environmental regulations become increasingly stringent, environmentally proactive manufacturers will have a market advantage. Manufacturers are also becoming more environmentally conscious as a result of the SuperFund law, which states the generator of a waste is permanently liable for costs of cleaning up that waste or its consequences, even if the waste disposal practice was legal in the past. In addition, more and more manufacturers are being fined for noncompliance with environmental regulations, and they are having to pay for associated legal fees and cleanup costs. Finally, there is a growing consumer demand for goods and services that are environmentally benign, or "green."

Environmentally conscious manufacturing is often an indication of a quality enterprise, since a truly green company is unlikely to be poorly managed. At a minimum, these companies put the most senior person possible in charge of environmental policy and make this policy available to the public. The policy includes specific goals with targeted dates, regular environmental assessments, and a plan for measuring results. Following are some proactive manufacturers' success stories:

Corporate environmental policy. The 3M Company, which has pioneered corporate pollution prevention, introduced its "Pollution Prevention Pays" (3P) program in 1975. 3M reports that in more than 3,000 projects, it cut wastewater by 1 billion gallons, air pollutants by 120,000 tons, and solid waste by more than 400,000 tons. Savings as of 1993 exceeded $530 million.

Chemical substitution. Intertox America, a paper processor, developed a way to bleach paper using hydrogen peroxide, thereby eliminating its use of toxic chlorine bleaching agents. Intertox now consults with other paper processing companies to help them switch to the new process.

Cleaner production. When American Cyanamid modified the manufacturing process of a yellow dye, it was able to eliminate the use of a toxic solvent. Equipment modification that allowed recycling of a substitute solvent cost $100,000 and saved the company $200,000 a year in reduced disposal and energy costs.

Equipment change. Rhone-Poulenc spent $10,000 to install in-line condensers for its salicylaldehyde process to avoid product loss during the drying stage. Product yield rose 0.5 percent. The plant saved $30,000 in the first year.

Industry cooperatives. Northern Telecom (Canada) and AT&T joined forces in 1989 to found the Industry Cooperative for Ozone Layer Protection. The co-op's members provide information and technical support to companies trying to make the transition to substances free of ozone-depleting compounds.

GENERAL MANUFACTURING PROCESS DESCRIPTION

In reality, there is no such thing as a "general process" for manufacturing. Manufacturing includes making a whole range of products—metal items, wood or plastic furniture, chemicals, books, and newspapers to name a few. For every type of manufacturing, each company has a slightly different process for making its product. Nonetheless, common process steps used by many manufacturers can be described.

First, the raw materials required to make the finished product are acquired. Some manufacturers start from scratch, such as a glass manufacturer that uses sand and other chemicals as starting material. Some manufacturers purchase materials that are already prepared for processing, such as printers who buy large rolls of paper ready to go to press. In either case, incoming materials are the beginning of the manufacturing process.

Next, these raw materials may undergo preparation for the following primary process steps. Many manufacturers have quality-control programs that reject any incoming materials not meeting specified standards, such as material purity. Depending on the raw materials used and how sensitive the primary process steps are, cleaning or stripping operations may take place or a chemical coating may be applied. Some raw materials, such as wood or sheets of metal, may need to be machined or formed.

The primary process then takes these prepared materials and turns them into a finished product. Depending on what product is being made and which materials and equipment are used, a variety of operations may take place. Prepared pieces may be welded, glued, or joined by mechanical means. They may be painted or imprinted, electroplated or chemically transformed.

Finishing operations often include cleaning, polishing, surface coating, sizing, and packaging operations. Not all of these steps may be needed for the finished product to function correctly, but they are necessary to meet quality and other specifications demanded by the manufacturer and customers.

MATERIAL BALANCES

Environmentally conscious manufacturers must regularly perform environmental assessments. These assessments identify and track all materials being used in the manufacturing facility and provide the basis for establishing company environmental goals and priorities. In addition, environmental assessments often bring to light new information and opportunities in other areas important to a successful company, such as quality, inventory control, hidden costs, and purchasing.

The foundation of a successful environmental assessment is material balances. In its simplest form, a **material balance** shows that the amount by weight of all raw materials coming in must be equal to the total amount of material leaving a manufacturing facility. In other words, any raw material that does not end up in the finished product must be consumed during the process or emitted as byproduct or waste. For example, solvent evaporation from a centrifuge process can be estimated as the difference between feed put into the centrifuge and the solids and liquid discharged from the centrifuge (Figure 2–1).

FIGURE 2–1
Centrifuge process.

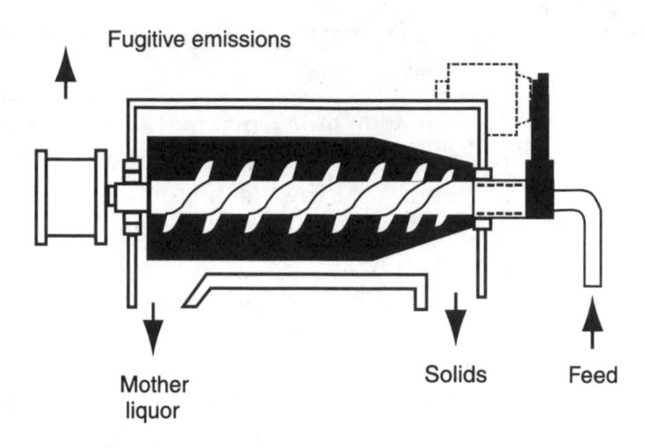

The first step in preparing a materials balance is to draw a process flow diagram, which is a visual means of organizing data on energy and materials flows, including material composition. This diagram shows the system boundaries and all streams entering and leaving the system, including points where wastes are emitted. Material balances are particularly useful if there are points in the production process where it is difficult or uneconomical to collect or analyze samples.

Material balances result in an organized system of accounting for the flow, generation, consumption, and accumulation of mass and energy in the manufacturing process. They are useful for organizing and extending environmental data and should be used wherever possible. However, the user should recognize that most balance diagrams will be incomplete, approximate, or both. Most manufacturers have numerous process streams and the exact composition of many streams is unknown and cannot be easily analyzed. Also, manufacturing operations or products may change frequently. Despite the limitations, material balances are essential to organize data, identify gaps, and estimate missing information.

HAZARDOUS WASTE STREAM IDENTIFICATION AND CHARACTERIZATION

While the amount of hazardous waste emitted by manufacturers may be very small compared to other emissions, these wastes represent a far more dangerous type of pollution. Toxic chemicals and materials in particular can be extremely hazardous to human health, plants and wildlife, or destructive to the environment if improperly managed. Therefore, Congress and its various entities have established several governing bodies and legislation to oversee environmental protection (see Figure 2–2).

Hazardous wastes are defined by the Environmental Protection Agency (EPA) under the authority of the Resource Conservation and Recovery Act (RCRA). To meet the definition of a hazardous waste, a material must first meet the definition for a solid waste. A **solid waste** is any material that is no longer used, including material being recycled or stored before disposal. Solid waste can be a solid, a liquid, semisolid, or even a contained gas.

Resource Conservation and Recovery Act (RCRA)

Definition of solid waste
Definition of hazardous waste
Generator requirements
Hazardous waste transportation requirements
Treatment, storage, and disposal facilities
- Land management units
- Incinerators
- Chemical, physical, and biotreatment units

Underground storage tanks
Household hazardous waste

Comprehensive Environmental Response, Compensation, and Liability Act (CERCLA)

SuperFund
Emergency Planning and Community Right-To-Know (SARA Title III)
The National Oil and Hazardous Substances Pollution Contingency Plan

Clean Air Act

National ambient air quality standards
Ground level ozone
Motor vehicles
Hazardous air pollutants
Accidental releases
Acid rain
Stratospheric ozone
Permits
State small business assistance programs

Clean Water Act

Water quality planning and management
Oil pollution
Industrial waste water
Storm water runoff
Safe Drinking Water Act
Underground injection
Waste water treatment sludge
Nonpoint source pollution
Marine sanitation devices

Occupational Safety and Health Act

Emergency plans and fire prevention plans
Flammable and combustible liquids

FIGURE 2–2
Major environmental legislation.

Hazard communication
Chemical exposure limits
Personal protective equipment (PPE)
Process safety management of highly hazardous chemicals
Hazardous waste operations and emergency response
Noise
Confined space
Lockout/tagout
Bloodborne pathogens
Ionizing radiation

The Marine Protection Research and Sanctuaries Act

Oil discharges
Marine sanitation devices
Benzene
Ozone depleting substances

Hazardous Materials Transportation Act

Hazard classes
Shipping requirements

Toxic Substance Control Act

Premanufacture notification
Manufacturers' requirements
Asbestos
Polychlorinated biphenyls

FIGURE 2–2
(continued)

A solid waste becomes a hazardous waste when it meets any one of the following conditions:

- It is on one of four EPA lists of hazardous wastes.
- It is a mixture of a solid waste and a hazardous waste on one of the lists.
- It is derived from treatment, storage, or disposal of a hazardous waste on one of the lists.
- It exhibits one or more of the four characteristics of a hazardous waste.

The burden of proving a waste stream is nonhazardous always falls on the manufacturer, also known as the waste generator.

Listed Wastes

The four EPA hazardous waste lists are keyed by specific letters—F, K, P, and U—which refer to types of waste regulated under RCRA regardless of the concentrations of chemical contamination. If a listed hazardous waste is mixed with nonhazardous solid waste, all of the waste may qualify as hazardous waste. Therefore, it is good practice to keep tight control of hazardous materials as a separate waste stream. In addition, segregation increases the potential for reuse, recycling, or treatment.

The F list is titled "Hazardous Waste from Non-Specific Sources" and includes generic industrial process wastes resulting from degreasing, solvent usage, electroplating, and heat treating, along with certain dioxin-contaminated wastes from the production of organic chemicals.

The K list, titled "Hazardous Waste from Specific Sources," is organized by industry and deals with commercial chemical products, manufacturing chemical intermediates, and off-specification species. Waste materials generated by listed industries must be compared to listing descriptions within the applicable industrial category.

The P list consists of generic chemical names and their Chemical Abstracts Service Registry Numbers. Titled "Discarded Acutely Toxic Commercial Chemical Products, Off-Specification Species, Container Residues, and Spills Thereof," it is a list of acutely hazardous materials. Acutely hazardous wastes have additional regulatory implications and any "empty" containers and liners of these materials must be triple rinsed or cleaned by an equivalent method.

The U list also consists of generic chemical names and their Chemical Abstracts Service Registry Numbers. It is called "Discarded Commercial Chemical Products, Off-Specification Species, Container Residues, and Spills Thereof."

Characteristic Wastes

Solid wastes not determined to be hazardous with respect to the F, K, P, and U listing descriptions must next be evaluated with respect to the four characteristic properties that qualify a waste as hazardous:

- Ignitability
- Corrosivity
- Reactivity
- Toxicity

An ignitable waste presents a fire hazard during routine management. This includes liquids with a flashpoint less than 140° F; solids capable of causing fire through friction, absorption of moisture, or spontaneous chemical changes; and ignitable compressed gases and oxidizers.

A corrosive waste is able to deteriorate standard containers, damage human tissue, and/or dissolve toxic components of other wastes. Corrosive wastes have a pH of less than 2 or greater than 12.5, or can corrode SAE 1020 steel at a rate greater than 0.25 inches per year at 130°F.

A reactive waste has a tendency to become chemically unstable. Reactive wastes may explode, generate pressure during handling, generate toxic gases, or react violently when exposed to air or water.

Toxic waste has the potential to poison or otherwise harm humans, fish, or wildlife. Toxic wastes are identified by using the Toxicity Characteristic Leaching Procedure (TCLP) test, designed to identify wastes likely to leach relatively large concentrations. These include heavy metals, pesticides, and solvents.

REGULATORY IMPACT

Environmental managers are faced with the overwhelmingly difficult task of complying with a myriad of regulations administered by several different agencies, with no sign of reprieve. The Clean Air Act Amendments of 1990 is the most complex and lengthy piece of environmental legislation to date, and future environmental legislation is expected to be similar.

Federal regulatory agencies are charged with interpreting the intent of laws passed by Congress and developing appropriate regulations to meet those intentions. Many times, state agencies choose to implement the regulatory program and develop state regulations that are as stringent, or more stringent, than federal regulations. Local agencies may also have additional requirements for their jurisdictions, and enforcement officers may need to make initial determinations of compliance status in certain situations. In addition, new regulations are including language that allows private citizens to call in reports of suspected violations.

Regulations can be overlapping, contradictory, and difficult to understand. The same chemical is often regulated by more than one agency under several different regulations. However, ignorance or misunderstanding of these regulations, which results in noncompliance, is not acceptable. Noncompliance is being met with increasingly more stringent enforcement, ranging from monetary fines to criminal charges. Penalties may be as much as $50,000 per day or imprisonment for up to fifteen years, or both, per violation. Fines are determined usually after consideration of three elements: seriousness of the violation, economic benefits gained from noncompliance, and history of the offender.

In addition to penalties for instances of noncompliance outlined previously, companies are also potentially liable for all cleanup costs and damages to natural resources from wastes, particularly those classified as hazardous. Liability under the Comprehensive Environmental Response, Compensation, and Liability Act (CERCLA), also known as SuperFund, is both strict and retroactive. This means it does not matter how carefully the waste generator acted or if the generator took all reasonable precautions against the release. It also does not matter if the generator complied with all laws in effect at the time of disposal.

Compliance with environmental regulations can be overwhelming, but it is necessary when compared to penalties associated with noncompliance, bad publicity that may ensue, and the fact that, once caught, a company will be required to come into compliance anyway. Also, compliance is the first step to being proactive, which saves everyone time, energy, and money. Many facilities are beginning to expand their compliance programs and look at pollution prevention or source reduction and hazardous waste minimization plans as key components in their ongoing compliance programs. These plans focus on eliminating or reducing use of materials and waste streams, which reduces potential liability and can sometimes decrease a company's regulatory burden by allowing it to drop down into a less-regulated class.

Regulated Materials

The federal government's approach to regulating hazardous materials is based on a history of a phase-specific direction. Hazardous contaminants entering the environment are regulated according to whether they are emitted to the air or to water, are utilized in food, or are involved with constituencies such as the agricultural industry. For example, the CWA regulates emissions to surface waters, the CAA emissions to the outside atmosphere. This approach has worked for several decades, particularly following World War II, but is currently under evaluation.

It has become apparent that regulations that are "phase-specific" do not integrate well with business practices and may not accomplish the objective of protecting public health, safety, and the environment. It is often difficult to determine for a particular situation all the applicable laws and regulating agencies. For example, volatile organic compounds (VOCs) can be regulated by the EPA, OSHA, the National Fire Protection Agency (NFPA), and the local water-treatment facility under multiple laws and regulations. Usually, the phase or situation with the least stringent regulation becomes the ultimate receptor of pollutants. Also, many situations fall under their own specific regulation, such as asbestos and PCBs, which are regulated under the Toxic Substances and Control Act (TSCA), and oil production facilities that must comply with Spill Prevention Control and Countermeasure (SPCC) plans.

In summary, it cannot be overemphasized that most materials are not regulated by a single agency or regulation and that the specific situation must be known to determine all applicable requirements. Major classifications of regulated materials are described below, and major federal legislation is noted when warranted. A brief outline of the major environmental regulations is also provided at the end of this chapter.

Air

An **air pollutant** is an airborne gas, aerosol, or particulate that occurs in concentrations that may threaten the well-being of organisms or disrupt the natural processes upon which they depend. Many regulated air pollutants are natural constituents of the atmosphere that are hazardous at large concentrations. Other air pollutants may be hazardous in any concentration.

Regulated air pollution can be broken into six major constituents—suspended particles, smog, sulfur dioxide, carbon monoxide, toxic air pollutants, and ozone depleting substances. In general, air pollutants emitted from a facility to the outside atmosphere are regulated by the EPA under the CAA. Air emissions within a facility are primarily regulated by OSHA.

Suspended particulates are actual pieces of ash, smoke, soot, dust, or liquid droplets released to the air by the burning of fuel, industrial processes, agricultural practices, or a number of natural processes. The smallest of these particles remain suspended in the air and, if inhaled, can lodge in the lungs and contribute to respiratory disease. Regulations focus on particles 10 microns in size, since these are most likely to get trapped in the lungs and create health problems.

Smog, or ground-level ozone, is the single-most serious air quality challenge for most urban areas and, in many instances, falls under local regulation in addition to state and federal regulation where it is a problem. Ground-level ozone forms when nitrogen oxides (NO_x) and VOCs react in the presence of sunlight and heat. Ozone, because of its reactivity, is harmful to most living organisms and can cause eye irritation and respiratory problems in humans. Regulations focus on limiting NO_x and VOC emissions.

Sulfur dioxide (SO_2) is a gas released when sulfur-containing fuels such as coal and oil are burned. Sulfur dioxide can cause respiratory problems and may also be converted to atmospheric acid, a component of acid rain. Some scientists theorize acid rain contributes to forest damage and that over time acid rain could increase the acidity of lakes, streams, and soils, which could change local ecosystems. Regulations focus on limiting sulfur dioxide emissions from coal-burning utility plants.

Carbon monoxide (CO) is a product of incomplete combustion. A variety of health effects have been associated with CO. Once inhaled, it interferes with the body's ability to absorb oxygen and causes drowsiness, headache, and even death, especially in those suffering from heart disease or respiratory problems. Regulations focus on fuel-combustion processes, including emissions from utility plants, automobiles, and incinerators.

The volume of **toxic air pollutants** emitted is very small compared to other forms of air pollution, but the potential impact on human health and the environment is a cause for concern. Toxics are chemicals that are known to or suspected of causing cancer or other serious health effects, including damage to the respiratory or nervous systems, birth defects, and reproductive effects. Some can cause death or serious injury if accidentally released in large amounts. Regulations focus on specific chemicals listed by the EPA and the Occupational Health and Safety Administration (OSHA). In addition, specific industries and practices or processes are often targeted. The Clean Air Act Amendments of 1990 extensively regulate toxic and other hazardous air pollutants.

Ozone depleting substances, commonly referred to as **chlorofluorocarbons** (CFCs), are regulated because of their potential to deplete the stratospheric ozone layer 12 to 22 miles above the earth's surface. This layer of gas screens the earth from the sun's powerful ultraviolet (UV-B) radiation. Increased UV-B radiation could, over the long run, cause a rise in cases of skin cancer and cataracts and damage important food crops and marine ecosystems. Under the Montreal Protocol, the United States and over 110 nations agreed to phase out production of CFCs by January 1, 1996. In addition, the Clean Air Act Amendments of 1990 require recycling and ban the release of refrigerants during the service, maintenance, and disposal of air conditioning and refrigeration equipment.

Water

Water pollution can be divided into seven areas—pathogens, oxygen-demanding wastes, plant nutrients, sediments, toxics, petroleum, and thermal pollution. Which agency or law regulates each of these is dependent on a variety of factors, including where the pollution is being discharged to and the intended final use of the water being discharged to. Discharges to inland surface waters fall under the CWA, with stormwater discharges falling under separate regulation. Groundwater discharges are regulated by RCRA, marine discharges under the Marine Protection, Research and Sanctuaries Act, drinking water under the Safe Drinking Water Act (SDWA), etc.

Pathogens, which include viruses, bacteria, protozoa, and parasitic worms, are primarily regulated under the SDWA and can cause such diseases as cholera, dysentery, gastroenteritis, hepatitis, typhoid fever, and other serious illnesses. Pathogens enter the water mainly through the feces and urine of infected people and animals. Regulations focus on measuring total coliform bacteria. Although coliform bacteria are not always harmful in themselves, they can indicate that a water system has been contaminated. Systems that exceed the standard more than 5 percent of the time may be declared in violation.

Oxygen-demanding wastes are controlled because of their potential to harm fish and other aquatic life, which cannot survive without oxygen for cellular respiration. Dissolved oxygen is removed from water primarily by organic waste decomposers. In most natural ecosystems, the quantity of organic waste is small; therefore, the amount of dissolved oxygen that decomposers use is also small. Industrial and municipal wastewater, however, typically contains high concentrations of organic wastes. This spurs the growth of decomposer populations, which consume large quantities of dissolved oxygen. Regulations focus on treating wastes with high biochemical oxygen demand (BOD), or the amount of dissolved oxygen that decomposers require to break down the material in a given volume of water.

Plant nutrients, primarily nitrogen and phosphates, are present in small amounts in natural water. High nitrate levels are a serious threat to children between six months and one year of age. These chemicals can react with oxygen-carrying hemoglobin in the blood to cause an anemic condition known as **blue baby syndrome.** In general, larger concentrations of plant nutrients accelerate the natural process of eutrophication. This process can cause dense growth of aquatic plants, blue-green algae blooms, and changes in the makeup of fish and aquatic life populations. Regulations focus on control of phosphate and nitrate levels in drinking water and discharge into treatment facilities. Nonpoint sources of plant nutrients, such as agricultural runoff, remain largely uncontrolled.

Sediments are particles of soil, sand, and minerals that are washed from the land into the water. Large amounts of sediments can become a problem because they may eventually fill stream channels, harbors, and reservoirs. As reservoirs behind dams fill with sediment, the capacity to generate electricity diminishes greatly and the reservoirs are less suitable for recreation. Sediments also impede navigation, cover bottom-dwelling organisms, eliminate fish-spawning grounds, and reduce the light penetration necessary for photosynthesis in aquatic plants. In addition, soils that are eroded from farmlands carry adhered nutrients. Thus, most lakes and reservoirs that receive heavy sediment loads are usually eutrophic.

Toxics are those substances posing a direct threat to fish, other aquatic life, and humans. Accidental discharge of highly toxic wastes may result in massive die-offs of aquatic organisms. Dumping these chemicals into sewage treatment plants can eliminate the microbial cultures that are essential in sewage treatment. Regulations focus on specific chemicals listed by the EPA.

Petroleum, or oil, is a threat to surface water quality, especially where oil is extracted or transported. Oil contains a large insoluble fraction, which when spilled on water gradually spreads and thins to form an ever-widening oil slick on the water's surface, the ecological consequences of which are not easy to assess. Near-shore oil spills that do not disperse rapidly are particularly damaging because they affect many vulnerable marine organisms, including aquatic birds and animals. Regulations focus on oil spill prevention and emergency response planning, with facilities engaged in oil production and transport required to prepare a SPCC plan.

Thermal pollution, or the raising of water temperature to levels that harm aquatic organisms, is primarily associated with the large-scale generation of electricity. Heat causes an increased oxygen demand for respiration by aquatic organisms such as fish, oysters, and clams, and at the same time reduces the capacity of water to absorb oxygen. This causes the organisms to suffer oxygen stress and can increase BOD levels. Regulations focus on controlling water discharge temperature.

Land

RCRA is the primary law implementing regulations that protect the land. Requirements focus on protection of groundwater and limiting what wastes can go into landfills, and landfill construction and closing specifications.

Pollutants are divided into nonhazardous and hazardous wastes. Nonhazardous wastes include paper and wood, yard wastes, glass, etc. Most of these wastes end up in a sanitary landfill. The remainder is either incinerated or recycled. Hazardous wastes include flammable, reactive, toxic, and corrosive materials. Materials are regulated either because of their potential for immediate threat to human health and the environment, or because of the waste's potential persistence in the environment.

Radioactive wastes present a unique hazard, and are managed separately by the Nuclear Regulatory Commission.

Regulatory Management

Regulatory management is often the most burdensome part of environmental, health, and safety programs. It is estimated that as much as 80 percent of a program's costs stems from regulatory (not technical) requirements, such as training and time spent on reporting requirements.

Unfortunately, there is not a single management standard that can be used to meet the environmental regulations, for each regulatory agency has its own requirements. Since many regulatory requirements are overlapping, one way to approach regulatory management is to divide requirements into four basic areas—accidental releases or emergency situations, routine use and emissions, reporting, and training requirements. OSHA regulations are a good place to start, particularly with regard to training.

Accidental Releases

Emergency action plans are required under various OSHA and EPA standards, particularly for hazardous materials. These written plans should address emergencies that may reasonably be expected, such as fires, toxic chemical releases, and natural disasters. Prevention plans to eliminate or reduce the possibility of emergency situations may also be required. These plans include installation, location, and maintenance schedules for alarms, communication systems, control, emergency, decontamination, and personal protective equipment. Names and phone numbers of emergency coordinators and local agencies may also be required. In some cases, detailed contingency plans containing a preformulated set of responses to a given emergency may be required.

Routine Use and Emissions

Monitoring of routinely used substances and suspected problem areas within a facility is required by OSHA for over twenty toxic or hazardous substances and over 400 toxic air contaminants. Emissions from a facility into the outside environment may require monitoring by the EPA. For example, continuous emissions monitoring is required for certain pollutants regulated under the CAA.

Exactly what type of monitoring is required is dependent on many factors, including the chemical and area being monitored. Monitoring of these materials is used to determine if further action is needed, such as engineering controls or personal protective equipment.

Routine emissions may also need to be tested, usually for specific listed or characteristic materials. This type of testing is used to determine if control equipment is needed or how to properly dispose of waste materials.

Transportation of hazardous materials is regulated by the Department of Transportation (DOT) under the Hazardous Materials Transportation Act (HMTA). Regulations require that these materials be properly classified, described, packaged, marked, labeled, and manifested.

Reporting

All environmental regulations have reporting and/or recordkeeping requirements, particularly for the industrial community and for hazardous materials. Reports and similar documents must be kept on file for a certain period of time, which varies according to regulation. Reporting is required to show compliance and provide information required under right-to-know laws, such as the Emergency Planning and Community Right-To-Know Act (EPCRA).

Training

OSHA requires hazard communication training for all employees in workplaces where hazardous materials are in use. In addition, environmental regulations may require specific training related to that regulation, such as CFC-certification training for refrigeration technicians. Training requirements range from awareness training to training for specific duties, and therefore the curriculum required varies from general to detailed information and specific topics that must be included. Some training must be given by certified instructors, while other training can be given by any "qualified" individual. It is important to focus on the intent of the training when complying with individual regulations.

FEDERAL REGISTER AND THE CODE OF FEDERAL REGULATIONS

All federal regulations, proposed regulations, and notices are published in the ***Federal Register.*** Printed every working day, the *Federal Register* serves as a source of information regarding current activities of all federal agencies.

New regulations and amendments to existing regulations published in the *Federal Register* are codified annually in the Code of Federal Regulations (CFR). The CFR is divided into 50 titles by federal agency. For example, Title 40 of the CFR is where EPA regulations are found, Title 29 deals with OSHA regulations, and Title 49 with DOT regulations. Each title is divided into chapters, parts, subparts, and sections. A part consists of a unified body of regulations devoted to a specific subject.

TREATMENT AND CONTROL TECHNOLOGIES

Starting with RCRA in 1976, Congress began passing environmental legislation at a rapid pace. RCRA gave the EPA responsibility for regulating the disposal of industrial hazardous waste. Since 10 percent of the nation's businesses generated over 95 percent of the waste, the EPA decided to begin by concentrating on these large entities. It developed standards, promulgated regulations and enforced the law using a command and control approach with an emphasis on end-of-pipe emissions.

To meet these environmental standards, manufacturers have developed a number of treatment and control technologies that protect human health and the environment. Treatment may also be needed to meet quality requirements (for example, secondary treatment of drinking water to meet aesthetic standards for taste, color, and odor).

TABLE 2–1
Examples of common treatment processes.

Physical Treatment	Chemical Treatment	Biological Treatment
Waste concentration	Solvent degreasing	Wastewater treatment
Distillation	Neutralization	Bioremediation
Dewatering	Incineration	

Treatment processes can be divided into three types—physical treatment, chemical treatment, and biological treatment (Table 2–1). Physical treatment processes change the state or structure of a material without changing its chemistry (for example, concentrating a sludge by removing excess solvent through evaporation or with a filter press). Chemical treatment, such as neutralization or incineration, actually changes the chemistry of the material through displacement, reduction, oxidation, etc. Biological treatment also changes the material, but relies on biological organisms and mechanisms rather than chemicals. The microbes react with or "eat" the material and store it or change it into a less hazardous form.

Quality Assurance and Quality Control

Quality assurance programs often reflect how well a company is doing environmentally, and they can be used as yet another "control technology." Since these programs focus on setting quality standards and making sure these standards are met as cost effectively as possible using quality control, quality assurance also helps guarantee that manufacturing processes are efficient. In other words, the best product is produced using the fewest resources possible.

By making environmental standards part of the quality assurance program, manufacturers can expect to reap advantages over competitors who lack such vision. In addition, environmental quality control is a useful tool for compliance since actual performance is compared to environmental standards. This allows the manufacturer to act on any differences as needed.

The Ford Motor Company believes quality excellence can best be achieved by preventing problems rather than by detecting and correcting them after they occur. Ford has developed a management model for its internal pollution prevention program that utilizes Total Quality Management (TQM) principles. Using this model, the Sheldon Road Plant in Plymouth, Michigan, successfully replaced its trichloroethylene (TCE) vapor degreasing system with an aqueous degreasing system, resulting in a superior process, lower costs, improved plant environment, and reduced environmental impact. Ford found that TQM and pollution prevention are complementary principles.

POLLUTION PREVENTION

Pollution control has long relied on familiar end-of-pipe technologies such as scrubbers, incinerators, and filters to control or reduce the toxicity of pollutants. This traditional command-and-control approach has resulted in significant environmental progress. Recently, however, we have begun to understand that even strong regulation and vigorous enforcement of end-of-pipe requirements do not effectively solve all environmental problems.

The result of this understanding is a radical but very sensible new approach, known as pollution prevention (P_2) or source reduction. P_2 relies on more efficient processes that reduce wastes while producing more product per unit of raw material. As can be seen in the Ford Motor Company example given earlier, P_2 can be a very useful tool in accomplishing manufacturing excellence. Increasingly, environmentally conscious manufacturers are making pollution prevention, or source reduction, their first choice in managing environmental issues as opposed to recycling or treatment.

Source Reduction

Source reduction includes product, material, process, and operational changes that eliminate or reduce sources of pollution.

Product and material changes include changes in material composition and substitution of intermediate and end products to reduce waste. Examples include switching to water-based solvents for cleaning and using lower VOC paints. When substitutions of hazardous materials cannot be made, materials often can be purified to reduce toxicity.

Process changes include changing process layout, increasing automation, and modifying process equipment. Process changes may include minimizing product handling, redesigning equipment and piping to reduce the volume of material in the system, and converting from batch to continuous operations. Also, adding drip pans and splash guards to equipment and installing speed controls on motors have been used to reduce and eliminate pollution sources.

Operational changes include inventory control, waste segregation, and employee training. Examples of operational changes include using one material for as many jobs as possible, buying only as much material as is needed, separating hazardous from nonhazardous waste, and actively involving employees in pollution prevention. Typically, operational changes are easily implemented and less expensive than product and process changes.

Pollution Prevention Barriers

Although there are strong incentives for setting up a pollution prevention program, many barriers must be identified and overcome. Some pollution prevention projects have large start-up costs, including equipment addition and replacement, employee training, and alternative raw material purchases. These initial costs may overwhelm management and doom a project from the beginning. Care should be taken to focus on the long-term positive economic impact a pollution prevention program brings a company.

Technical barriers may also stand in the way. Potential barriers include limited availability of information on P_2 techniques, especially for small- and medium-sized companies, limited flexibility in manufacturing, and rigorous product quality and customer specifications. Companies can overcome the information barrier by contacting state and federal regulatory agencies for assistance, encouraging employees to watch for P_2 information in technical journals and newspapers, and talking with other professionals in the technical and environmental field. Companies can also hire a consultant to work on specific pollution prevention projects.

Many companies have limited flexibility in their manufacturing process. Therefore, implementing pollution prevention measures within the process may cause slow-downs or complete shut-downs in production. The flexibility barrier may be overcome by involving design and production personnel in the initial stages of setting up pollution prevention projects. Additionally, implementing proven technologies and starting with pilot operations will help overcome the flexibility barrier.

Product quality and customer specifications are always a major concern in the manufacturing process. While many P_2 techniques improve quality, some may harm product quality and not meet customer specifications. To avoid this, a company must verify customer expectations and test product and process quality before establishing prevention projects.

Finally, state and federal regulations may hinder or complicate some pollution prevention programs. For example, changing the feed to a less hazardous material may require changing existing facility permits. By consulting with state and federal agencies during the initial pollution prevention design stages, this barrier may be lessened. Additionally, working with regulatory agencies in the long-range planning of the plant design may overcome this obstacle.

ENVIRONMENTAL HEALTH AND SAFETY ISSUES

OSHA sets safety and health standards in the workplace. The law requires employers to provide a place of employment free of hazards that can cause death or serious physical harm, and deals with issues ranging from general occupational safety to hazardous materials, noise, and energy. The goal of an effective environmental, health, and safety program is to protect workers from health hazards and ensure that people are capable of performing their jobs efficiently and safely. Some of the issues covered by OSHA are explained here.

Emergency Plans

Written emergency action plans are required under many OSHA standards. These plans address emergencies that reasonably may be expected in the workplace—for instance fires and toxic chemical releases. The plans must include reporting procedures, emergency escape routes and procedures, and lists of equipment or systems to be used during an emergency. OSHA requires emergency action plans to contain the following elements, at a minimum:

- Procedures for reporting fires and other emergencies
- Emergency escape procedures and routes
- Employee procedures for those who remain to operate critical plant operations until their evacuation becomes absolutely necessary
- Procedures to account for all employees after emergency evacuation has been completed
- Rescue and medical duties to be performed and by whom
- Contacts for further information or explanation of the plan

Hazard Communication

Chemical manufacturers and importers must determine the hazards of each product and convey this information to employers by means of labels on containers and material safety data sheets (MSDS).

All employers are required to have a written hazard communication (Right-To-Know) program. This program must include identification and listing of all hazardous chemicals at the workplace, container labeling, MSDSs, and employee training. OSHA also requires employers to maintain records, track diseases, and set standards for regulating substances that may cause cancer.

Chemical Exposure Limits

OSHA has established permissible exposure limits (PELs) for over twenty toxic or hazardous substances and over 400 toxic air contaminants. Exposure limits for extremely hazardous substances are given as ceiling values, and they may not be exceeded at any time. Other substances have limits based on a 40-hour-week period of exposure and must not be exceeded during an eight-hour work shift. OSHA also sets exposure limits for PCBs, asbestos, and ionizing radiation under separate regulation.

Because regulatory limits may not be changed without new legislation, many manufacturers also use exposure limit recommendations set by the American Conference of Government Industrial Hygienists (ACGIH) when determining what actions to take in protecting employee health and safety. This is why one often finds both OSHA and ACGIH permissible exposure limits given on a MSDS. General industrial practice is to adhere to the more stringent limit.

Personal Protective Equipment

Manufacturers must provide their employees with personal protective equipment, such as gloves and respirators, wherever necessary and train them in how to use and properly maintain this equipment. Respirator fit testing and training is particularly involved. Training in the proper use of this equipment is the most important activity of an environmental, health, and safety program because ultimately safety depends on the individual worker.

Process Safety Management of Highly Hazardous Chemicals

Safe operation of chemical processes requires a thorough analysis of safety hazards and for this reason OSHA requires manufacturers to compile written safety information on the chemical hazards, technology, and equipment of processes and then perform a process hazard analysis. For simple operations, a chemical hazard survey with workable plans to avoid fire or accident may be sufficient. For more complex operations, a detailed analysis must identify, evaluate, and control all highly hazardous chemicals, and written operating procedures and emergency action plans must be developed (see Figure 2–3).

Hazardous Waste Operations and Emergency Response

Employers involved in hazardous waste cleanup, hazardous waste treatment, storage, and disposal, and hazardous materials emergency response operations

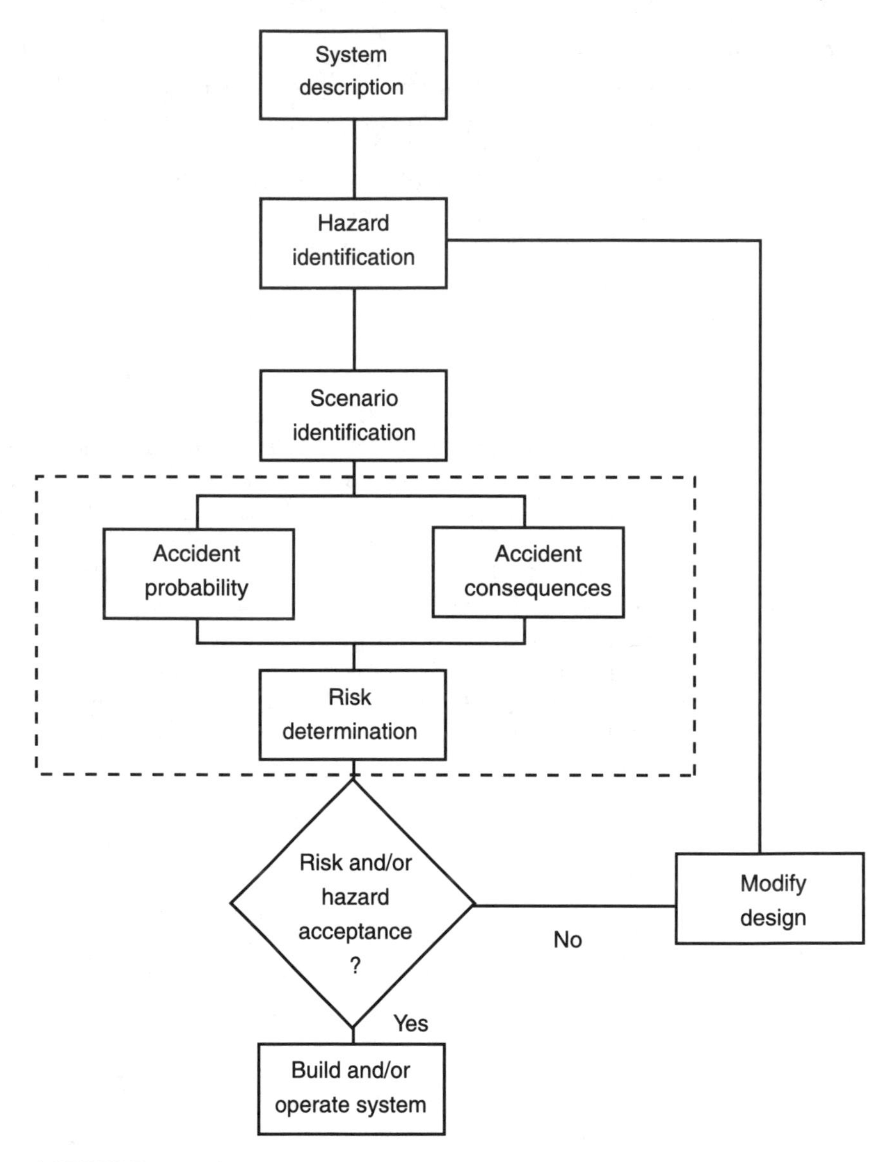

FIGURE 2–3
Preliminary hazard review decision diagram. Adapted from *Guidelines for Hazard Evaluation Procedures,* AICHE, 1987, p. 1–9.

must have a written safety and health program. This program should identify specific hazards and determine appropriate safety and health control procedures needed to protect employees and control exposure to hazardous substances. Employers must also institute a medical surveillance program and engineering controls, work practices, and personal protective equipment for employee protection.

All employees must receive training before engaging in operations that could expose them to hazardous substances, safety, or health hazards. Annual refresher training is also required. Training is given at five levels, based on employee duties and function.

SUMMARY

- Environmental regulations are becoming increasingly stringent and complex. Emissions to the air, land, and water as well as energy use are regulated at all points of the manufacturing process. Generally speaking, emissions within a manufacturing facility are regulated by OSHA and emissions to the environment outside of a manufacturing facility are regulated by the EPA.
- Manufacturers are practicing pollution prevention and environmentally conscious manufacturing to alleviate the burden of environmental regulations and to increase manufacturing efficiency. Both practices focus on reducing risks to human health and the environment in a manner that allows the manufacturer to remain competitive.

QUESTIONS FOR REVIEW

1. The chemical methyl ethyl ketone is regulated by several different agencies under different laws. Under which law and agency is it regulated in each of the following instances?
   - As a degreasing solvent used in a closed-loop degreasing tank.
   - As a degreasing solvent that is vented to the outside of the facility.
   - As a paint solvent in which overspray is captured by solid filters sent to the local landfill.
   - As a paint solvent in which the paint booth uses a waterwash system to capture overspray.
   - As a solvent stored on-site.
2. How is environmentally conscious manufacturing different than pollution prevention? How is it the same?
3. A mid-sized manufacturer of metal parts uses a traditional wet spray paint to paint all finished pieces and is considering changing to an electrostatic powder coating. Draw material balance diagrams for each process.
   - Based on the material balance diagrams alone, which process do you think the manufacturer should use?
   - Adding in estimated costs for capital equipment and materials, which process should be used?
4. What is the difference between a hazardous waste, a hazardous material, and a hazardous substance? A listed waste and characteristic waste?
5. Which environmental law has retroactive liability and what does this mean?
6. What are the advantages and limitations of biological treatment?
7. What relationship is there between quality assurance/quality control and waste generation?
8. Give an example of source reduction, as opposed to recycling or treatment.
9. What is the key element of a good environmental, health, and safety program?
10. Where are federal regulations first published? At what point(s) in the rule-making process does a manufacturer have input?

ACTIVITIES

1. Draw a material balance diagram for a real process. To do this, you can visit a local manufacturer, draw it for a manufacturing process familiar to you, or use an example from your daily life such as making coffee. Whatever you use, be sure to set appropriate boundaries and label all material streams with their direction, concentration, and rate of generation.
2. Pretend you have just won the lottery and are the proud new owner of a small manufacturing business of your choice. Using any method you can think of, get as much information as you

can regarding environmental issues and regulations that affect your business and summarize it. At the end of this summary, describe your information-gathering experience.

3. For the same business used in Activity 2, list all material streams you can think of. Next to each list, rate each material as to its importance on a scale of 1 to 10 (1 being not very important, 10 being extremely important) with regard to quantity, toxicity, regulatory burden, and other concerns such as company policy. Total your ratings for each listed material. Table 2–2 shows an example.
Come up with possible alternatives for the three materials having the highest totals. Using the same system, how do these alternatives compare?

TABLE 2–2

Material	Quantity	Toxicity	Regulations	Other	Total
Paper	10	1	1	8	20
Inks	2	7	8	8	25
Blanket cleaner	6	7	9	6	28
Roller cleaner	4	7	9	4	24

REFERENCES

Center for Chemical Process Safety of the American Institute of Chemical Engineers. 1995. *Guidelines for Process Safety Fundamentals in General Plant Operations.*

Chatfield-Taylor, R.F. 1992. *Environmental Compliance: Negotiating the Regulatory Maze.*

Enviroene at http://es.inel.gov/index.html

Epstein, S.S., M.D., L.O. Brown, and C. Pope. 1982. *Hazardous Waste in America.* San Francisco: Sierra Club Books.

Griffin, R.D. 1988. *Principles of Hazardous Materials Management,* Chelsea, MI: Lewis Publishers, Inc.

James M. Montgomery Consulting Engineers. 1985. *Water Treatment Principles and Design.* New York: John Wiley and Sons.

Lindgren, G.F. 1989. *Managing Industrial Hazardous Waste,* Lewis Publishers.

Phifer, R.W., and W.R. McTigue Jr. 1988. *Handbook of Hazardous Waste Management for Small Quantity Generators.* Chelsea, MI: Lewis Publishers.

Pollution Prevention Institute. 1995. “Compliance with Environmental Regulations–Selected Summaries.” Manhattan, KS: Kansas State University.

Pollution Prevention Institute. 1995. “Pollution Prevention.” Manhattan, KS: Kansas State University.

U.S. Environmental Protection Agency. 1991. “Pollution Prevention 1991: Progress on Reducing Industrial Pollutants.”

U.S. Environmental Protection Agency. 1992. *Facility Pollution Prevention Guide* (EPA/600/R-92/088).

World Resources Institute. 1993. *The 1993 Information Please Environmental Almanac.* New York: Houghton Mifflin Company.

3

Metal-Finishing Wastes

William Christie

Upon completion of this chapter, you will be able to meet the following objectives:

- Describe metal-finishing manufacturing general processes.
- Describe the materials balance of a metal-finishing facility.
- Know the hazardous waste streams generated by metal-finishing operations.
- Understand the current environmental regulatory impact on the industrial category.
- Describe common treatment and control technologies for metal-finishing facilities.
- Discuss available pollution prevention technology option.
- Understand the environmental, health, and safety issues inherent to facility operations.

INTRODUCTION

Metal finishing as an industrial category is broad and varied and can be addressed by any one of several approaches. The U.S. Environmental Protection Agency (EPA) currently defines thirteen separate metal industry categories, one of which is metal finishing. These thirteen categories fall into three basic manufacturing areas. Table 3–1 provides a summary of the EPA metal industry areas and categories.

TABLE 3–1
Summary of the metals industry.

Manufacturing Area	Industry Category
Metal and metal alloy manufacturing	Iron and steel manufacturing Nonferrous metals manufacturing Ferroalloy manufacturing
Metal forming	Iron and steel manufacturing Metal molding and casting Aluminum forming Copper forming Nonferrous metals forming and metal powders
Component finishing	Electroplating Iron and steel manufacturing Metal finishing Battery manufacturing Coil coating Porcelain enameling Electrical and electronic component manufacturing

As Table 3–1 indicates, metal finishing as an industrial category falls into the third area—component finishing. Separate sections in this chapter are dedicated to other related manufacturing areas and industrial categories (i.e., primary metals smelting, electroplating, and printed circuitboard production). The discussion in this section emphasizes the metal-finishing industry as a separate and distinct point source category as determined by the EPA and its effluent limitations guidelines and standards. Another industry, metal products and machinery (MP&M), will also be discussed to a limited degree. At the time of this writing, the EPA defined MP&M as a new industrial point source category and issued proposed technology-based effluent limitations guidelines for new and existing direct dischargers and pretreatment standards for new and existing indirect dischargers.

GENERAL PROCESS DESCRIPTIONS

The metal-finishing industry as defined by the EPA has six basic operations comprising forty individual process operations. These six basic operations are:

1. anodizing
2. electroless plating
3. chemical etching and milling
4. electroplating
5. coating
6. printed circuit board manufacturing

TABLE 3–2
Metal-finishing industry process operations.

Abrasive blasting	Flame spraying	Salt bath descaling
Abrasive jet machining	Grinding	Shearing
Assembly	Heat treating	Sintering
Brazing	Hot dip coating	Soldering
Burnishing	Impact deformation	Solvent degreasing
Calibration	Laminating	Sputtering
Cleaning	Laser beam machining	Testing
Coating stripping	Machining	Thermal cutting
Electric discharge machining	Mechanical plating	Thermal infusion
Electrochemical machining	Painting	Tumbling
Electron beam machining	Plasma arc machining	Ultrasonic machining
Electropainting	Polishing	Vacuum metalizing
Electrostatic painting	Pressure deformation	Vapor plating
		Welding

Table 3–2 itemizes, in alphabetic order, the forty individual process operations that comprise the six basic operations of the metal-finishing industry.

This EPA categorization scheme of six basic operations and forty individual process operations is the manner by which water quality regulations and effluent limitations guidelines are applied against those metal finishing industrial sites that discharge process wastewaters.

This section focuses on only four of the six basic metal-finishing operations: anodizing, chemical etching and milling, coating, and electroless plating. The other two EPA basic operations, electroplating and printed circuitboard manufacturing, are discussed in separate sections in this chapter.

Perhaps a more logical way to discuss this industrial category is to combine and rearrange the EPA basic and process operations into six unit operations with the individual process operations grouped accordingly. These six unit operations are:

1. Form-Cut-Join
2. Machining
3. Cleaning
4. Finishing
5. Coating
6. Miscellaneous

This approach allows a functional flow that follows the industrial processes that typically occur in the metal-finishing industry. With this in mind, the industry's basic and process operations are regrouped into these six unit operations. Table 3–3 details the six unit operations and the corresponding process operations.

The newly proposed MP&M industry category has basic unit operations and individual process operations that are very similar to the metal-finishing industry. MP&M basic unit operations include metal shaping, surface preparation, organic deposition, surface finishing, and assembly operations.

TABLE 3–3
Metal-finishing industry unit operations and process operations.

Form-Cut-Join	Machining	Cleaning
Abrasive jet machining	Abrasive blasting	Coating stripping
Brazing	Electrical disch. machining	Salt bath descaling
Impact deformation	Electrochem. machining	Solvent degreasing
Laminating	Electron beam machining	
Pressure deformation	Grinding	
Shearing	Laser beam machining	
Sintering	Plasma arc machining	
Soldering	Ultrasonic machining	
Thermal cutting		
Welding		
Finishing	**Coating**	**Miscellaneous**
Burnishing	Anodizing*	Assembly
Chemical etching/milling*	Coating*	Calibration
Polishing	Electroless plating*	Heat treating
Tumbling	Electropainting	Printed circuit board*
	Electroplating*	Testing
	Electrostatic painting	
	Flame spraying	
	Hot dip coating	
	Mechanical plating	
	Painting	
	Sputtering	
	Thermal infusion	
	Vacuum metalizing	
	Vapor plating	

*EPA basic operation

Metal finishing is an intermediate step in the fabrication of goods and commodities formed or fabricated from primary and secondary metals and metallic alloys. Ores are first mined and then smelted. Subsequent industrial processes then yield the feed stock for the products that will undergo the various metal-finishing processes. Feed stock includes metals and alloys in the form of rod, bar, sheet, plate, tube, or ingot.

Form-Cut-Join

The first unit process in the metal-finishing industry is to take the feed stock material and perform a gross forming, cutting, or joining action on it. These gross actions include:

- Abrasive jet machining
- Brazing
- Impact deformation

- Laminating
- Pressure deformation
- Shearing
- Sintering
- Soldering
- Thermal cutting
- Welding

Abrasive jet machining is a mechanical process in which a controlled stream of abrasive-bearing fluid is directed at the part for the purpose of cutting or metal removal. This process is typically used for cutting hard, brittle metal substrates and polymer composites.

Brazing is the joining of metals by flowing a thin layer of nonferrous filler metal into the space between two stock materials. Bonding results when high temperature, greater than 800° F, dissolves the base metal in the molten filler metal, but without fusion of the base metals.

Impact deformation is the permanent shaping of the stock material through the application of an impact force. Impact deformation operations include coining, forging, heading, high-energy rate forming, peening, shot peening, and stamping.

Laminating is the adhesive bonding of metal layers to form a part. Adhesive bonding is a joining action that occurs by applying a thermoset, thermoplastic, or chemically reactive cure adhesive to the contact surfaces of the parts.

Pressure deformation is the permanent shaping of the stock material through the application of pressure or an impact force (at a slower rate than impact deformation). Pressure deformation operations include bending, crimping, drawing, embossing, flaring, forming, necking, and rolling.

Shearing is the severing or cutting of the stock material by forcing a sharp edge or opposed sharp edges into the workpiece, stressing the material to the point of shear failure and separation.

Sintering is the formation of a part from powdered metal whereby particles are fused together under heat and pressure.

Soldering is the joining of metals by flowing a thin layer of nonferrous filler metal into the space between the two stock materials. Bonding results when high temperature, less than 800° F, dissolves the base metal in the molten filler metal, but without fusion of the base metals.

Thermal cutting is a cutting, slotting, or piercing action using an oxyacetylene oxygen lance or an electric arc cutting tool.

Welding is the joining of two or more stock materials by applying heat, pressure, or both, with or without filler material that produces a bond through fusion or recrystallization. Welding processes include arc, cold, electron beam, gas, laser beam, and resistance welding.

Machining

Once the stock material has been formed, cut, or joined, the machining processes provide finer and more detailed finishing to the workpiece. Machining is the general process of removing stock from the workpiece by forcing a cutting tool through the workpiece and removing a chip of the base material. Machining processes include:

- Abrasive blasting
- Electrical discharge
- Electrochemical
- Electron beam
- Grinding
- Laser beam
- Plasma arc
- Ultrasonic

Abrasive blasting is a pneumatic impingement of abrasive grains upon the workpiece. The abrasive media includes glass or plastic beads, sand, metal shot, grit, slag, silica, pumice, or other materials such as walnut shells.

Electrical discharge machining is a metal-removing action caused by the formation of an electrical spark between an electrode, shaped to the required contour, and the part. Electrical discharge machining is also known as **spark machining** or **electronic erosion**.

Electrochemical machining is a controlled electrolytic metal-removing action whereby the metal workpiece is the anode and the shaped tool is the cathode (the opposite of electroplating). By controlling the current distribution, the applied potential causes a selective dissolution of metal from the part in those areas directly adjacent to the tool.

Electron beam machining is a thermoelectric process whereby the material is removed from a metal part by a high-energy electron beam. At the point where the energy of the high-velocity electrons is focused, sufficient thermal energy is generated to vaporize the metal substrate.

Grinding is the action of removing stock material by using a tool consisting of abrasive grains held by a rigid or semirigid binder. The most common grinding tool abrasives are aluminum oxide, silicon carbide, and diamond.

Laser beam machining is the thermoelectric removal of material by using a highly focused monochromatic collimated beam of light targeted on the workpiece. Metal removal is largely accomplished by evaporation, although some material is removed in the liquid state at high velocity.

Plasma arc machining is the removal of material by a high-velocity jet of high temperature (30,000° F) ionized gas that melts and displaces the material. Nitrogen, argon, or hydrogen gas is passed through an electric arc, causing the gas to become ionized, and can be used on almost any metal substrate.

Ultrasonic machining is a mechanical process designed to effectively machine hard, brittle materials. This process removes material using abrasive grains. The grains are carried in a liquid between the tool and the part and bombard the work surface at high velocity.

Cleaning

At various stages in the metal-finishing processes, the stock material needs to be cleaned after some other process activity. Grease, oil, scale, or other surface residues are removed in preparation of a subsequent process. Cleaning solutions are specific to the residue targeted for removal. Cleaning processes include:

- Coating stripping
- Salt bath descaling
- Solvent degreasing

Coating stripping is the removal of metallic or organic coatings from the metal workpiece. Metallic coating stripping most often uses chemical baths. Organic coating stripping employs thermal, mechanical, and chemical means. Chemical stripping solutions include caustics, acids, solvents, or molten salt.

Salt bath descaling is the removal of surface oxides or scale from the workpiece by immersing the workpiece into a molten salt bath or a hot salt solution. Salt bath descaling solutions can contain molten salts, caustic soda, sodium hydride, and other chemical additives.

Solvent degreasing is the removal of oil and grease from the workpiece using organic solvents including aliphatic petroleum, aromatics, oxygenated hydrocarbons, and halogenated hydrocarbons. Solvent cleaning can be accomplished in either the liquid or vapor phase, or a combination of both. Figure 3–1 provides simplified schematics for solvent degreasing operations.

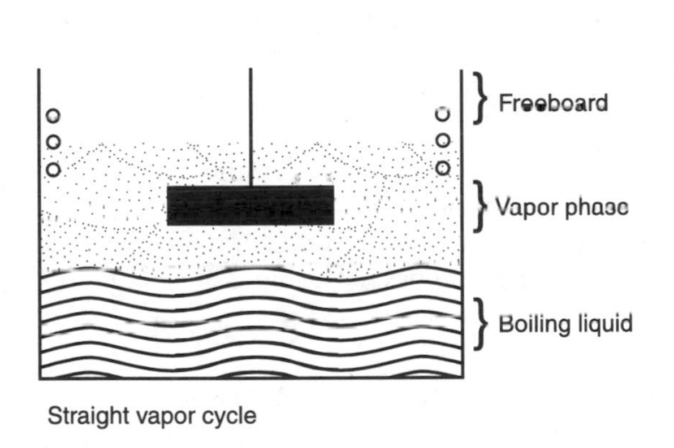

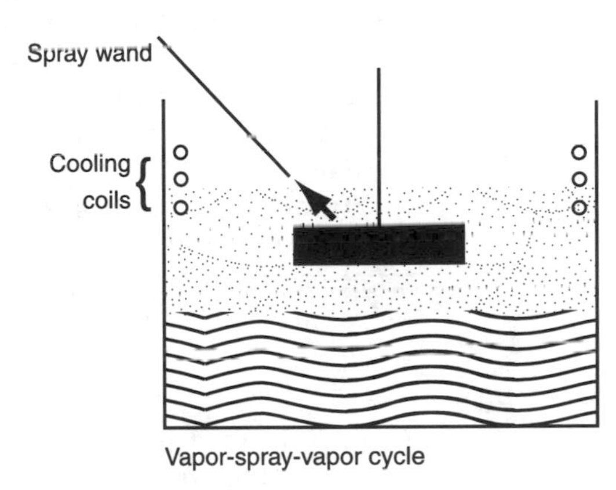

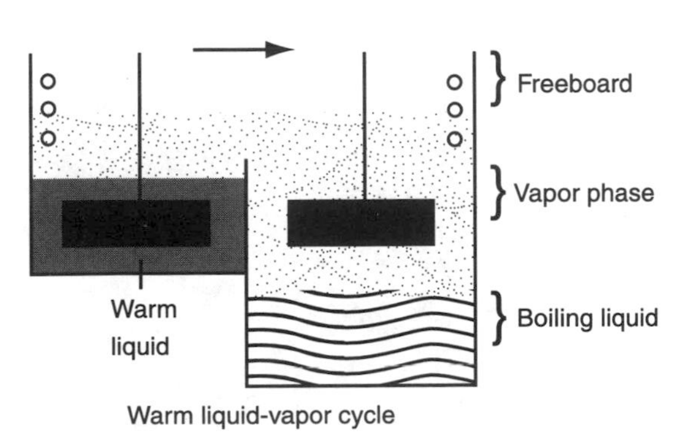

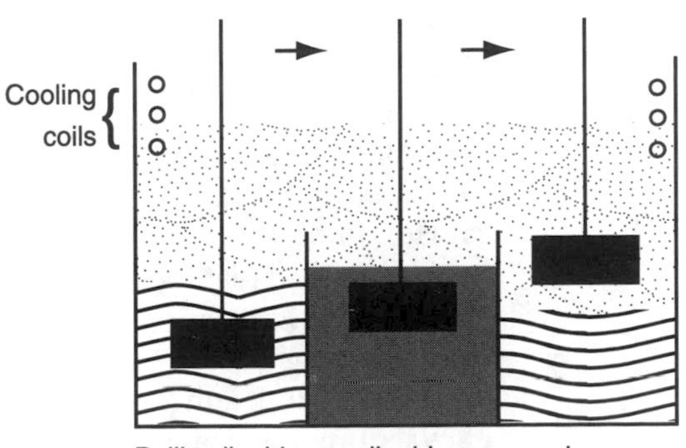

FIGURE 3–1
Schematics of solvent degreasing operations.

Finishing

After the cleaning of the workpiece, the finishing process puts on the final touch before the last process of coating. Finishing processes include:

- Burnishing
- Chemical etching and milling
- Polishing
- Tumbling

Burnishing is the action of finish sizing or smooth finishing the workpiece by displacement, rather than removal, of minute surface irregularities. Burnishing is accomplished with a smooth point or line-contact and with fixed or rotating tools that are cooled with lubricants or soap solutions.

Chemical etching and milling are actions that produce specific design configurations or surface appearances on the workpiece and are a controlled dissolution using chemical reagents or etchants (acidic or alkaline). Etching and milling are also used to eliminate scratches, nicks, and other surface imperfections.

Polishing is an abrading action used to remove or smooth out any surface defects that adversely affect the appearance or function of the workpiece. Polishing is usually performed with either a belt or wheel that is precoated with an abrasive compound. Lubricants are applied to wheels and belts to extend their life and to improve the finish.

Tumbling, or barrel finishing, is a controlled action of processing the workpiece to remove burrs, scale, flash, and oxides as well as to improve the final surface finish. A rotating barrel or vibrating unit is used with an abrasive media, water, and chemical additives to obtain a uniform surface finish typically not possible by hand finishing.

Coating

Coating the workpiece is the final process and usually involves the application of a protective film for the purposes of corrosion resistance or to produce a decorative finish. Coating processes include:

- Anodizing
- Electroless plating
- Electropainting
- Electroplating
- Electrostatic painting
- Flame spraying
- Hot dipping
- Mechanical plating
- Painting
- Sputtering
- Thermal infusion
- Vacuum metalizing
- Vapor plating

Anodizing is an electrochemical process that converts the metal surface to a coating of an insoluble oxide by making the workpiece anodic in an acidic solution.

The oxide layer begins forming on the surface, and as the reaction proceeds, the oxide layer thickens as more metal is oxidized.

Electroless plating is the chemical deposition and application of a metal coating on the workpiece by immersion in a plating solution without the use of electricity. The process depends on the catalytic reduction of a metallic ion in an aqueous solution containing a reducing agent.

Electropainting is the coating of the workpiece by making it either anodic or cathodic in a bath of the emulsion coating material. It is used primarily for corrosion-resistant primer coatings. The electrodeposition bath contains stabilized resin, disperse pigment, surfactants, and sometimes organic solvents in water.

Electroplating is the production of a thin surface coating of one metal upon another by electrodeposition where the workpiece is the cathode. Metal ions in acid, alkaline, or neutral solutions are reduced on the cathodic surfaces of the part being plated. (Electroplating is a separate section in this chapter.)

Electrostatic painting is the application of charged paint particles onto the oppositely charged workpiece and is followed by thermal fusing to form a cohesive film. By charging the part differently than the particles, the paint is attracted to the part surface, resulting in improved coverage.

Flame spraying is the application of a metal coating onto the workpiece using finely powdered fragments of wire, together with suitable fluxes, projected through a flame.

Hot-dip coating is the coating of the metallic workpiece with another metal by immersion in a molten bath. This process leaves a thin layer of molten metal on the part surface and an alloy is formed at the interface of the two metals. Galvanizing is a hot-dip process that uses molten zinc.

Mechanical plating is the process of depositing a metal coating onto the workpiece using a tumbling barrel in which the plating metal is applied to the surface as a result of contact with solid metal powder. Metal powder, impacting media, and the part are tumbled together to obtain the desired coating.

Painting is the process of applying an organic coating onto the workpiece. This operation includes applying paint varnish, lacquer, shellac, and plastics by processes such as spraying, dipping, brushing, roll coating, lithography, and wiping.

Sputtering is a vacuum-coating process in which a metallic or nonmetallic part is coated with thin metallic films. The covering of the workpiece with a thin film of metal occurs by bombarding the part surfaces with positive ions in a low-pressure gas discharge tube.

Thermal infusion is the process of applying a fused metal coating, typically zinc or cadmium, onto a ferrous workpiece using metal powder or dust in the presence of heat. A high-temperature fusion agent with a small amount of boron or silicon is deposited on the base metal and fused to the surface.

Vacuum metalizing is the process of coating the workpiece with metal by flash heating metal vapor in a high-vacuum chamber. The vapor condenses on all exposed surfaces of the workpiece. Aluminum is the primary metal used in this process for decorative coatings as well as corrosion protection.

Vapor plating is the process of metal deposition upon the heated surface of the workpiece by reduction or decomposition of a volatile compound. The reduction is accomplished by thermal dissociation or reaction with the base metal.

Miscellaneous

Miscellaneous processes are those that do not fall into the other five unit operations and include:

- Assembly
- Calibration
- Testing
- Heat treating
- Printed circuitboard manufacturing

Assembly involves the fitting together of the finished components into a complete part or product. **Calibration** is the application of thermal, electrical, or mechanical energy to the finished component to set or establish reference points for operational and quality tolerances. **Testing** also involves the application of these same energies to the part to determine the suitability of functionality of the component.

Heat treating is the modification of the physical properties of the workpiece through the application of controlled heating and cooling cycles. **Heat treating** includes aging, annealing, austempering, austenitizing, carburizing, cyaniding, malleabilizing, martempering, nitriding, normalizing, siliconizing, and tempering.

Printed circuitboard manufacturing is the formation of an electrical circuit pattern of conductive metal on nonconductive board material. The printed circuitboard manufacturing category is a separate industry and is a separate section in this chapter.

MATERIALS BALANCE

Metal-finishing facilities use a diverse number of raw materials for their feed stock. As determined by their level of manufacturing and fabrication operations, the facilities may be involved with primary, secondary, or tertiary tiers of component finishing. Primary facilities fabricate their end-products from the raw material feed stock. This includes metals, alloys, or nonmetals in the form of rod, bar, sheet, plate, tube, or ingot. A secondary facility will take a partially finished workpiece from a primary facility as its feed stock and continue the component finishing processes for that particular workpiece. Finally, a tertiary facility will take a partially finished workpiece from a secondary facility as its feed stock and continue the component finishing processes to produce the finished product. Any metal-finishing facility may involve any or all of these varying tiers of component finishing.

Inputs to a metal-finishing facility include energy sources, raw materials, chemicals, and water. Energy sources may include electricity, natural gas, heating oil, diesel fuel, etc. Raw materials include the feed stock for the workpieces, i.e., primary and secondary metals, alloys, and nonmetal base materials. Metals and alloys include aluminum, brass, bronze, copper, iron, steel, titanium, etc. Nonmetal base materials include various types and forms of plastics and composites. Chemicals include coolants, lubricants, oils, paints, and the various components of the process baths and solutions. These processes primarily involve acids (chromic, nitric, sulfuric, etc.) and caustics (sodium hydroxide, potassium hydroxide, etc.).

Outputs from a metal-finishing facility include air emissions, hazardous wastes, solid wastes, wastewater discharges, and, of course, the finished products. Air emissions include gases, particulates, and vapors, some of which are controlled waste streams and some of which may be uncontrolled or fugitive emissions. Hazardous waste vary widely and may include spent process bath solutions and sludges, heavy metal contaminated debris, spent acids and caustics, spent solvents and degreasers, etc. Solid wastes include the typical trash and garbage that are nonhazardous and cannot be recycled that result from routine business operations. Wastewater discharges include sanitary effluent, along with pretreated and untreated effluents that are generated by the manufacturing operations with the metal-finishing facility.

There exists numerous recycling opportunities in the metal-finishing industry. These range from the simple and routine paper and cardboard recycling programs to exotic chemical recovery and regeneration technologies. Specific examples of recycling include metal-working fluids (coolants, oils, lubricants), spent blasting abrasives, metals recovery, rinse waters, and spent acid solutions. Additional detail on recycling is provided in the discussion on pollution prevention found later in this section.

Perhaps the most obvious recycling opportunity is the actual raw material entering the initial metal-finishing unit process that forms, cuts, or joins the primary workpiece. An example of this is the primary aluminum sheet and bar stock that is used in aircraft component manufacturing. For every 3 pounds of aluminum that enters the factory, 2 pounds ends up on the floor as a result of cutting, drilling, boring, and forming operations that are performed on the feed stock. All of this scrap aluminum, primarily in the form of chips, is collected and compressed in briquettes and recycled.

Finished products from the metal-finishing industry include almost any conceivable commodity that is made of metal or coated with metal. These commodities range from kitchen appliances and aerospace components to automobile bumpers and tie tacks. A broad view of materials balance in the metal-finishing industry is depicted in Figure 3–2.

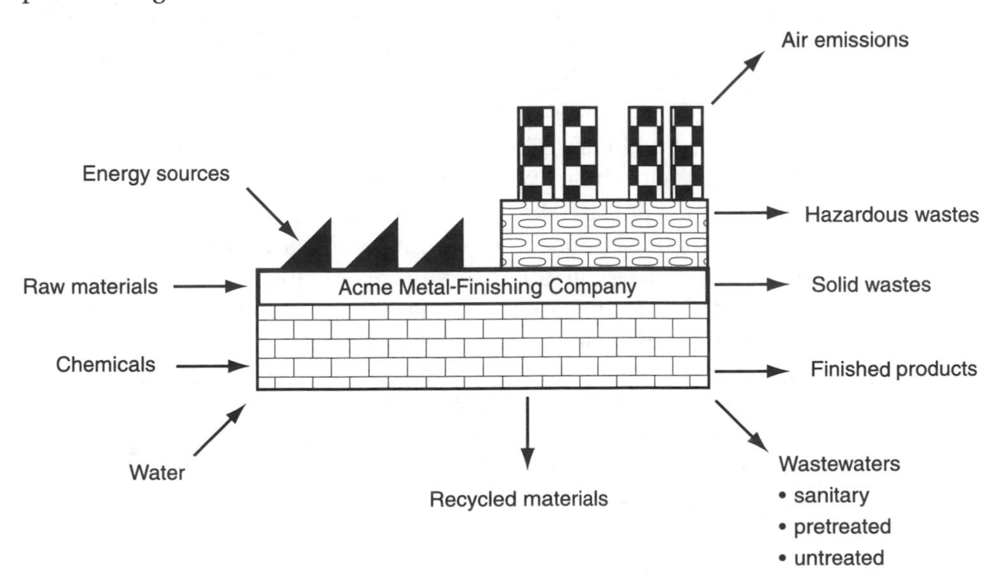

FIGURE 3–2
Materials balance in the metal-finishing industry.

HAZARDOUS WASTE GENERATION

As the metal-finishing industry processes and operations are broad and varied, so too are the waste streams generated. Typical waste streams from the metal-finishing industry include oils and greases, acidic rinses, alkaline rinses, heavy metals, toxic organic solvents, cyanide, and sludges. Of primary concern is the generation of hexavalent chromium (Cr^{+6}) and cyanide waste streams, both of which are environmentally hazardous in low concentrations.

As a whole, the metal-finishing industry generates wastewater effluents, hazardous solid wastes, and airborne emissions. Wastewater effluents are typically generated from processes involving baths, solutions, and rinsing operations. Hazardous solid wastes are typically generated from the form-cut-join and machining unit operations and also from processes involving bath and solution sludges. Airborne emissions typically result in the form of gases, fumes, mists, and vapors generated in the processes using high-temperature or air-agitated baths and solutions. On an individual basis, several processes are typically zero-discharge operations. These include sintering (form-cut-join); electron and laser beam machining, and plasma arc and ultrasonic machining (machining); sputtering, thermal infusion, vacuum metalizing, and vapor plating (coating); and calibration (miscellaneous).

The primary waste streams generated from the form-cut-join unit operation are heavy metals and oil and grease. Heavy metal wastes typically result from the gross physical actions on the workpieces. Much of the oil and grease is generated when the workpieces are quenched, cooled, or annealed. This quenching, cooling, or annealing action usually involves various water-based solutions or emulsified oils.

The primary waste streams generated from the machining unit operation are also heavy metals and oil and grease, plus cyanide. Heavy metal wastes result from the stock material removal actions during machining, grinding, etc. Concurrently, a number of these processes involve oil-based cooling and lubrication, thus resulting in the generation of oil and grease wastes. The cyanide waste stream is generated from the electrochemical machining process where the workpieces are immersed in a cyanide- and metal-based etching solution.

The cleaning unit operation generates all of the individual waste streams common to the metal-finishing industry. Cleaning the workpieces involves various acid- and alkaline-based baths and their associated rinse waters. These baths, along with wetting agents or detergents, remove oil, grease, dirt, oxides, and scale from the workpiece surfaces. Solvent degreasing generates toxic organic wastes from the organic solvent used in the degreasing operations. Paint stripping and salt bath descaling also generate toxic organics along with heavy metals and oily wastes.

The finishing unit operation also generates all of the individual waste streams common to the metal-finishing industry. Finishing the workpiece is similar to cleaning, but with a finer level of detail in preparation of the final step of coating the workpiece. Typically, wastes are generated by final rinsings and from the spent process solutions. Cyanide wastes are typically generated in the burnishing process where it is used as a wetting agent and rust inhibitor.

The coating unit operation generates all of the typical metal-finishing industry waste streams except for oil and grease. A number of the chemical coating processes involve heavy metal-based acidic plating solutions and their associated rinse operations. Also, a number of coating processes involve quenching, cooling, and sealing steps.

The miscellaneous unit operation processes do not generate significant quantities of hazardous waste except for heat treating and printed circuitboard manufacturing. Heat-treating process wastes result from rinses and bath discharges. Printed circuitboard manufacturing wastes are discussed in a separate section of this chapter.

Table 3–4 summarizes the various waste streams generated by the six unit operations.

Of the three typical industrial waste streams (wastewaters, hazardous solid wastes, and airborne emissions) the wastewater discharges are the predominant waste stream resulting from the various metal-finishing unit processes.

A majority of the unit processes in the form-cut-join operation are typically dry processes but may involve a water-based cooling, quenching, or rinse step. The exceptions to these mostly dry processes are abrasive jet machining and shearing. The media carrier in abrasive jet machining is typically an alkaline solution or an emulsified oil. When it is to be recharged, the spent solution is usually dumped as a wastewater. Waste streams generated by the shearing process include metal-working fluids and rinses.

Approximately half of the unit processes in the machining operation are dry processes. Abrasive blasting generates spent abrasive, and water as a media carrier must be replenished. Electrical discharge machining results in the dielectric fluids being discharged as a wastewater. The spent electrolytes along with rinses from electro-chemical machining are discharged. The grinding process involves spent coolants as well as rinse waters.

All of the individual unit processes in the cleaning and finishing operations result in wastewater discharges. These include spent process solutions, rinse waters, and quenching solutions. A majority of these processes use caustics, acids, solvents, and other organic chemicals in their solutions.

Several of the unit processes in the coating operation are dry processes. The other processes, primarily anodizing, plating, and painting, generate wastewater discharges in the form of spent process solutions, wash waters, and rinse waters.

A majority of the unit processes in the miscellaneous operations are typically dry processes. Assembly may involve cooling or rinse waters. Heat treating will generate spent bath solutions along with some quench and rinse waters. Testing may include some wet tests that use fluids for dye penetrant, hydraulic, or salt spray operations.

TABLE 3–4
Metal-finishing unit operations waste streams.

Metal-Finishing Unit Operation	Heavy Metals	Hex. (+6) Chrome	Cyanide	Oil and Grease	Toxic Organics
Form-cut-join	X			X	
Machining	X		X	X	X
Cleaning	X	X	X	X	X
Finishing	X	X	X	X	X
Coating	X	X	X		X
Miscellaneous	X		X	X	X

REGULATORY IMPACT

The metal-finishing industry is regulated by the federal acts controlling water pollution, hazardous waste generation, and air pollution. Water pollution control requirements are codified at 40 CFR 433. Hazardous waste generation provisions are codified at 40 CFR 260–280. Air pollution control requirements that may apply are codified beginning at 40 CFR 50. This discussion emphasizes the water quality regulations.

Water pollution control requirements include effluent limitations guidelines as well as both general pretreatment regulations (40 CFR 403) and categorical pretreatment standards (40 CFR 433). Effluent limitations are authorized in the Clean Water Act (CWA) under the provisions of ξ304 and 306. Pretreatment standards are authorized in ξ307 (b) and (c) to develop national pretreatment standards for new and existing dischargers to publicly owned treatment works (POTWs). Overall, the CWA was adopted to restore and maintain the chemical, physical, and biological integrity of the nation's waters.

Metal-finishing facilities that send their effluent to a POTW are considered **indirect dischargers**. Those facilities that discharge their effluent directly to the environment without POTW treatment are considered **direct dischargers**. Wastewater effluents from direct discharging facilities are controlled by a National Pollutant Discharge Elimination System (NPDES) permit.

General pretreatment standards (40 CFR 403) establish administrative mechanisms that require POTWs to develop local pretreatment programs to enforce the general discharge prohibitions as well as the specific categorical pretreatment standards. Categorical pretreatment standards for the metal-finishing industry (40 CFR 433) are designed to prevent the discharge of pollutants that pass through, interfere with, or are otherwise incompatible with the operation of the receiving POTW. These standards are technology-based for the removal of toxic pollutants. Also, these standards contain specific numerical limitations based on an evaluation of specific technologies for the particular industrial category.

Originally, when categorical pretreatment standards were first established in 1974, they were applied to the electroplating industry (40 CFR 413). Later, successful petitions to the EPA resulted in a separate industrial category for the metal-finishing industry with regulations promulgated in 1983 (40 CFR 433). In many cases, the electroplating industry must also comply with the metal-finishing categorical pretreatment standards. This dual category compliance requirement is determined by the industrial activities involving the six basic operations and the forty process operations discussed earlier in this section. Electroplating is a separate industrial category and is discussed in a separate section in this chapter.

More recently, the EPA is proposed additional regulations that concern a new point source category known as MP&M. When finalized, these regulations will be codified at 40 CFR 438 and will provide effluent limitation guidelines, pretreatment standards, and new source performance standards. MP&M requirements apply to process wastewater discharges from sites engaged in manufacturing, rebuilding, or maintaining finished metal parts, products, or machines.

Regulatory standards for MP&M are to be phased in across two categories of industrial sites. The MP&M Phase I point source category applies to the following seven industrial sectors:

- Aerospace
- Aircraft
- Electronic equipment
- Hardware
- Mobile industrial equipment
- Ordnance
- Stationary industrial equipment

MP&M Phase II will be proposed and promulgated approximately three years after MP&M Phase I and intends to cover the following eight industrial sectors:

- Bus and truck
- Household equipment
- Instruments
- Motor vehicle
- Office machine
- Precious and nonprecious metals
- Railroad
- Ships and boats

Discussion here concentrates on the categorical effluent limitations and pretreatment standards for the metal-finishing point source category, excluding electroplating and printed circuitboard manufacturing. Control technologies levied against the industry include best practicable control technology currently available (BPT), best available technology economically achievable (BAT), and best conventional pollutant control technology (BCT). Additionally, new sources must abide by performance standards (NSPS). Pretreatment standards (PS) are divided into two groups: existing sources (PSES) and new sources (PSNS).

BPT effluent limitation guidelines are generally based on the average of the best existing performance by plants of various sizes, ages, and unit processes within the category or subcategory for control of pollutants. Considerations include the total cost of achieving effluent reductions in relation to the effluent reduction benefits, the age of the equipment and facilities involved, the processes employed, process change required, engineering aspects of the control technologies, non-water quality environmental impacts (i.e., energy), and other factors deemed appropriate.

BAT effluent limitations represent the best existing economically achievable performance of plants in the category or subcategory. BAT is established as the principal national means of controlling the direct discharge of toxic pollutants and nonconventional pollutants to the nation's waters. Factors considered in assessing BAT include the age of equipment and facilities involved, the process employed, potential process changes, and non-water quality environmental impacts.

BCT was established for discharges of conventional pollutants from existing industrial point sources. Conventional pollutants include biochemical oxygen demanding pollutants (BOD), total suspended solids (TSS), fecal coliform, pH, oil and grease, and others. BCT replaces BAT for the control of conventional pollutants at certain facilities. BCT limitations are established with a two-part cost reasonableness test.

NSPS are based on the best available demonstrated treatment technology. New facilities have the opportunity to install the best and most efficient production processes and waste-treatment technologies. As a result, NSPS represents the most stringent numerical values attainable through the application of the best available control technology for all pollutants (i.e., conventional, nonconventional, and toxic pollutants). Considerations include the cost of achieving the effluent reduction and any non-water quality environmental impacts.

PSES are designed to prevent the discharge of pollutants that pass through, interfere with, or are otherwise incompatible with the operation of POTWs. Pollutants must be pretreated that pass through or interfere with the POTW treatment processes or sludge-disposal methods. PSES are technology-based and analogous to BAT-effluent limitation guidelines. In determining whether to promulgate national category-wide pretreatment standards, the EPA generally determines that there is a pass-through of a pollutant and, thus, a need for categorical standards if the nationwide average percentage of a pollutant removed by well-operated POTWs achieving secondary treatment is less than the percentage removed by the BAT model treatment system.

PSNS, like PSES, are designed to prevent the discharge of pollutants that pass through, interfere with, or are otherwise incompatible with POTW operations. PSNS are to be issued at the same time as NSPS. New indirect dischargers, like the new direct dischargers, have the opportunity to incorporate into their facilities the best available demonstrated technologies. The same factors are considered in promulgating PSNS as for NSPS.

For metal-finishing facilities, BPT-effluent limitations are applied for seven heavy metals, cyanides, total toxic organics (TTO), oil and grease, total suspended solids (TSS), and pH. BAT-effluent limitations are applied for seven heavy metals, cyanides, and TTO. The seven heavy metals are: cadmium, chromium, copper, lead, nickel, silver, and zinc. All metals are measured as total portions (T), as compared to dissolved or suspended portions. Cyanide is also measured as total unless the facility has a cyanide treatment process, then the cyanide is measured as that portion amenable (A) to alkaline chlorination. In all pretreatment schemes, augmenting the use of process wastewater or other diluting techniques to achieve compliance with effluent limitations is forbidden. (Dilution is not the solution to pollution.)

The numerical values of BAT-effluent limitations are identical to those of BPT-effluent limitations. Oil and grease, TSS, and pH are not included as pollutants for BAT. For both BPT and BAT, the alternative cyanide (A) effluent limitations are 0.86 milligrams per liter (mg/l) for any one-day maximum, and shall not exceed 0.32 mg/l for the monthly average.

PSES in the metal-finishing industry point source category are identical to those for BPT-effluent limitations. Oil and grease, TSS, and pH are not included as pollutants for PSES. PSNS require a lower cadmium limit but are otherwise identical for BPT-effluent limitations. (The lower limits for cadmium are 0.11 mg/l for any one-day maximum, and shall not exceed 0.07 mg/l for the monthly average.) Oil and grease, TSS, and pH are not included as pollutants for PSNS.

Table 3–5 provides the current BTP-based effluent limitations for the metal-finishing industry point source category (40 CFR 433).

NSPS numerical values are identical to those for BPT-effluent limitations, but require a lower cadmium limit equivalent to PSNS. For PSES, PSNS, and NSPS, the alternative cyanide (A) effluent limitations are 0.86 mg/l for any one-day maximum, and shall not exceed 0.32 mg/l for the monthly average.

TABLE 3–5
Metal-finishing industry BPT-effluent limitations.

Pollutant or Pollutant Property	Maximum for any One Day (mg/l)	Monthly Average Shall Not Exceed (mg/l)
Cadmium (T)	0.69	0.26
Chromium (T)	2.77	1.71
Copper (T)	3.38	2.07
Lead (T)	0.69	0.43
Nickel (T)	3.98	2.38
Silver (T)	0.43	0.24
Zinc (T)	2.61	1.48
Cyanide (T)	1.20	0.65
TTO	2.13	–
Oil and grease	52	26
TSS	60	31
pH*	6.0–9.0	6.0–9.0

*Standard units

Slight differences exist with the proposed effluent limitations for the newly defined MP&M industrial category. Aluminum and iron have been added to the list of pollutant parameters; while lead, silver, and total toxic organics (TTO) have been deleted. All of the metals plus cyanide are measured as total portions (T) as they are in the effluent limitations for the metal-finishing industry category. The concentration limits for all the metals and cyanide apply to BPT, BAT, NSPS, PSES, and PSNS. The oil and grease parameter is proposed as an indicator for organic pollutants and the concentration limits for oil and grease also apply to all levels of regulatory control. The concentration limits for total suspended solids apply only to BPT, BCT, and NSPS.

Table 3–6 provides the proposed concentration-based effluent limitations for the metal products and machinery (Phase I) point source category (40 CFR 438).

As indicated earlier in Table 3–1, the EPA has established effluent limitations guidelines and standards for thirteen metals industries. With the recently proposed MP&M requirements, it is anticipated that facilities currently regulated by existing regulations will now have to abide by the new and more stringent MP&M standards. The EPA is proposing that MP&M Phase I limitations replace the existing metal-finishing effluent guidelines for those industrial facilities that meet and satisfy the MP&M Phase I criteria.

The hazardous waste regulations that are applied to the metal-finishing industry are not source-specific or category-specific, as are the water pollution control and effluent limitation requirements. Should any metal-finishing facility generate hazardous waste, those activities and operations are regulated by the RCRA. RCRA controls the generation of hazardous wastes, along with the transportation, treatment, storage, and disposal of those hazardous wastes.

As discussed earlier, all six metal-finishing industry unit operations result in the generation of hazardous wastes. Hazardous solid wastes are typically generated from the form-cut-join and machining unit operations, and also from any processes involving bath and solution sludges.

TABLE 3–6
MP&M Phase I concentration-based effluent limitations (proposed).

Pollutant or Pollutant Property	Maximum for any One Day (mg/l)	Monthly Average Shall Not Exceed (mg/l)
Aluminum (T)	1.40	1.00
Cadmium (T)	0.70	0.30
Chromium (T)	0.30	0.20
Copper (T)	1.30	0.60
Iron (T)	2.40	1.30
Nickel (T)	1.10	0.50
Zinc (T)	0.80	0.40
Cyanide (T)	0.03	0.02
Oil and grease	35	17
TSS	73	36
pH*	6.0–9.0	6.0–9.0

*Standard units

Air pollution control requirements are diverse and scattered widely across the regulations. National primary and secondary ambient air quality standards are found at 40 CFR 50. These standards regulate sulfur oxides (SO_x), particulate matter, carbon monoxide (CO), ozone, nitrous oxides (NO_x), and lead. Standards of performance for new stationary sources are found at 40 CFR 60. Several subparts within 40 CFR 60 address requirements for specific industries that fit the criteria to be considered metal-finishing industry categories. The performance standards for these industry categories that are related to the metal-finishing industry include the following:

Subpart EE	Surface Coating of Metal Furniture
Subpart MM	Automobile and Light-duty Truck Surface Coating Operations
Subpart SS	Industrial Surface Coating-Large Appliances
Subpart TT	Metal Coil Surface Coating

National emission standards for hazardous air pollutants (NESHAPs) are found at 40 CFR 61. The original NESHAP requirements address asbestos, benzene, beryllium, coke oven emissions, inorganic arsenic, mercury, radionuclides, and vinyl chloride. More specific requirements are found at 40 CFR 63, NESHAPs for Source Categories. The NESHAP for chromium emissions from hard and decorative chromium electroplating and chromium anodizing tanks is found at 40 CFR 63 Subpart N. The NESHAP for solvent cleaning machines is found at 40 CFR Subpart T. The NESHAP final rule for the aerospace manufacturing and rework industry is codified in 40 CFR Subpart GG. The hazardous air pollutants (HAPs) of concern for the aerospace industry category include cadmium, chromium, ethylene glycol, glycol ethers, methylene chloride, methyl ethyl ketone, toluene, and xylene.

Metal-finishing industry facilities need to address these source categories with a broad approach because of the ever-increasing promulgation of the 189 hazardous air pollutant (HAPs) standards. The EPA is proposing both source-specific as well as pollutant-specific requirements as part of the industry category specific regulations to be phased in across the next several years.

TREATMENT AND CONTROL TECHNOLOGY

Wastewater treatment for metal-finishing effluents involves technologies for common metal wastes, complexed metal wastes, precious metal wastes, hexavalent chromium, cyanide wastes, toxic organics, oily wastes, and sludges. The technologies for these predominant waste streams address intermediate steps as well as end-of-pipe controls. Figure 3–3 provides an overview of wastewater treatment processes for a typical metal-finishing facility.

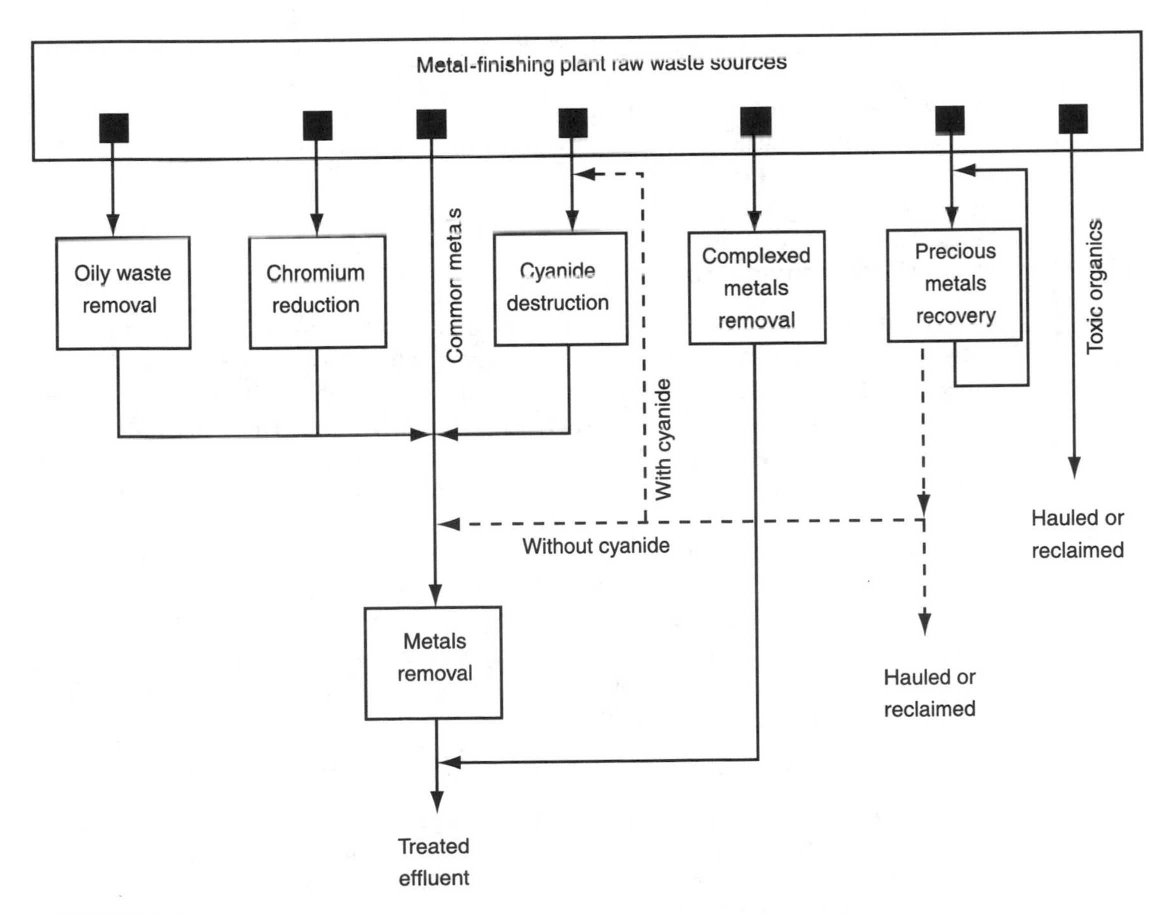

FIGURE 3–3
Metal-finishing industry wastewater treatment.

Also, in-process control technologies are available that reduce water usage and drag out pollutants from process baths, and facilitate the efficient handling of process wastes. In-process control technologies include:

- Flow reduction through efficient rinsing
- Waste oil segregation
- Countercurrent and static rinsing
- Process bath segregation
- Process bath conservation
- Process modification
- Integrated waste treatment
- Cutting fluid cleaning

Common end-of-pipe treatment technologies employed in the metal-finishing industry include the following:

- Chemical precipitation and sedimentation
- Oil skimming
- Chemical reduction of hexavalent chromium
- Neutralization
- Alkaline chlorination for cyanide destruction
- Multimedia filtration
- Chemical emulsion breaking
- Gravity settling of wastewater
- Pressure filtration of sludge
- Gravity thickening of sludge

Treatment of common metal wastes consists of hydroxide precipitation followed by sedimentation. This technique precipitates the dissolved metals that are then removed by sedimentation or filtration. It is not effective for hexavalent chromium and cyanide tends to interfere with the process. Treatment of complex metal wastes involves high pH precipitation that dissociates the metals, resulting in free metal ions that can then be removed by standard hydroxide precipitation and sedimentation.

Precious metal wastes are treated similar to the common metal wastes, with evaporation, ion exchange, or electrolytic recovery techniques used to recover the precious metal content of the waste stream. Hexavalent chromium (Cr^{+6}) must be reduced to trivalent chromium (Cr^{+3}). Sodium metabisulfite, sodium bisulfite, and sulfur dioxide are the most widely used reducing agents. The reaction in these processes is illustrated for the following sulfur dioxide reaction.

$$2H_2CrO_4 + 3SO_2 \longrightarrow Cr_2(SO_4)_3 + 2H_2O$$

Reduction using other reagents is chemically similar to the previous illustration. Once reduced, the chromium is separated from solution in conjunction with other metal salts by alkaline precipitation. Figure 3–4 provides an overview of the treatment technologies for the reduction and precipitation of chromium-bearing wastes typically found in metal-finishing facilities.

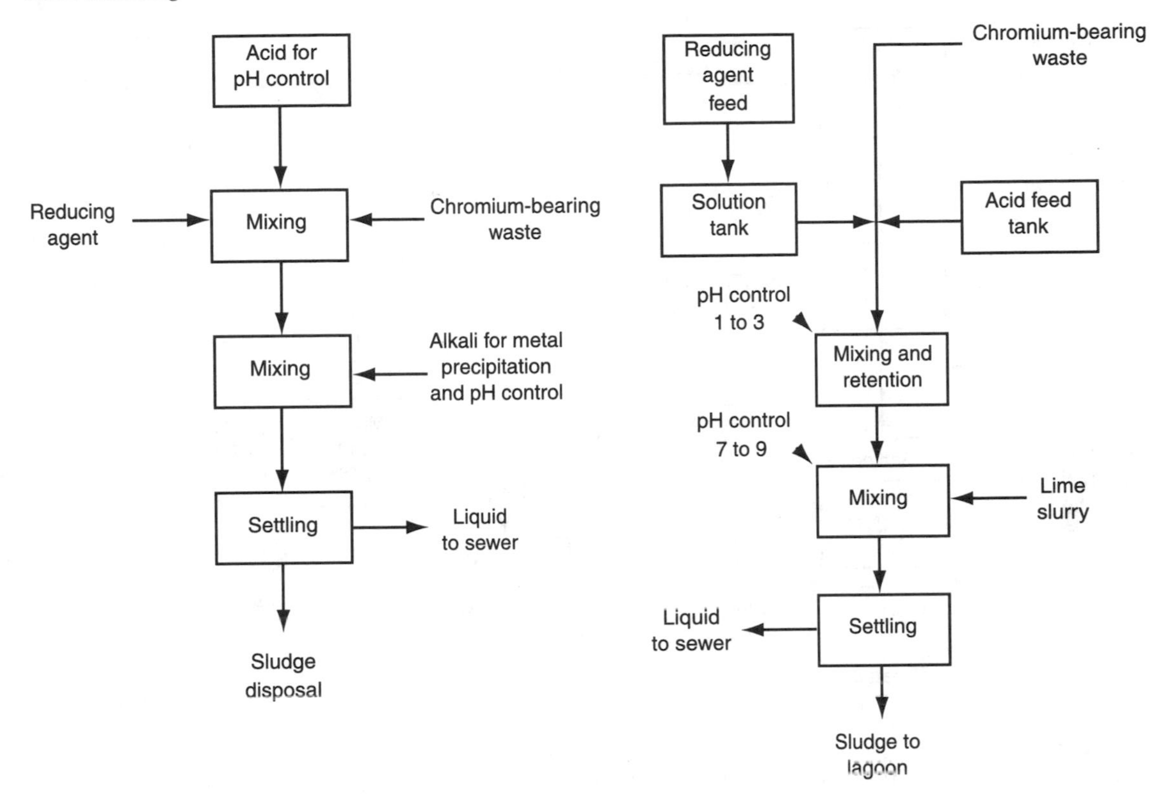

FIGURE 3–4
Reduction and precipitation of chromium-bearing wastes.

Treatment of cyanide wastes is typically performed by alkaline chlorination that oxidizes the cyanide, which is then amenable to chlorination. Chlorine is typically used as either chlorine gas or sodium hypochlorite. This process can be illustrated by the following two-step chemical reaction:

$$Cl_2 + NaCN + 2NaOH \longrightarrow NaCNO + 2NaCl + H_2O$$
$$3Cl_2 + 4NaOH + 2NaCNO \longrightarrow 2CO_2 + N_2 + 6NaCl + 2H_2O$$

The cyanide destruction process results in carbon dioxide, nitrogen products, salt, and water. Also, complexed cyanides can be precipitated with ferrous sulfate. Cyanide waste streams can be treated in a continuous process or in a batch process. Figure 3–5 provides an overview of the cyanide treatment processes typically found in metal-finishing facilities.

Treatment for toxic organics usually involves in-plant control to prevent their entering any waste stream. This is accomplished by proper storage, waste stream segregation, and other appropriate management. Treatment of oily wastes commonly involves skimming, coalescing, emulsion breaking, flotation, centrifugation, ultrafiltration, or reverse osmosis. Typically sludge-treatment techniques include thickening, pressure filtration, or bed drying. The sludge is then managed as hazardous waste and is disposed of accordingly.

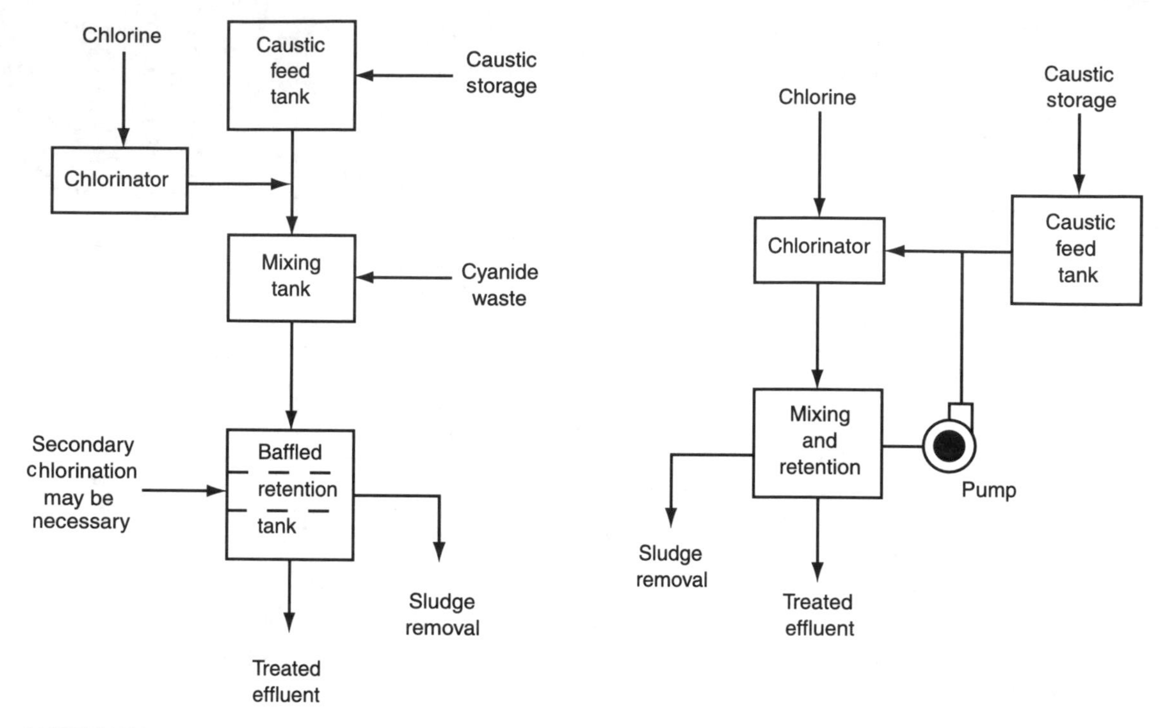

FIGURE 3–5
Cyanide chlorination-continuous and batch processes.

Air pollution control technologies for the metal-finishing industry are associated with specific process emissions. By the nature of the industry, it is necessary to apply technologies against all unit and process operations to control gaseous emissions and particulate matter. Particulate matter is a concern especially for the form-cut-join and machining unit operations. Gaseous emissions along with vapors and fumes are a concern in any unit or process operation that involves chemicals (cleaning, finishing, coating, etc.).

Air pollution control technology for particulate matter includes: settling chamber, cyclone, bag filter, electrostatic precipitator, and wet scrubber. Choice of technology is determined by flow rate, particle size, collection efficiency, pressure drop, etc. The settling chamber removes particulates by slowing the velocity of the air flow so that the particulates drop out of the air stream. The cyclone uses a centrifugal motion on the air flow to force the particulates to the outer wall, where they fall into a collection device. Bag filters pump the air stream through a filtering medium that traps the particulates. Electrostatic precipitators use an electrical current to attract the charged particles in the air stream. Wet scrubbers direct a water spray into the air stream that knocks down the particulate matter. Figures 3–6a–6e provide simplified schematics of the various air pollution control devices used in the metal-finishing industrial category.

In all cases of air pollution control technology for particulate matter, the particles must be removed from the control devices on a recurring basis and frequency.

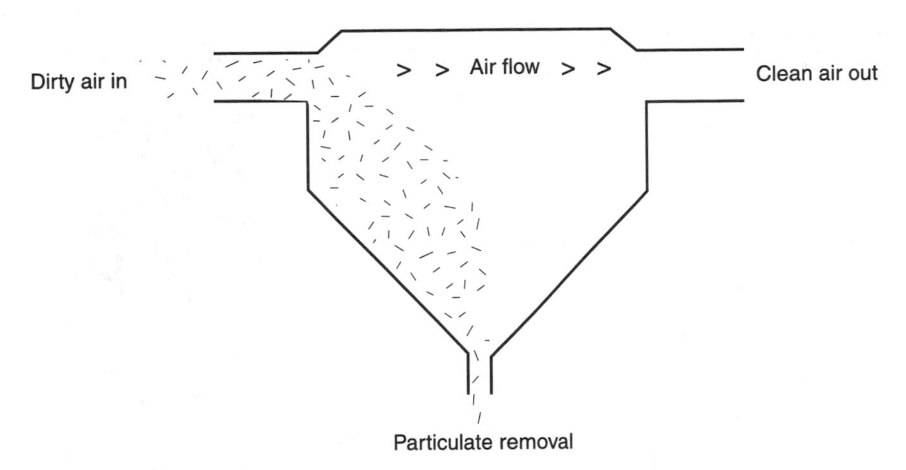

FIGURE 3–6A
Settling chamber.

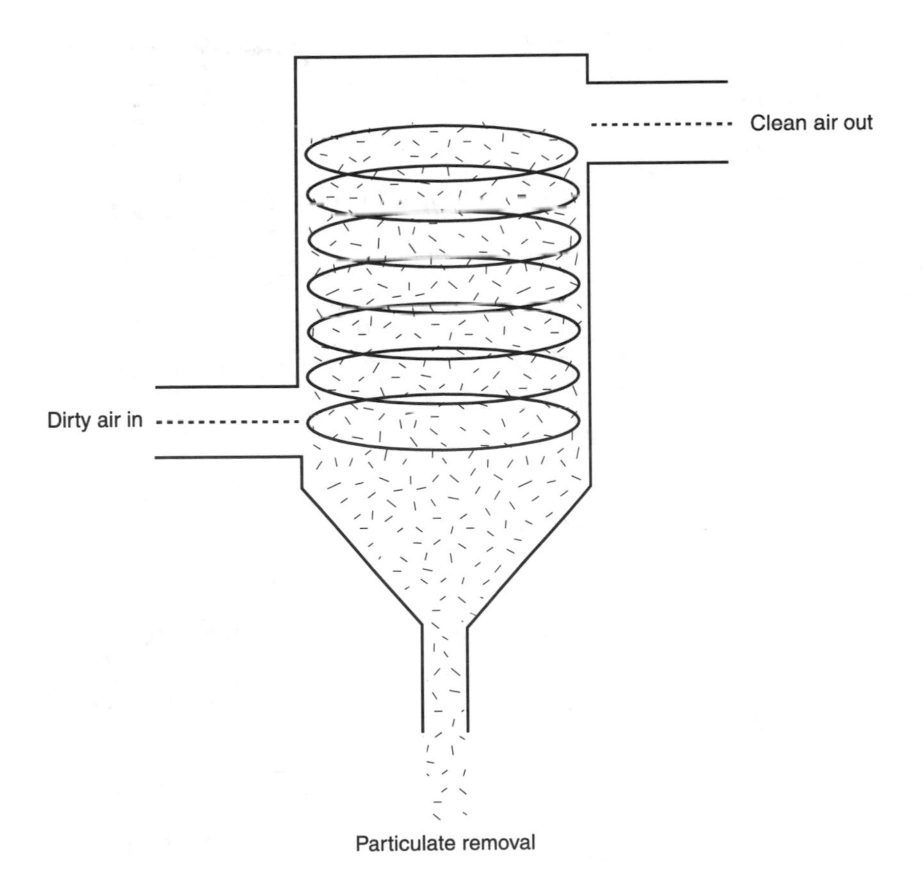

FIGURE 3–6B
Cyclone.

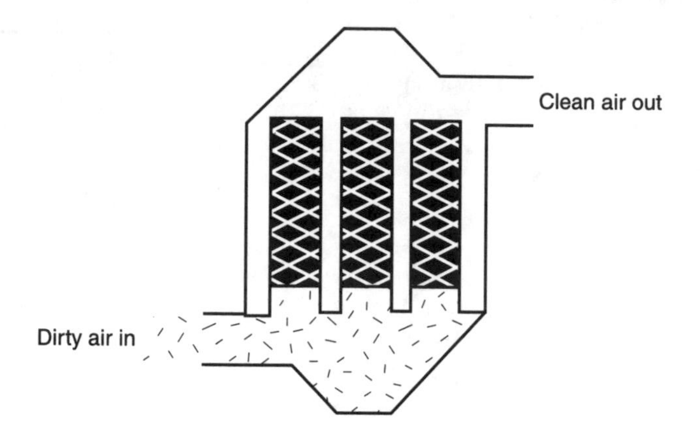

FIGURE 3–6C
Bag filter.

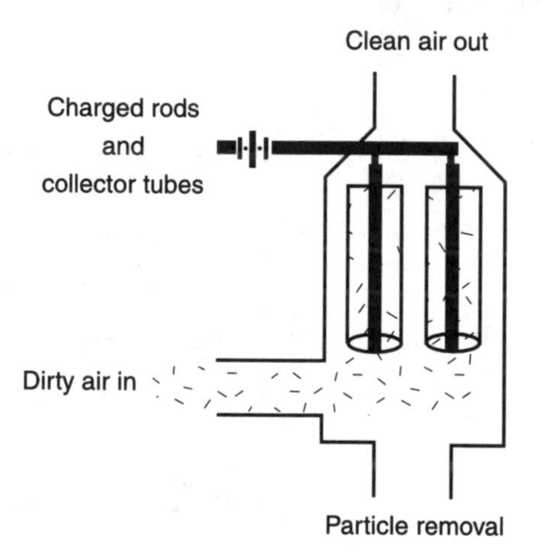

FIGURE 3–6D
Electrostatic precipitator.

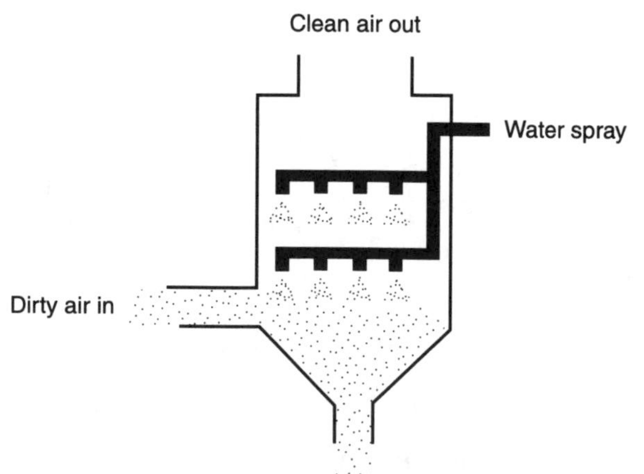

FIGURE 3–6E
Wet scrubber.

This collected residue must then be disposed of in an appropriate manner. As determined by the metal-finishing unit or process operation, this residue may be considered to be a hazardous waste by a regulatory agency.

Air pollution control technology for gaseous emissions includes: sorption, oxidation, and wet scrubbers. Whereas the discussion for particulate matter centered on specific types of control devices and their respective mechanical operations, control of gaseous emissions must be discussed in terms of process chemistry. Sorption includes both adsorption and absorption, and involves both physical and chemical reactions using a sorptive medium such as activated carbon. Oxidation includes both catalytic and thermal techniques to remove VOCs. Wet scrubbers use water or other liquid solutions to dilute, dissolve, or react with the chemical components of the gaseous emission.

The unit operations for cleaning, finishing, and coating all involve chemicals in a majority of their process operations (stripping, descaling, anodizing, etc.). These processes involve baths and rinses that contain acidic and alkaline solutions. Air pollution control technology is required because of the agitation and hot temperatures of the chemical process baths. Typically, wet scrubbers can be used to control and neutralize the gaseous emissions, thereby protecting the environment and human health. The vapors and mists released from an acid process bath will be scrubbed with an alkaline solution and an alkaline bath will be scrubbed with an acidic solution.

Most frequently, wet scrubbers are applied to electroplating, cleaning, and coating processes. Two common scrubbing devices are the wet-packed scrubber and the composite mesh pad mist eliminator. The **wet-packed scrubber** consists of a spray chamber filled with packing material. Process water is continuously sprayed onto the packing and the air stream is pulled through the packing by a fan. Contaminants in the air stream are absorbed by the water spray and the air is released to the atmosphere. Composite mesh pads are most frequently used to control emissions from hard chromium electroplating and chromic acid anodizing. The **composite mesh pad mist eliminator** removes chromic acid from the air stream by slowing the velocity of the air and causing the entrained chromic acid droplets to impinge on the fiber pads.

Other gaseous emissions from metal-finishing operations include organic solvents and other VOCs. Much of these emissions result from evaporation and volatilization of chemicals used in coating process operations and also from solvent degreasing.

POLLUTION PREVENTION

Three groups of technology options exist for controlling pollution resulting from the operational processes found in the metal-finishing industry. These three are: source reduction, recycling, and end-of-pipe treatment. Pollution prevention specifically addresses the first two options. Also, the EPA has assigned the highest priority to source reduction (including in-process source reduction), and the second highest priority to recycling. As always, end-of-pipe treatment (or disposal) receives the lowest priority in the beneficial environmental management hierarchy.

Broad concepts and techniques for source reduction and recycling include production planning and sequencing, process or equipment modification, raw material substitution, waste segregation and separation, and loss prevention and housekeeping. Specifically, the predominant technology options for source reduction and recycling available to the metal-finishing industry include the following:

- Centrifugation and pasteurization of machining coolants
- Ion exchange
- Centrifugation and recycling of painting water curtains
- Reverse osmosis
- Supported liquid membrane separation
- Electrolytic recovery
- Flow reduction for rinses and baths
- Countercurrent cascade rinsing

Machining coolants used in the metal-finishing industry typically consist of water-soluble oils dispersed in water. The coolant is pumped through a coolant sump, over the machining tool and the part during machining, and returned to the sump. The coolant will become spent over time due to several reasons, including build-up of suspended solids, build-up of nonemulsified tramp oils, excess microbial growth, and additive consumption. Centrifugation and pasteurization can be used along with oil skimming and biocide addition to reduce coolant discharge and pollution generation at the source. These technologies result in reducing the amount of coolant and wastewater requiring treatment or disposal, and in reducing the amount of new coolant that must be purchased.

Paint spray booths use either a dry filter mechanism or a water curtain to trap paint overspray. The water curtain method allows the water to be continuously recirculated until the solids content in the wastewater necessitates treatment and recycling or disposal. Centrifugation of these wash waters removes the solids, recycles the water, and eliminates the need for discharging the water. Wastewater from paint spray booths commonly contains organic pollutants as well as certain metals. Eliminating the discharge of this wastewater may eliminate the need for end-of-pipe treatment for organic pollutants at certain metal-finishing facilities.

Supported liquid membrane technology uses a liquid membrane transport process for fluid and particle separation and filtration. Supported liquid membranes can be used for metal ion separation, solvent extraction, VOC removal, and other applications. The liquid membrane aspect combines solvent extraction and a stripping operation in one process. For metal-finishing operations, this membrane technology can be used for the recovery and separation of metals from aqueous process solutions. For metals recovery, the supported liquid membrane technology is usually more cost-effective than alkali chlorination or ion exchange. Figure 3–7 shows a typical metals plating line using this membrane system for recovering the metal salt.

Flow reduction for process rinses and process baths involves both rinse water reduction as well as bath maintenance and regeneration. Rinse water reduction is implemented through flow restrictors, rinse timers, or conductivity controllers. Bath maintenance and regeneration is necessary as process baths become contaminated with impurities that affect their performance. Options include filtration, carbon treatment, electrolysis, carbonate freezing, and chemical precipitation. These tech-

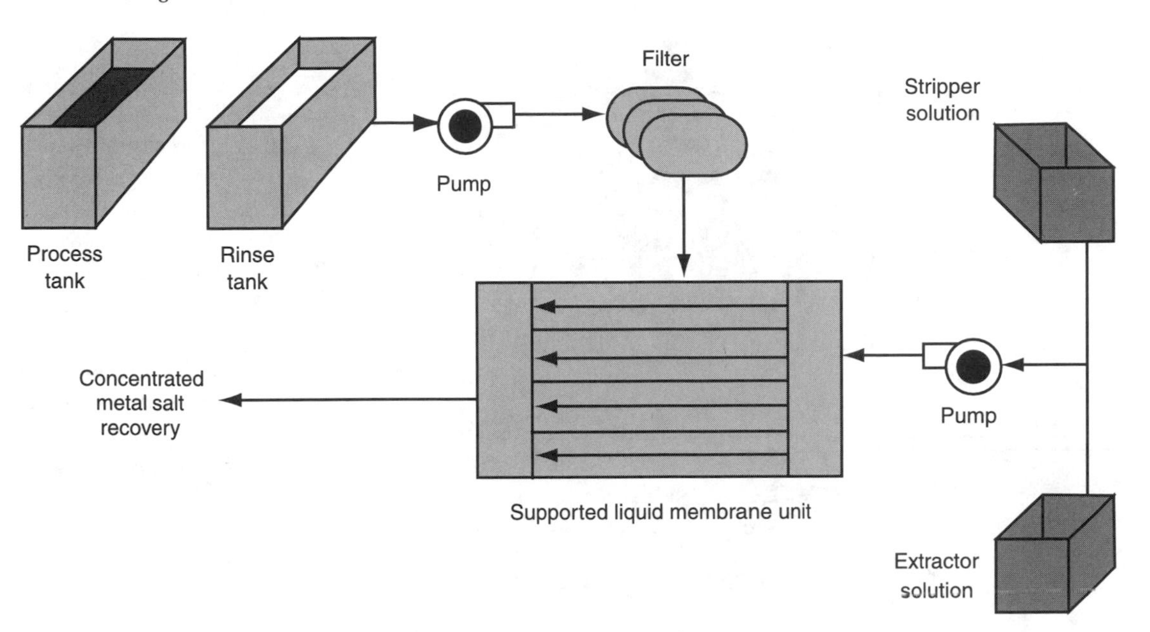

FIGURE 3–7
Metal recovery using a supported liquid membrane system.

nologies keep the baths in good operating condition and result in a more consistent production quality. Plant loadings to wastewater treatment systems are reduced, resulting in decreased purchases of treatment chemicals and sludge disposal costs.

In-process ion exchange is a reversible chemical reaction that exchanges ions in a feed stream for ions of a like charge on the surface of an ion exchange resin. Typical cation resins exchange H^+ for other cations, while anion resins exchange OH^- for other anions. This technology is used for water recycling or metals recovery. In water recycling, deionizing removes all cations and anions from a relatively dilute rinse stream and recycles the deionized water back to the rinsing process. For metals recovery, metal scavenging recovers only the metal content from the drag-out. Scavenging also provides a highly concentrated regenerant, which is particularly suitable for electrolytic recovery.

Reverse osmosis is a membrane-separation technology used for chemical recovery. The membrane separates the feed stream into a reject stream and a permeate. The reject stream contains most of the dissolved solids and is deflected by the membrane while the permeate passes through. The permeate stream is usually of sufficient quality to be recycled as rinse water. Reverse osmosis membranes reject organics and nonionic dissolved solids as well as more than 99 percent of multivalent ions and up to 96 percent of monovalent ions. This technology is most applicable to electroplating rinse waters for nickel and cyanide processes. Figure 3–8 shows the conceptual theory of reverse osmosis technology.

Electrolytic recovery is an electrochemical process used to recover metals from process baths and solutions. Metal ions are removed from the waste stream by passing the waste stream through an electrolytic cell involving closely spaced anodes and cathodes. Drag-out recovery rinses and ion-exchange regenerant are common

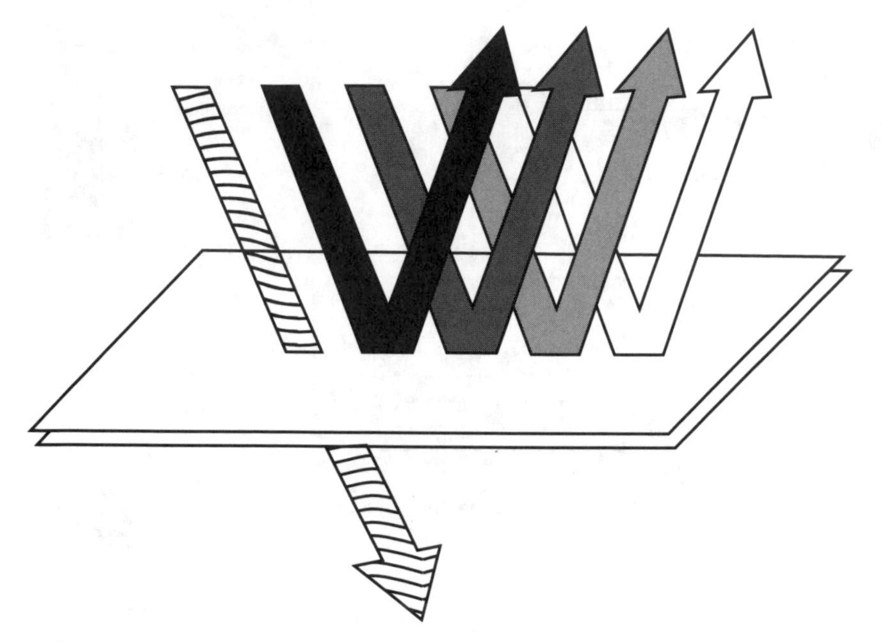

The porosity of the reverse osmosis membrane filter media allows water to permeate through, but rejects salts, low- and high-molecular weight compounds, as well as suspended colloids and particles.

FIGURE 3–8
Conceptual theory of reverse osmosis.

sources of solutions that are processed using electrolytic recovery. This recovery technology works well in the metal-finishing industry for process solutions containing cadmium, copper, nickel, and precious metals. Process solutions containing chromium or aluminum are poor candidates for electrolytic recovery.

Rinsing operations are included in numerous metal-finishing processes to dilute and remove dirt, oil, or chemicals that remain on the parts and part racks. Rinsing removes the drag-out from the part surfaces, improves the surface finishing process, and prevents the contamination of subsequent process operations. Countercurrent cascade rinsing is widely used to reduce the discharge rate of rinse waters at metal-finishing facilities. Over time, the first rinse becomes contaminated with drag-out and reaches a stabile pollutant concentration. Subsequent rinses stabilize at lower and lower concentrations. The more rinse tanks used in the rinsing process, the less water needed to adequately remove the process solution.

ENVIRONMENTAL, HEALTH, AND SAFETY ISSUES

As the metal-finishing industry operations are broad and varied, so too are the potential resultant environmental, health, and safety issues. All of this industrial category's unit operations and individual process operations can present potential hazards to the environment as well as to the health and safety of each facility's employees. All metal-finishing unit operations have the potential to generate waste streams that are hazardous to the environment.

As indicated earlier in Table 3–4, all six of the industry's unit operations generate heavy metal wastes; three also generate hexavalent chromium as a waste. Five of the six unit operations generate cyanide, oil and grease, and toxic organic waste streams. From an environmental standpoint, all of these waste streams must be treated, neutralized, or disposed of in a legal and acceptable manner to minimize their impact.

As with any industrial process or operation, numerous occupational health and safety hazards provide opportunity to injure the employee. Appropriate personal protective clothing and equipment must be used. These include head, hand, eye, ear, and foot protection; respiratory protection; and chemical exposure protection. In addition to clothing and equipment, administrative controls and engineering controls are available to afford protection to the employee. Administrative controls include such things as training and education, and worker rotation. Engineering controls include ventilation, enclosing or isolating the process, less-hazardous chemical substitution, etc.

The predominant occupation safety and health hazards in the metal-finishing industry can be classified as follows:

- Kinetic or mechanical
- Thermal
- Electrical
- Chemical
- Acoustic

Kinetic or mechanical hazards include slip-trip-fall injuries along with struck-by and striking injuries. Thermal hazards include temperature extremes as well as fires and explosions. Electrical hazards are normally present in any industrial facility. Chemical hazards are present because many of the process operations use chemical baths and solutions. Acoustic hazards are also normally present in an industrial facility. Table 3–7 summarizes the occupational health and safety hazards present in the various unit operations found in metal-finishing facilities.

As Table 3–7 indicates, all six metal-finishing unit operations provide opportunity for employees to be exposed to various health and safety hazards. Of special note is the potential for chemical exposure hazards in each one of the unit operations.

TABLE 3–7
Metal-finishing unit operations health and safety hazards.

Metal-Finishing Unit Operation	Kinetic/ Mechanical	Thermal	Electrical	Chemical	Acoustic
Form-cut-join	X	X	X	X	X
Machining	X	X	X	X	X
Cleaning	X	X		X	
Finishing	X	X		X	X
Coating		X	X	X	
Miscellaneous	X	X	X	X	

TABLE 3–8
Metal-finishing unit operations chemical hazards.

Metal-Finishing Unit Operation	Typical Exposure Scenario	Health Hazard or Target Organ
Form-cut-join	Heavy-metal dusts Welding fumes	Kidney and lung damage Nuisance particulates Eye, nose, throat irritation Systemic toxicity
Machining	Heavy-metal dusts	Kidney and lung damage Nuisance particulates Eye, nose, throat irritation Systemic toxicity
Cleaning	Salt bath chemicals Degreasing chemicals Solvents	Chemical burns, dermatitis Eye, nose, throat irritation Narcosis, systemic toxicity
Finishing	Corrosive chemicals	Chemical burns, dermatitis Eye, nose, throat irritation
Coating	Corrosive chemicals Volatile organics	Chemical burns, dermatitis Eye, nose, throat irritation Narcosis
Miscellaneous	Heat treat chemicals	Eye, nose, throat irritation Dermatitis

Table 3–8 provides information on selected potential exposures and health hazards from the various chemicals used in metal-finishing unit operations.

SUMMARY

- The metal-finishing industry comprises six basic unit operations consisting of forty individual processes. The basic unit operations include shaping, machining, cleaning, finishing, and coating. Individual processes range from abrasive blasting to hot dip coating to vacuum metalizing. As an industrial category, the metal-finishing industry is closely related to several other industry categories including electroplating, electrical and electronic components manufacturing, and coil coating.
- Inputs to metal-finishing facilities include metals, alloys, and nonmetallic materials as feed stock along with numerous hazardous chemicals. These chemicals are used in a majority of the process operations that are performed on the workpieces. Output from metal-finishing facilities includes a wide array of finished products as well as air emissions, hazardous wastes, and wastewater discharges. Hazardous wastes generated by the industrial operations include heavy metals, cyanide, solvents, oil and grease, and toxic organics. The predominant waste stream from metal-finishing facilities is the discharge of wastewaters.

- The metal-finishing industry is regulated by a number of regulations that control air pollution, water pollution, and hazardous wastes. The Clean Air Act, its amendments, along with source-specific and category-specific emission standards for hazardous air pollutants, regulates the air emissions. The Clean Water Act with its category-specific pretreatment standards and effluent limitations regulates the wastewater discharges. The Resource Conservation and Recovery Act with its far-reaching hazardous waste regulations controls the generation, treatment, storage, and disposal of solid hazardous wastes. In addition to these environmental regulations, there are occupational safety and health requirements in place that protect the workers at metal-finishing facilities.
- Treatment and control technologies, in combination with pollution prevention programs, range from the most simple techniques to very complex technological innovations. As with any industry that uses hazardous chemicals in its manufacturing processes, the metal-finishing industry has its share of environmental issues along with worker health and safety concerns.

QUESTIONS FOR REVIEW

1. List the six basic operations of the metal-finishing industry as determined by the EPA industrial category definition.
2. List the six unit operations for metal-finishing manufacturing operations; for each unit operation, list and briefly describe two process operations.
3. Briefly describe the general materials balance for the metal-finishing industry.
4. List the five primary hazardous waste streams generated by the various process operations in the metal-finishing industrial category.
5. Of the six unit operations, which two unit operations generate hazardous wastes in all five primary waste streams?
6. Which hazardous waste stream is generated by all six unit operations?
7. List the twelve pollutant categories regulated by the BPT-effluent limitations for the metal-finishing industrial category.
8. List four in-process control technologies and four end-of-pipe treatment technologies for metal-finishing operations.
9. Briefly describe three air pollution control technologies used in the metal-finishing industry.
10. Briefly describe three pollution prevention technology options for source reduction or recycling for the metal-finishing industry.
11. List the five categories of occupational health and safety hazards that exist due to metal-finishing manufacturing operations.
12. Briefly discuss three chemical hazard exposure scenarios in metal-finishing unit operations and their specific systemic health hazards or target organs.

ACTIVITIES

1. Visit a metal-plating facility and discuss plant operations and pollution prevention with the plant manager.
2. Visit a local sewage treatment plant that receives industrial waste discharges and discuss treatment technology for those industrial wastewaters.
3. Visit with local/regional environmental protection agency personnel and discuss MP&M regulations and/or NESHAP regulations.

REFERENCES

American Society for Metals. 1982. *Metals Handbook, 9th Ed. Volume 5: Surface Cleaning, Finishing, and Coating.*

Buonicore, A.J., and W.T. Davis, Eds. 1992. *Air Pollution Engineering Manual.* New York: Van Nostrand Reinhold.

Burgess, W.A. 1981. *Recognition of Health Hazards in Industry—A Review of Materials and Processes.* New York: John Wiley & Sons.

Clayton, G.D., and F.E. Clayton, Eds. 1991. *Patty's Industrial Hygiene and Toxicology, 4th Ed., Volume I, Part A-General Principles.* New York: John Wiley & Sons.

National Safety Council, Occupational Safety and Health Data Sheets (various). Chicago: National Safety Council.

Nemerow, N.L., and A. Dasgupta. 1991. *Industrial and Hazardous Waste Treatment.* New York: Van Nostrand Reinhold.

Plog, B.A., Ed. 1988. *Fundamentals of Industrial Hygiene, 3rd Ed.* Chicago: National Safety Council.

United States Environmental Protection Agency. 1995. "Development Document for the Proposed Effluent Limitations Guidelines and Standards for the Metal Products and Machinery Phase I Point Source Category." Washington, D.C.

United States Environmental Protection Agency. 1984. *Guidance Manual for Electroplating and Metal Finishing Pretreatment Standards.* Washington, D.C.

United States Environmental Protection Agency. 1983. Title 40, Code of Federal Regulations, Part 433—Metal Finishing Point Source Category. Washington, D.C.

4

Construction Wastes

Brian F. Goetz

Upon completion of this chapter, you will be able to meet the following objectives:

- Know the basic principles for identifying, managing, handling, and disposing of construction and demolition wastes.
- Emphasize Pollution Prevention (P_2), since waste reduction and recycling are becoming an evermore important aspect of construction and demolition projects due to increased disposal costs and long-term liabilities. (Lesser emphasis has been placed on safety issues since they are quite broad in and of themselves.)

INTRODUCTION

Construction waste disposal is an often overlooked constituent of the items disposed of in the United States and is regulated on the state and local level rather than by the federal government. Therefore, the guidelines for disposing construction wastes vary from state to state and there is no way to summarize all 50 states in a single chapter. However, a basic understanding can be derived from studying the methods of construction waste disposal as well as the alternatives such as waste reduction and recycling.

The EPA defines construction waste as:

> *waste building materials, dredging materials, tree stumps, and rubble resulting from construction remodeling, repair, and demolition of homes, commercial buildings, other structures, and pavements. This debris may contain lead, asbestos, and other hazardous substances.*

◗ **PHOTO 4–1**
Housing construction generates a good portion of the construction waste generated and disposed of today.

Construction wastes used to be disposed of in various manners, from burning to burial onsite. However, these practices have been widely abandoned and in most states made illegal. Today most of the construction waste generated in North America is deposited in dumpsters and hauled either to a permitted construction and demolition (C&D) waste landfill or a municipal solid waste (MSW) facility. A 1995 report from the EPA estimated that there were 1,800 C&D landfills operating in the United States at that time.

Another source of disposal for C&D waste is on-site facilities. Typically, these sites are used only for the disposal of C&D waste generated at that site and are closed after the construction work is completed. Because these sites are on privately owned land and receive only waste generated at that site, there is not much data on them available nationwide.

MATERIAL COMPOSITION OF CONSTRUCTION WASTE DEBRIS

The amount of construction waste disposed of in a given year is often unknown. There have been numerous studies over the years classifying the amount and type of trash disposed of in municipal landfills. However, since construction and waste debris are not considered to be municipal wastes, they are usually left out of these equations even though they often end up in municipal landfills. Secondly, where separate construction and demolition landfills are permitted these operations tend to be privately, rather than municipally, operated and their owners do not have to thoroughly report the quantity and types of material they receive. Commonly, when required by individual state laws small demolition disposal sites only have to provide a registration or notification of operations and to maintain simple records of the quantity and origin of wastes disposed.

The EPA report *Characterization of Municipal Solid Waste in the United States, 1960 to 2000* estimated that C&D waste accounted for approximately 24 percent of the municipal solid waste disposed of in sanitary landfills in 1984. Unlike municipal wastes, which has established and predictable generation rates, C&D waste generation is variable because it is dependent on a number of factors. The state of the economy will affect the amount of building that takes place—the better the economy, the more building and more wastes. Catastrophic events such as earthquakes, tornadoes, floods, and hurricanes will create a lot of C&D wastes. C&D wastes will vary with the seasons and regional climates. Public projects such as roadway and utility construction also add to the C&D waste stream.

Studies focused on construction wastes have revealed various estimates of how much of the nation's total waste stream comes from C&D materials, ranging anywhere from 10 to 30 percent. The Garbage Project at the University of Arizona has performed extensive studies on the type and amount of waste disposed of throughout the United States. It estimates that construction and demolition debris accounts for 12 percent by volume of a typical landfill's contents. However, when compared by weight with other garbage, construction debris accounts for 28 percent of the landfill's contents. Table 4–1 shows the results of a study performed by Toronto, Canada, that reveals the average composition of construction wastes deposited in its metropolitan facilities.

Another often overlooked item of waste being disposed of is road construction debris. Wastes from road building and reconstruction include fill dirt, concrete, asphalt, vegetation, and woody debris. Most of these materials are buried onsite or are given away to landowners needing them. Grading ordinances, such as the one in King County, Washington, have limited the activities of fill material work that needs a permit. County residents used to be able to accept up to 500 cubic yards of fill material without a permit. In 1991, an ordinance was passed that limited the amount to 100 cubic yards (about ten dump truck loads) in sensitive areas. Therefore, local regulations have put limits on the activities of disposing of these wastes and have focused on alternatives to land disposal.

Construction and demolition activities can generally be placed into one of five categories: roadwork, excavation, building demolition, construction/renovation, and site clearance. The following is a list of the typical components of each category:

- Roadwork material—asphalt, concrete, and earth fill
- Excavated material—earth, sand, stones, and miscellaneous cleared materials
- Building demolition—mixed rubble, concrete, steel beams and pipes, brick, glass, and wood

TABLE 4–1
Average composition of construction wastes.

Percentage	Material
34.8	Wood
26.9	Glass, ceramic, rubble, aggregate
16.6	Building materials
7.8	Paper and paperboard
7.3	Steel
2.5	Plastic
3.7	Miscellaneous

PHOTO 4–2
Building demolition will generate waste materials such as concrete, steel and wood.

- Construction/renovation—wood, roofing, fixtures, wall board, insulation, ductwork, pipes, metal, glass, paint supplies, plastic, cardboard shipping containers, pallets and carpet
- Site clearance—dirt, trees, roots, brush, concrete, rubble, sand and steel

HAZARDOUS WASTE IDENTIFICATION AND DISPOSAL

On-site handling and disposal of hazardous wastes is not regulated unless a facility generates over 200 pounds of any one hazardous waste in a given year. The EPA has defined generators of less than 100 kg per month as **conditionally exempt small-quantity generators.** The Resource Conservation and Recovery Act (RCRA) governs present and future activities that generate solid wastes. Under RCRA, generators are responsible for determining if their wastes are hazardous; in most cases, the builder is considered the generator. A generator is considered "conditionally exempt" if less than 220 pounds of hazardous waste are generated in one calendar month and less than 2,200 pounds of hazardous waste are stored onsite.

Given the general contractor's responsibility under federal laws, it is wise to fully document methods of disposal to ensure that such disposal is lawful. The Comprehensive Environmental Response, Compensation, and Liability Act (CERCLA) covers provisions of already existing hazardous waste sites. Under CERCLA, a contractor could be liable at a later date as a generator or transporter of any amount of potentially hazardous substance whether it is defined as a product or a waste. Examples of potentially hazardous materials include: adhesives, sealers, paint stripper, asphalt, paints, resins and epoxies, antifreeze, solvents, caulking, and waterproofing agents.

In general, construction debris is not considered toxic waste, unless it consists of asbestos or lead. These materials are encountered for the most part in demolition and renovation projects of properties containing asbestos ceilings, insulation, flooring or shingles, and lead-based paint products and require proper identification, handling, and disposal.

Asbestos

Asbestos is a widely used, mineral-based material that is resistant to heat and corrosive chemicals. Typically, asbestos appears as a whitish, fibrous material that may release fibers that range in texture from coarse to silky. Airborne fibers that can cause health damage may be too small to see with the naked eye. Exposure to asbestos can cause respiratory problems and cancer. Federal regulations regarding the handling and disposal of asbestos include:

- 40 CFR Part 763: Asbestos Containing Materials in Schools
- 40 CFR Part 61: National Emissions Standards for Hazardous Air Pollutants (NESHAP)
- 40 CFR Part 763: Asbestos Model Accreditation Plan Rule
- 40 CFR Part 763: Asbestos Abatement Projects; Worker Protection

The following materials may contain asbestos:

- Resilient floor covering tiles
- Asphalt roofing products
- Asbestos-cement products commonly used for duct insulation, pipes and siding

Many buildings constructed between 1950 and 1970 contain asbestos-sprayed ceilings and walls. The Environmental Protection Agency (EPA) estimates that there are asbestos-containing materials in most of the nation's approximately 107,000 primary and secondary schools and 733,000 public and commercial buildings. Intact and undamaged asbestos material can be handled with little risk. In fact, the EPA used to recommend removing any asbestos found in buildings and has since changed its policy to recommend that asbestos be managed and maintained until a building is torn down. Asbestos is only dangerous when it becomes airborne and is inhaled; therefore, precautions must be taken to guard against the risk of exposure to these particles.

While it is often possible to determine that a material or product contains asbestos by visually examining it, the material or product must be analyzed by instrument to be certain. Until a product is tested, it is best to assume that it contains asbestos, unless the label or manufacturer verifies that it does not. The EPA requires that the asbestos content of suspect materials be determined by collecting bulk samples and analyzing them by polarized light microscopy. This technique determines both the percent and type of asbestos in the bulk material. EPA regional offices can provide information about laboratories that test for asbestos.

If asbestos is present at a job site, it is the employer's responsibility to monitor the workplace for potential asbestos hazards, including:

- Workplace exposure to asbestos must be limited to 0.2 fibers per cubic centimeter of air (f/cc), averaged over an eight-hour work shift.
- Daily monitoring must be continued until the exposure drops below an action level of 0.1 f/cc. Daily monitoring is not required where employees are using supplied-air respirators operated in the positive-pressure mode.
- Warning signs must be posted at each regulated area and caution labels placed on all raw materials, mixtures, scrap, waste, debris, and other products containing asbestos fiber.

Generators and handlers of asbestos wastes are responsible for ensuring that all shipments of asbestos-containing waste material be accompanied by a **Waste Shipment Record** (WSR). The WSR documents the movement and ultimate disposition of asbestos waste. The WSR form stays with the waste and requires signatures from each of these responsible parties that handles the waste: the generator, the transporter, and the disposal site operator.

The generator is responsible for initiating the WSR paperwork, the hauler for making sure that the paperwork stays with the material, and the site operator for knowing when the waste was received and where it was ultimately disposed of. Waste disposal site operators are not expected to open bags or other containers to verify that the material is asbestos, as long as the WSR shows sufficient verification of the material's origination and transport. The disposal site operator is then responsible for returning the WSR to the generator. If the generator does not receive a copy of this document within forty-five days, the disposal site operator must file a discrepancy report.

Waste generators must retain copies of all WSRs, including WSRs signed by the owner or operator of the waste disposal site where the waste was deposited for at least two years. Waste disposal site operators must also keep copies of these records for two years. Additionally, disposal site operators must maintain up-to-date records that indicate the location, depth, area, and quantity of asbestos-containing waste material within the disposal site on a map or diagram of the disposal area.

Lead

Exposure to lead can cause health problems ranging from delays in neurological and physical development, to nervous and reproductive system disorders. Lead exposure occurs mainly in older buildings that contain lead-based paint. Like asbestos, lead-based materials are not harmful if they remain intact and undisturbed. It is only when lead peels and is ingested, or is made airborne and inhaled that it becomes a problem. Activities such as scraping, sanding, or using a heat gun on surfaces that contain lead-based paint can release large amounts of lead dust and fumes. Federal regulations regarding the handling and disposal of lead include:

- Toxic Substance Control Act, Title IV
- Lead-Based Paint Poisoning Prevention Act of 1971, 42 U.S.C. 4822
- Lead Contamination Control Act of 1988, 42 U.S.C. 201
- Residential Lead-Based Paint Hazard Reduction Act of 1992 (Public Law 102-550). This act is better known as Title X (of the Housing and Community Development Act).

Various agencies set standards for evaluation and management of lead, including:

- The Department of Housing and Urban Development (HUD). This department evaluates and promotes efforts to reduce lead hazards in privately owned housing. It also provides grants to communities to reduce lead hazards in housing.
- The Centers for Disease Control and Prevention (CDC). The CDC promotes state and local screening efforts and develops improved treatments for lead exposure.
- The Occupational Safety and Health Administration (OSHA). OSHA develops work standards and exposure limits to protect workers from occupational lead exposure.
- The Consumer Product Safety Commission (CPSC). This commission identifies and regulates sources of lead exposure in consumer products.

When question able materials are found onsite, a certified environmental site assessor who can perform a toxic inventory site assessment and recommend handling and disposal procedures if hazardous wastes are encountered should be brought in. His on-site visit and assessment should include:

- Visual inspection of paint condition and location
- Lab tests of paint samples
- Surface dust tests

To permanently remove lead hazards, a lead-abatement contractor must be hired. Abatement methods include removing, sealing, or enclosing lead-based paint with special materials. Certified contractors will employ qualified workers and follow strict safety rules set by state and federal government. They will also know and have proper avenues for disposing of lead-based wastes.

Lead contained in used batteries is also a concern for the construction industry. In fact, according to the EPA, batteries account for over 60 percent of all lead disposed of in landfills. Applicability and requirements for the handling and disposal of used batteries are found in 40 CFR ξ 266.80. Persons who generate, transport, or collect spent batteries, or who store spent batteries but do not reclaim them, are not subject to regulation under ξ 3010 of RCRA. Owners and operators of facilities that store spent batteries before reclaiming them also are subject to regulation under this section.

Other Hazardous Waste Materials

The Garbage Project has looked at the quantity of hazardous waste discarded in everyday household garbage. It estimates that nearly 1 percent of household waste is toxic. A study in Marin County, California, yielded an annual total of 64,700 pounds of toxic chemicals going into the county's landfills. Construction waste may constitute similar quantities when everything is added up—paint thinners, paint cans, adhesive containers and tubes, empty caulking tubes, batteries, and used oil. A study performed in 1993 by the North Carolina Department of Environment, Health and Natural Resources identified the amount of hazardous waste disposed of to be about 0.4 percent of construction waste by weight. A Portland, Oregon, study identified a total of 46 pounds of hazardous materials from the construction of a typical 1,850 square-foot house—15 pounds of sealer/caulking tubes, 15 pounds of aerosol cans,

10 pounds of joint compound, 5 pounds of adhesives, and 1 pound of resins. Additionally, many C&D wastes contain inseparable hazardous constituents, such as carpeting that can leach formaldehyde and treated wood products.

The EPA has summarized and divided the types of hazardous wastes that may be present in the C&D waste stream into four categories:

- Excess materials used in construction and their containers, including adhesives, adhesive containers, solvent containers, silicone containers, plastic laminant, leftover paint and paint containers, caulking containers, excess roofing cement, and roofing cement cans
- Waste oils, grease, and fluids, such as machinery lubricants, brake fluid, driveway sealers, oils, and engine oils
- Batteries, fluorescent bulbs, appliances
- Inseparable constituents of bulk items, such as formaldehyde present in carpet and treated or coated wood

Studies have documented what often happens to hazardous wastes at construction sites. A 1995 EPA Office of Solid Waste report titled *Construction and Demolition Waste Landfills* identified some of the disposal methods:

> *In some cases construction workers dump leftover paints or solvents on the ground. Others may use sawdust, kitty litter, or masking tape to "dry" up empty paint cans and solvent containers. "Hazardous" wastes may be disposed of in a dumpster, left at the construction site for a cleanup contractor, self-hauled to a landfill, or returned to the shop.*

A good practice for building and demolition contractors is to provide a separate collection of these materials and locate an approved disposal facility to haul them to, no matter what the volumes are.

The RCRA Hotline

(800) 424-9346

8:30 a.m. to 7:30 p.m. (EST), Monday–Friday.

The EPA's RCRA hotline provides quick response to regulatory and other questions concerning waste management and disposal. Assistance is offered to anyone who calls the toll-free number from concerned citizens and state and local government officials, to consultants, manufacturers, lawyers, and EPA and other federal employees. In addition to answering questions, the hotline staff can help identify other sources of information and can take orders for Office of Solid Waste publications. Questions or requests for information that are beyond the scope of the hotline's services are generally referred to the appropriate office (the EPA, or other federal, state or local agencies) for followup.

REGULATORY IMPACT

The1987, Subtitle D, set regulations for the permitting, control, and operation of municipal and hazardous waste landfills in the United States. The EPA promulgated rules regarding the construction and operation of these facilities. Requirements included installing liners to protect the groundwater, pipes to collect leachate, and monitoring wells to detect groundwater contamination. However, at that time the agency did not set any standards for construction and demolition landfills, leaving those regulations up to state and local lawmakers.

A nationwide survey performed by Gershin, Brickner and Bratton, Inc., in the early 1990s found that over forty of the fifty states have put regulations in place for demolition landfills. It also found that state regulations were less stringent than those for municipal solid waste because most states considered demolition wastes to be relatively inert. A study of C&D regulations performed by the EPA in 1995 summarizes the varying degree with which states regulate these wastes:

- Eleven states required all C&D landfills to meet sanitary landfill requirements.
- Twenty-four states had separate requirements for C&D landfills.
- Twenty-four states specifically prohibited any disposal of hazardous wastes at off-site C&D landfills.
- A total of seven states exempted all on-site C&D landfills from regulatory requirements.
- All fifty states had varying degrees of groundwater monitoring requirements for landfill leachate.

A lawsuit filed in the early 1990s against the EPA by the Sierra Club challenged this approach. The Sierra Club charged that C&D landfills were as potentially hazardous as municipal landfills. Their clubs research contended that C&D landfills posed a significant risk to the environment and public health, citing numerous cases of C&D facilities that accepted hazardous materials and were showing evidence of groundwater contamination. In the settlement of that lawsuit, the EPA agreed to promulgate revised criteria for nonmunicipal solid waste facilities.

In response, the National Association of Demolition Contractors (NADC) commissioned an independent study of C&D landfills and their disposal practices. Though it concurred with some of the Sierra Club's claims—high leachate levels of lead, manganese, ethyl benzene, chloride, iron and sulfates—it stressed that the majority of these offending sites were poorly managed landfills that had switched from accepting municipal solid waste to C&D-only disposal. It stated that many of these landfills did this to avoid the stricter guidelines set by the RCRA.

The NADC cited cases of well-managed C&D landfills that had good operating records and stressed that standards should be set with these operations in mind. It commented that:

> *State-of-the art demolition landfills pose no significant environmental risk. The NADC's studies show that demolition landfills, when monitored and operated consistent with industry guidelines, provide appropriate environmental safeguards. Leachate representative of such facilities meets National Primary Drinking Water Standards and can be managed so that it does not pose a significant environmental threat.*

However, as with most regulations, it is often the worst-case scenario that drives the regulation. At the time of this writing, and as a result of the January 1994 consent agreement reached in the Sierra Club suit, the EPA is reviewing its policy on C&D handling and disposal, especially as it relates to hazardous wastes. It will, in all probability, set standards that are at least as strict as those recommended by the NADC (Figure 4–1).

Responsible, trained personnel—appropriate supervision of facility operations; training requirements for all on-site employees

Routine procedures and protocols—*Plan-of-Operations Manual;* training in site safety/operational practices required of all staff

Defined listing of acceptable and unacceptable wastes—wastes allowable for receipt well-defined; personnel trained in identification

Inspection of all incoming waste loads—required disclosure of waste type and source; visual inspection of material when delivered

Siting—suitable site surface and subsurface conditions; compatible with adjacent land uses

Leachate containment—capacity to contain leachate either through native soil conditions, compaction of native soils, or other containment system

Groundwater monitoring—upgradient and downgradient groundwater monitoring for appropriate parameters, tested at least annually

Recordkeeping—maintenance of records of waste receipts and waste placement

Financial assurance—long-term funding for postclosure cover maintenance

Closure plan—design for installation and maintenance of final cover

Adapted from *The NADC Reports: Demolition Contractors Manage and Dispose of Waste Responsibly*

FIGURE 4–1
NADC recommended operating practices and design characteristics for state-of-the-art demolition landfills.

Other considerations in locating and operating a waste-disposal site include: the site (is it located on a floodplane or unstable ground); gas (are landfill gases monitored); covers (is the site covered to protect from wind blowing debris around); closure (is there a plan for monitoring the gas and groundwater after the facility has been closed). All of these things must be taken into account to assure a good operation.

Many builders are finding that disposing of construction waste can be problematic and expensive. In Toronto and other municipalities, cardboard, fine paper, clean wood, concrete and rubble, drywall, and scrap metal are banned from the landfill. More restrictions are being enacted in other urban areas as well. Elsewhere, builders are finding the costs of disposal increasing as landfill space decreases and tipping fees rise. Tipping fees rose an average of 30 percent from 1988 to 1990 and are expected to continue that way for the foreseeable future.

TREATMENT AND CONTROL TECHNOLOGIES

Many aspects of treatment and control technologies for construction wastes have and will be covered by other sections of this chapter. However, even though C&D wastes have been dealt with for years, added disposal costs and increased regulations will continue to provide the driving force for the evolution of new ideas and technologies.

One such technology to emerge has been large-scale mechanical sorting facilities. These facilities are designed to accept C&D wastes in large loads and then sort and separate the materials. They can also accept a wide range of loads, from large demolition projects to small remodeling job wastes. Incoming wastes are graded and weighed according to quality. Fees are based on the materials received and their quality. Contractors who separate and bring good materials are rewarded with lower disposal costs, while those who bring in poor quality or mixed materials must pay more.

Once the wastes are dropped off, workers pull out the large, valuable, and undesirable items and the rest is placed on conveyor belts. Grinders, screens, blowers, and magnets along the conveyor line mechanically sort items by size, density, and material composition. Ultimately, sorting facilities are designed to recover up to 80 percent of the wastes they receive. The other 20 percent has to be landfilled. However, with paybacks for recovered and recyclable materials, these facilities have been able to charge about 25 percent less than a typical landfill.

POLLUTION PREVENTION AND RECYCLING

In light of increased scrutiny focused on the handling and disposal of construction wastes, builders and communities are looking to alternatives such as waste reduction, reuse, and recycling. Many have discovered that not only can they avoid regulatory requirements and increased disposal costs, but they can save money by implementing more efficient building methods and selling recyclable materials.

A study by the Metropolitan Service District of Portland, Oregon, added up the waste disposed of during housing construction. It found that the average single-family detached house generates from 2–7 tons of waste during construction. The study described the largest component as woody material. The second largest category was drywall, followed by corrugated cardboard. Lesser amounts of metal, insulation, roofing, and concrete were also listed.

Another study, conducted by the National Association of Homebuilders (NAHB) in 1995, estimated that wood accounts for 40–50 percent of construction waste, and drywall about 15 percent. Next in line were cardboard and siding waste materials. Understandably, with this amount of waste there is incentive to reduce and recycle the amount of waste at a construction site; however, the cost and work necessary to do so may not outweigh the costs of simply disposing of it.

The NAHB study stated that the average disposal cost for a single-family home construction project's waste was $511 in 1994. Contractors looking to reduce and recycle will obviously compare their disposal costs with the alternatives. The same axioms that apply to municipal solid waste disposal apply to C&D wastes, most notably, the three Rs—*reduce, reuse,* and *recycle.*

- **Reduce.** Reducing the amount of waste produced at a construction site can save a contractor a considerable amount of money. Not only will it save on materials through more efficient use of the raw goods used to build a home or structure, it will also save on disposal costs. Fewer items will be thrown away, thus, fewer loads will go to the landfill. With tipping fee costs continuing to rise, C&D waste generators will choose to manage and reduce their own material discards more carefully. As more and more MSW landfills close and environmental regulations associated with C&D landfills become more stringent, alternative costs will act as the economic stimulus for waste reduction.

- **Reuse.** Sifting through a dumpster at a typical construction site, one may find a wide variety of usable items—from 2 by 4s to insulation, asphalt shingles to plywood, and heating duct to piping. Finding alternatives to disposal requires a little planning and creativity. Heavy waste materials such as concrete, block, and brick can be used as fill material. Centralized cutting areas for lumber and drywall can be used to save materials when a job requires a short length board or wall. A laborer can be designated to hunt and collect useful scraps of materials, gathering them in a central location. Insulation scraps can be used as sound deadeners in interior partition walls. Scrap wood can be given away for kindling fires. Chips from clean wood can be used as landscaping mulch and as bulking agents for composted sewage sludge. Large pieces of clean carpet and vinyl flooring can be saved for other projects, closets, or entryways. They can also be saved for the homeowner to store and use for replacement materials when the originals are damaged or worn. And paint supplies can be collected and consolidated for future use.
- **Recycling.** As much as 95 percent of waste generated on the job site is potentially recyclable, and most of the material, even in scrap form, is clean, uncontaminated, and unmixed with other waste materials. The key is to collect it in volumes large enough to make recycling economically viable. Most construction sites will generate modest volumes of wood, cardboard, and other materials; however, collecting them separately for recycling can be prohibitably expensive. Large-volume construction projects have more potential for establishing viable recycling operations than single projects. One builder, the Toll Brothers of Huntington Valley, Pennsylvania, saved $96,000 in landfill fees through recycling in 1995.

In Portland, Oregon, and Chicago, Illinois, job-site recycling rates had risen to nearly 50 percent in 1994. Much of this could be attributed to progressive policies and cooperative ventures by local waste haulers and building contractors. However, large increases in disposal costs may have provided as much or more of the impetus for the increase in recycling rates. Portland's tipping fee of $75 per ton of construction waste in 1994 was quite a bit above the national average of $30.

There are many uses for recycled C&D materials. As mentioned, wood scraps can be shredded and used for landscaping and sewage sludge. Gypsum, the main component of drywall, also has a high potential for recycling. It can be ground up and reused in the manufacture of wallboard or as a soil treatment for croplands. (Gypsum added to soil will act like limestone, adding sulfur.) Asphalt shingles can be recycled into road-paving and road-patching materials. And excess building materials can be collected and given to nonprofit organizations for distribution to needy families. In Gray, Maine, the Building Materials Bank collects everything from building materials to used appliances from remodeling jobs and sells them at a fraction of their original costs to low-income families that otherwise couldn't afford to maintain their homes. In 1994, the bank distributed over 90 tons of material to approximately 300 families.

Urban Ore, Inc., works in conjunction with the local waste management facility in Berkeley, California. It has a Building Materials Exchange, which deals in construction materials dropped off by individuals or businesses, or picked up at local job sites. Urban Ore pays cash for some materials and charges a fee to accept others. Reusable postconsumer materials are diverted from the landfills through this effort.

Wood Recycling, Inc., of Peabody, Massachusetts, collects and recycles wood products from construction sites. It uses the wood fiber to make composite wood products and particle-board fibers. Big City Forest, Inc., of the Bronx, New York, uses construction wood waste to make new pallets, furniture, and flooring. Its efforts led to the reclamation of 3 million board feet of lumber in 1995.

ReClaim, Inc., of Tampa, Florida, recycles asphalt roofing materials collected from C&D projects. ReClaim works with the community to identify sources and quantities of roofing debris and turns them into value-added products. American Reclamation Corporation of Southboro, Massachusetts, also recycles roofing shingles. It crushes these shingles and turns them into cold mix asphalt.

Finding space on the job site to locate separate recycling bins for each type of material can be a challenge. Materials usually have to be stored onsite until a full load is available for pickup or hauling to the recycling facility. However, when an on-site disposal system mixes a variety of discards in one waste bin, it cancels their recycling value by adding resorting costs and time. For example, recycling wood waste is much easier when it is not mixed with drywall, paint, and shingles, so if at all possible separate bins for recyclables is the best method to ensure your program is successful.

Opportunities for reducing waste start with a working knowledge of what is being discarded and who is responsible for that waste. First, a responsible crew member should be assigned to the task of auditing the construction site waste

PHOTO 4–3
A typical construction site will include a large dumpster like this one, which does little to encourage waste reduction, reuse, and recycling.

stream and developing alternatives to disposal. A person familiar with the building process and materials used will be able to coordinate better management practices to reduce these wastes. Routinely inspecting a construction waste site can reveal how efficient contractors and crews are at using their materials. Other workers can also be instructed to fill out a log whenever they deposit waste in the dumpster. Times, dates, materials, and volumes will provide valuable information on what, how much, and when things get thrown out. Once all of this information has been gathered, decisions can be made about the viability and options for waste reduction and recycling.

Secondly, general contractors should make their subcontractors responsible for their own waste. Most job sites have a single dumpster provided by the general contractor and used by everyone. If the general makes everyone responsible for their own wastes, then each sub will have an incentive to control and reduce what they discard. Guidelines for waste reduction can also be included in legal contracts for construction and demolition jobs.

In this age of environmental awareness, reducing construction and demolition wastes can be a positive marketing tool for contractors. By reducing and recycling materials, a contractor can market his or her services as environmentally responsible. Persons hiring a contractor may favor one who integrates sound waste management over one who doesn't, because a contractor who is concerned about the trash that leaves the job site will more than likely be conscientious about the quality of the work that goes into the construction.

ENVIRONMENTAL HEALTH AND SAFETY ISSUES

Worker safety and proper handling of construction wastes are an important aspect of any building job site. The regulations fall under OSHA guidelines and can be found in 29 CFR (Code of Federal Regulations), Part 1926. Subpart H addresses materials handling, storage use, and disposal. It also addresses worker safety when dealing with asbestos. Subpart T covers demolition activities.

Asbestos Disposal

Part 1926.58 describes the wastes that are generated by asbestos, including:

- Empty asbestos shipping containers
- Process wastes such as cuttings, trimmings, or reject materials
- Housekeeping wastes from sweeping or vacuuming
- Asbestos fireproofing or insulating material that is removed from buildings
- Asbestos-containing building products removed during building renovation or demolition
- Contaminated disposal protective equipment

The requirements for handling asbestos during demolition are also covered in Part 1926.252. Empty shipping bags can be flattened under exhaust hoods and packed into airtight containers for disposal. Empty shipping drums are difficult to clean and should be sealed. Vacuum bags or disposable paper filters should not be cleaned, but should be sprayed with a fine water mist and place into a labeled waste container. Process waste and housekeeping waste should be watered down or a mixture of water and surfactant used prior to packaging in disposable containers.

Any asbestos waste that is removed from buildings must be disposed of in leak-tight 6-mil.-thick plastic bags, plastic-lined cardboard containers, or plastic-lined metal containers. These wastes are to be wetted before removing to minimize the release of materials during handling. All handlers of asbestos waste must have the proper protective clothing and respiratory protection. Worker protection requirements are further defined in Part 1926.251.

As mentioned before, early regulations dealing with asbestos concentrated on its removal. However, asbestos that is in good condition does not pose a risk to the environment or health of individuals. It is only when asbestos particles become airborne and are at risk of being inhaled that they become a problem. Therefore, regulations dealing with asbestos concentrate on "managing" the material and leaving it in place if there is no risk of it becoming airborne.

Disposal of Waste Materials

Part 1926.252 covers the safety requirements of job sites dealing with waste materials. They include:

- Requiring a chute—a slide, closed in on all sides, through which materials are moved from a high place to a lower one—whenever materials are dropped more than 20 feet to any point lying outside the exterior walls of a building.
- When debris is dropped through holes in the floor without the use of chutes, the area onto which the material is dropped shall be completely enclosed with barricades not less than 42 inches high and not less than 6 feet back from the projected edge of the opening above. Signs warning of the hazard of falling materials should be posted at each level. Removal shall not be permitted in this lower area until debris handling ceases above.
- All scrap lumber, waste material, and rubbish shall be removed from the immediate work area as the work progresses.
- Disposal of waste material or debris by burning shall comply with local fire regulations.
- All solvent waste, oily rags, and flammable liquids shall be kept in fire-resistant covered containers until removed from the work site.

Demolition Site Regulations

Subpart T of Part 1926 lists the regulations pertaining to demolition sites. They include:

- Site preparation (ξ 1926.850). An inspection must be performed by an engineer or other competent person to determine the hazards that exist onsite. The integrity of the walls, floors, and other structures should be checked. If any of these structures pose a hazard to workers in the area, they must be braced. Water and electrical lines that are required on the job site must be rerouted and protected from damage. Hazardous chemicals and flammable materials must be removed prior to demolition. Where there is a potential for glass debris to create a hazard, it must be removed. Any opening that creates a falling hazard for workers must be protected to a height of 42 inches, and canopies must be constructed to protect employees from falling debris when entering the job site.
- Stairs, passageways, and ladders (ξ 1926.851). Only those stairways, passageways, and ladders designated as means of access to the structure of a building shall be

PHOTO 4–4
Covered walkways, like this one over a sidewalk next to the construction of a new skyscraper, are required to protect workers and the public from falling debris.

used. Other access ways shall be entirely closed at all times. These structures must be periodically inspected and maintained in a clean, safe condition.

- Chutes (ξ 1926.852). This section details the use and proper installation of chutes and when they are required on a job site.
- Removal of materials through floor openings (ξ 1926.853)
- Removal of walls, masonry sections, and chimneys (ξ 1926.854). This section describes procedures, load limits, and safety precautions required for the demolition and removal of walls, masonry, and chimneys.
- Manual removal of floors (ξ 1926.855)
- Removal of walls, floors, and material with equipment (ξ 1926.856)
- Storage of waste material (ξ 1926.857). This section describes the safe loading of waste material allowable on floors.
- Removal of steel construction (ξ 1926.858)
- Mechanical demolition (ξ 1926.859)
- Selective demolition by explosives (ξ 1926.860)

SUMMARY

- Construction waste disposal and landfilling regulations vary from state to state following the general guidelines set by the EPA. The EPA defines construction waste as "waste building materials, dredging materials, tree stumps, and rubble resulting from construction remodeling, repair, and demolition of homes, commercial buildings, other structures, and pavements. This debris may contain lead, asbestos, and other hazardous substances."

- The EPA estimated that approximately 24 percent of the municipal solid waste disposed of in sanitary landfills is construction waste. Of this waste, approximately 35 percent is wood, 27 percent is glass/ceramic/rubble/aggregate, 17 percent is building material, 8 percent is paper and paperboard, 7 percent is steel, 2.5 percent is plastic, and 4 percent is miscellaneous. Demolition activities like roadwork, excavation, building demolition, renovation, and site clearance also contribute a good deal to the construction waste stream.
- Most construction waste is not considered hazardous waste unless a facility generates over 200 pounds of any one hazardous waste in a given year. Under RCRA, generators (building contractors) are responsible for determining if their wastes are hazardous. In general, construction waste is not considered toxic waste, unless it contains asbestos or lead. If asbestos or lead are encountered, then these materials have to be disposed of according to the federal regulations regarding their handling and disposal. Other materials, such as paint cans, solvents, and glues, are often disposed of at construction sites. It is a good management practice for building and demolition contractors to provide a separate collection container for these materials and to locate an approved disposal facility to haul them to.
- The 1987 RCRA Act, Subtitle D, set regulations for the permitting, control and operation of municipal and hazardous waste landfills in the United States. However, it did not set any standards regulating the operations of construction and demolition landfills. A lawsuit filed by the Sierra Club challenged this approach, so the EPA is now in the process of reviewing its rules regarding the operation of these facilities, which will lead to stricter controls in the future.
- There are few treatment and control technologies for construction wastes other than mechanical sorting facilities that separate the wastes for reuse. Grinders, screens, blowers, and magnets separate the materials by size, density, and material composition. These sorting facilities are designed to recover 80 percent of the wastes they receive.
- As much as 95 percent of waste generated on a construction site is potentially recyclable. Wood, gypsum (from drywall), and cardboard are three of the main components of construction waste that can easily be recycled. Increasing landfill fees and environmental awareness is creating a market for this type of recycling. Numerous businesses have capitalized on this phenomena and are now working to divert this waste from landfills.
- Worker safety and health regulations are dealt with by OSHA under 29 CFR Part 1926. This regulation addresses materials handling, storage, use, and disposal activities. It also addresses the handling of asbestos and worker safety at demolition sites.

QUESTIONS FOR REVIEW

1. What does "conditionally exempt" mean?
2. Why do contractors want to keep good records that show they have properly identified and disposed of hazardous wastes?
3. What is a WSR?
4. Estimates show that C&D wastes account for about 12 percent of a landfill's volume, while they account for 28 percent of its weight. Why is this?

5. In the past, states have regulated C&D waste landfills less stringently than municipal solid waste landfills. Why?
6. Why are tipping fees going up so rapidly?
7. How much waste does an average home generate during construction?
8. Typically, what materials can be recycled from C&D wastes?
9. What are the benefits of an on-site C&D recycling program?
10. What factors affect the amount of C&D waste disposed of annually?

ACTIVITIES

- Contact your local solid waste management authority through the telephone book or the state department that handles solid waste and ask where the closest construction waste landfill for your community is located. What type of permit does this landfill have? How much does it charge for disposing of construction waste? How does this compare with other landfilling charges around the United States? How many years of life does this landfill have left? Is any construction waste recycling taking place at this landfill or in your community?
- Find a building under contruction and locate construction site's waste bin. Look in the bin and determine what kind of materials are being thrown out. Talk with the contractor and ask how much it costs to dispose of construction waste for the project. What percentage of the total building cost will go towards hauling this waste away? Does the contractor recycle any of this waste?
- Call the RCRA Hotline and ask them to send you the most recent information regarding the regulation of construction waste disposal in your state. What did they say? Were they helpful? What kind of information did they send you? Did they refer you to any other agencies?

RESOURCES

Center for Resourceful Building Technology
P.O. Box 100
Missoula, MT 59806
(406) 549-7678

Environmental Building News
RR1, P.O. Box 161
Brattleboro, VT 05301
(802) 257-7300

Municipal and Industrial Solid Waste Division—Office of Solid Waste
United States Environmental Protection Agency
401 M St., SW
Washington, DC 40460
(703) 308-7267
Internet: http://www.epa.gov/

The National Association of Demolition Contractors
16 North Franklin St.
Doylestown, PA 18901
(800) 541-2412
(215) 348-4949

Solid Waste Information Clearinghouse
The Solid Waste Association of North America
P.O. Box 7219
Silver Spring, MD 20907-7219
(301) 585-2898

READINGS

Brickner, R. 1995. "Construction Waste and Demolition Debris Get More Attention." *Resource Recycling.* August 1995, pp. 79–83.

Cassidy and Griffin. 1993. "Developing a County Road Construction Debris Recycling Program." *Resource Recycling.* April 1993, pp. 64–70.

Donnely, C. 1993. "Construction Waste." *Fine Homebuilding.* August 1993, pp. 70–75.

Filtz, Swanson, and Felker. 1993. "New C&D Processing Technologies." *Resource Recycling.* August 1993, pp. 28–34.

Goddard, J. 1995. "Promoting Building Industry Recycling: A How-to Guide." *Resource Recycling.* December 1995, pp. 25–26.

Holmes, H. 1995. "Where Does Construction Debris Go?" *Safety+Health.* April 1995, pp. 83–84.

McLeister, D. 1995. "Recycling Saves Toll Brothers $96,000." *Professional Builder.* October 1995, p. 33.

Mumma, T. 1995. *Guide to Resource Efficient Building Elements–5th Edition.* Missoula, MT: Center for Resourceful Building Technology.

National Association of Homebuilders Research Center, Inc. 1995. *Residential Construction Waste: From Disposal to Management–Interim Document.* April. Upper Marlboro, MD.

Rathje and Murphy. 1992. *Rubbish!* New York: HarperCollins Publishers, Inc.

Tesar, J. 1991. *The Waste Crisis.* New York: Facts on File, Inc.

The National Association of Demolition Contractors. 1995. *The NADC Reports—Demolition Contractors Manage and Dispose of Waste Responsibly.* A report based on research conducted by Gershman, Brickner & Bratton, Inc. February 1995. Doylestown, PA.

The Solid Waste Association of North America. 1993. *Construction Waste and Demolition Debris Recycling ... A Primer.* Silver Spring, MD.

U.S. Department of Labor. 1992. *Protection Against Asbestos.* OSHA Fact Sheet #92-06.

U.S. Environmental Protection Agency. 1989. *Decision Makers Guide to Solid Waste Management.* Document # EPA/530-SW-89-072. November 1989.

U.S. Environmental Protection Agency. 1990. *Field Guide for the National Emission Standards for Hazardous Air Pollutants* (NESHAP), Document # EPA-340/1-90-016. December 1990.

U.S. Environmental Protection Agency Office of Solid Waste. 1995. *Construction and Demolition Waste Landfills.* Document # EPA530-R-95-018. February 1995.

U.S. Environmental Protection Agency Office of Solid Waste. 1995. *The EPA Municipal Solid Waste Factbook,* Version 2.0.

U.S. Government Printing Office. 1987. *Title II—Solid Waste Disposal Act (RCRA) of 1987.* Washington.

5

Understanding and Preventing Pollution in the Printing and Publishing Industries

Kong S. Chiu

Upon completion of this chapter, you will be able to meet the following objectives:

- Describe the five most widely used printing processes.
- Identify the key chemicals and raw materials used in the printing industry as well as key sources of pollution.
- Explain how federal environmental laws impact the industry.
- Explain the principles of pollution prevention and apply the waste management hierarchy advocated under the Pollution Prevention Act to the printing industry.
- Identify opportunities for P_2 in the printing industry.

INTRODUCTION

The printed page is ubiquitous in today's society. Despite advances in video and computer technology, printed material continues to play an important role in the information age. The U.S. printing industry has seen steady growth in the last decade and current rates of growth are expected to continue into the next century. Annual growth between 4 and 5 percent has been projected for the industry.[1] In 1993, the printing and publishing industry produced $172 billion in goods and employed almost 1.5 million people.[2] In 1992, there were about 61,000 printing and publishing establishments across the country. Approximately 80 percent of these employed fewer than twenty people, making printing and publishing the largest collection of small manufacturing businesses in the country.[3]

The activities and processes described in this chapter fall under Standard Industrial Classification (SIC) Code Major Group 27 (Printing, Publishing and Allied Industries). The SIC Manual describes this group as one that "includes establishments engaged in printing by one or more common processes . . . and those establishments that perform services for the printing trade, such as bookbinding and platemaking. This major group also includes establishments engaged in publishing newspapers, books, and periodicals, regardless of whether or not they do their own printing."[4]

GENERAL PROCESS

In general, the printing process involves the application of ink (in the form of pictures and text) to a variety of surfaces that may include paper, plastic, cloth, and metal. Five major printing processes are commonly used in the United States. Each process uses a unique image carrier and different techniques to transfer the image to the printing surface. These five processes, in order of widest use, are: lithography, letterpress, flexography, gravure, and screen printing.

Printing and publishing activities can be roughly grouped into three stages: preparation, printing, and finishing. The three steps and differences between each of the major printing processes will be described in this section. Figure 5–1 shows a block diagram that outlines the entire printing process from prepress through postpress. Refer to this diagram as you read about the individual steps and processes used to produce a printed document.

Preparation (Prepress)

Prepress activities include everything that is done before the image is transferred to the printing surface. For text that is being printed, the manuscript must be prepared, the style and arrangement of the text must be chosen, and the text must be composed or typeset. For pictures and illustrations that are being printed, the art must be prepared and made camera-ready.

Once the text and pictures are ready, they are assembled together in the copy preparation process. (This process is also called "copy layout" or "copy assembly.") The purpose of copy preparation is to put together a camera-ready version of the page to be printed. The camera-ready copy is then photographed and a transparency is made. The transparency is a thin sheet of transparent acetate that carries positive images, like a slide. When light is shone through the transparency, a high-quality image of the page is produced.

The transparency is used to prepare the ***image carrier*** that transfers the image to the printing surface. Image carriers vary between the different printing processes. In lithography, flexography, and letterpress, different types of printing plates are used. In gravure, an etched copper cylinder is used while in screen printing the image carrier is a stencil placed over a fine mesh screen.

In general, photomechanic techniques are used to prepare the image carrier. The surface of the image carrier is covered with a light-sensitive coating and the image of

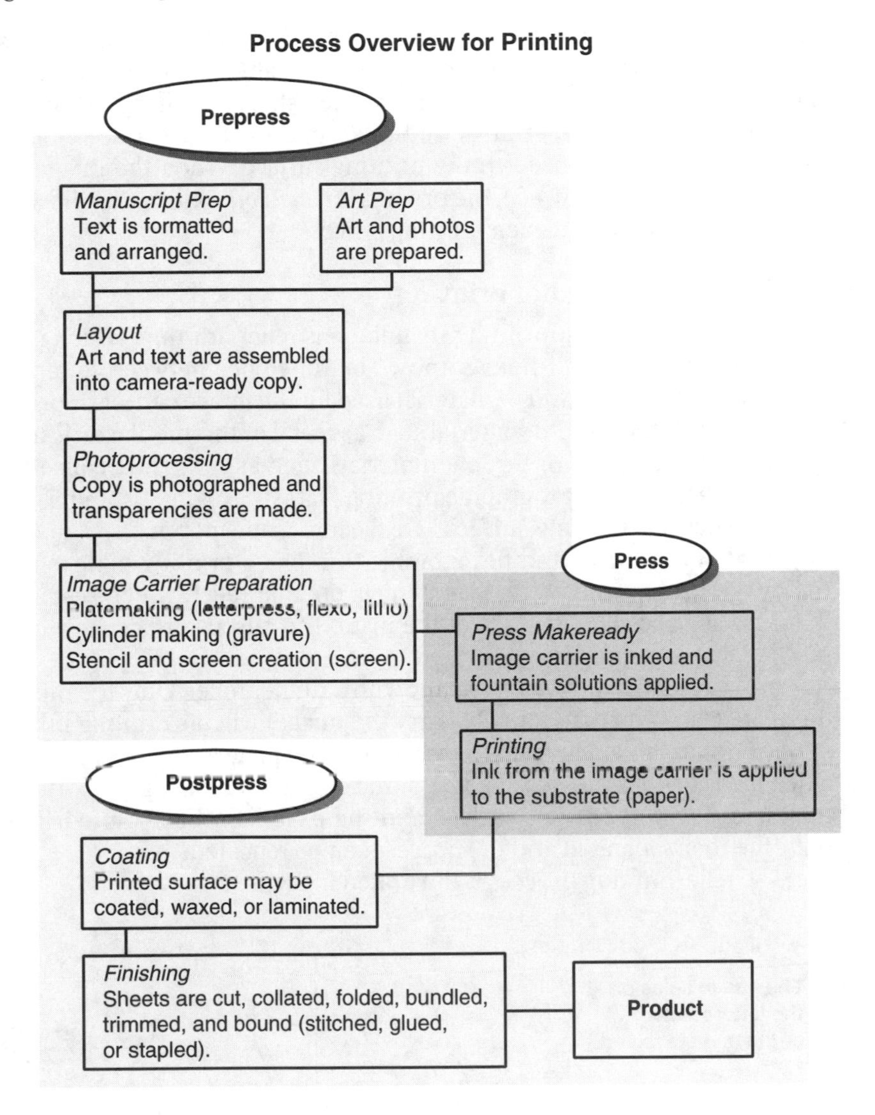

FIGURE 5–1

the page is projected onto the coating. Exposure to light (by the image being projected through the transparency) changes the characteristics of the light-sensitive coating, allowing either the exposed or unexposed portions to be removed with a solvent. Depending on the printing process, additional steps may be taken to produce the image carrier. When the image carrier is completed, the image is ready to go to press.

Printing (Press)

In the press stage, the image carrier is used to transfer the image to the printing surface. Press hardware is thoroughly cleaned between press runs or when colors are changed. The printing surface can be fed in a number of ways. In many printing operations, a continuous roll of paper or other substrate is fed into the printing assembly. This is known as ***web fed*** printing. Individual sheets or pieces are cut from the continuous roll after the press stage. An alternative to web fed printing is sheetfed printing, which uses individually fed sheets. In many instances, the image from the image carrier is transferred to an intermediary surface (such as a rubber roller) and then transferred to the printing surface. When the image is transferred with an intermediary surface, the process is called ***offset printing.*** Each aspect of the press stage is different in each printing process.

Letterpress and Flexography (Relief Printing)

The oldest and simplest printing technique is relief printing, which is used in both **letterpress printing** and **flexography.** The difference between letterpress printing and flexography is the kind of material used for the image carrier or printing plate. In letterpress printing, a hard material such as metal or thermoplastic is used. In flexography, plates are made of flexible materials such as rubber. Because of the porous and flexible nature of flexographic printing plates, flexography often is used in applications where the printing surfaces are flexible or unsmooth.

In both letterpress and flexographic printing, a printing plate is engraved with the image to be printed. The engraving is done in relief, which results in an image that is raised above the surface of the plate. The raised surfaces of a coin are examples of relief engraving.

Ink is applied to the plate, usually with a rubber roller. Only the raised surfaces of the plate (the parts that actually carry the image) will pick up the ink. The printing surface (for example paper) is then laid over the plate and pressed by another roller. Once the roller has pressed the two surfaces together, the ink and image have been transferred onto the paper surface and the page has been printed. The paper and plate are then separated and the process can be repeated.

The relief printing process is illustrated in Figure 5–2.

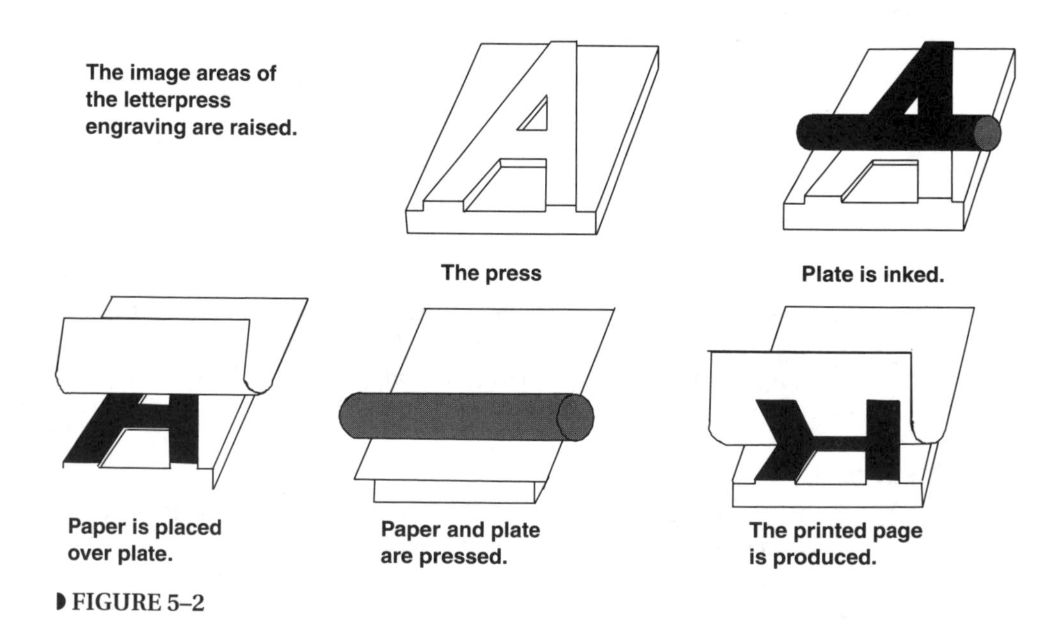

FIGURE 5–2

Lithography (Planographic Printing)

In planographic printing, the image area on the printing plate is in the same plane as the non-image area. Instead of using a raised image, planographic printing takes advantage of physical and chemical differences between the image and non-image areas. This is the technique used in **lithographic printing,** the most widely used printing process in America.

The physiochemical principle employed in lithography is the fact that oil and water do not mix. Using this principle, the image area of a lithographic plate can be rendered both ink receptive and water repellant.

A lithographic printing plate consists of a surface where the image area is covered by an oil-based coating. The oil-based coating accepts ink but repels water. Water is applied through a fountain solution to the printing plate and adheres only to the non-image area where there is no oil. When ink (oil-based) is applied to the plate, it is received by the oil-coated image area and repelled by the wet non-image area. The surface to be printed is then laid over the plate as in relief printing and the two surfaces are pressed together, transferring the ink and image to the paper. The surfaces are separated and the process can be repeated.

The lithography printing process is illustrated in Figure 5–3.

Gravure (Intaglio Printing)

A polished metal cylinder is used as the image carrier in **gravure printing**. Gravure cylinders are usually made of copper and have the printing image engraved on their surface. The image area is sunken (the opposite of relief printing) and is transferred to the cylinder through either chemical or electromechanical etching. After the image has been etched, the cylinder is usually electroplated with a thin coating of chrome to make it more durable.

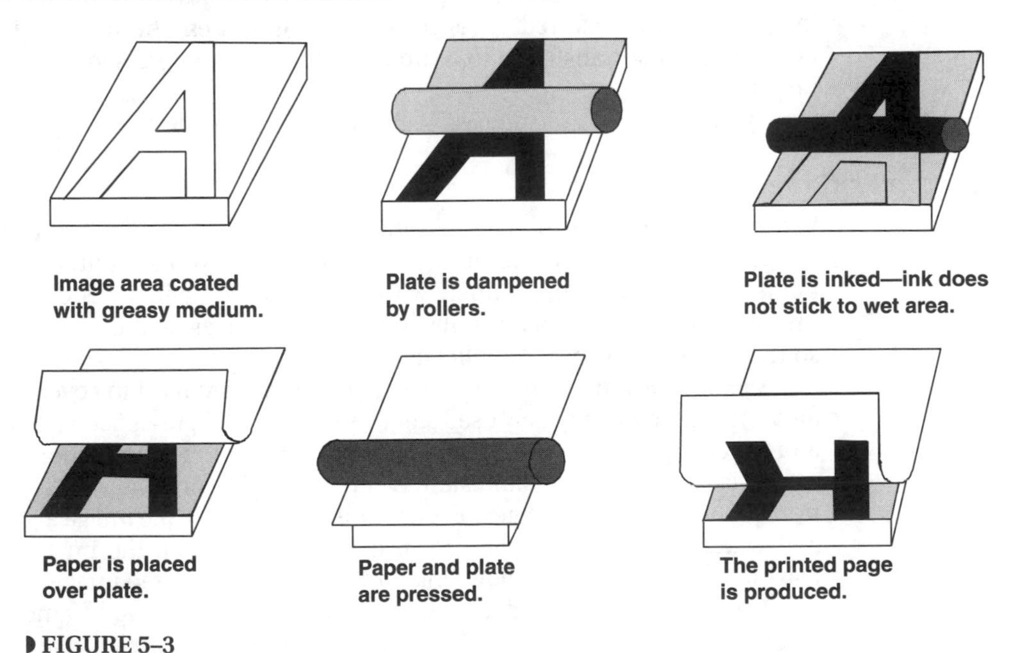

FIGURE 5–3

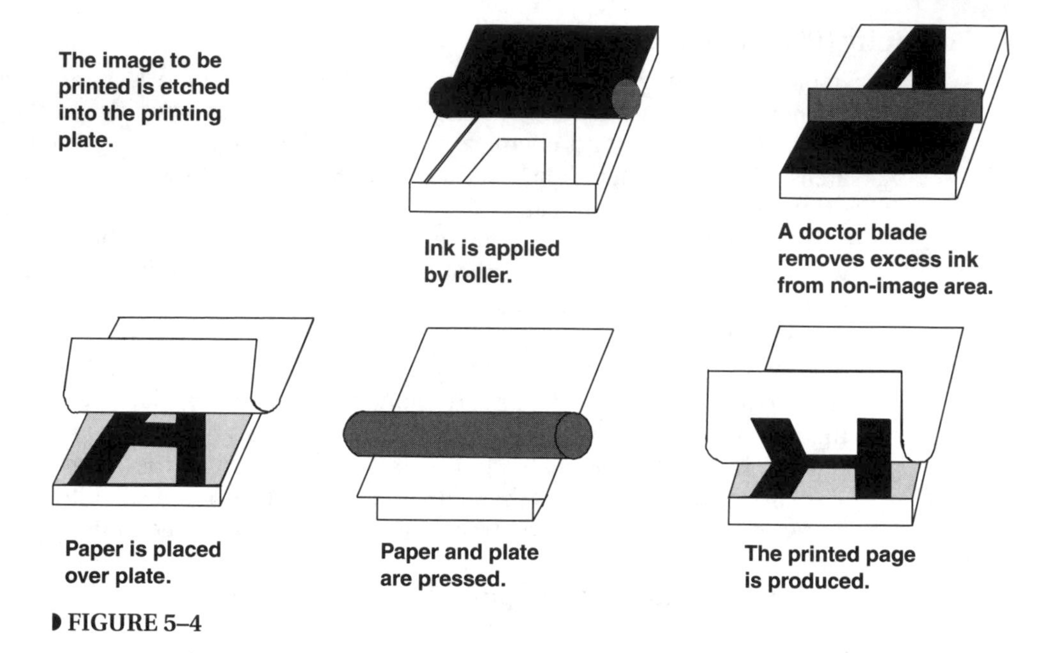

FIGURE 5–4

The cylinder is mounted on a press that rotates at high speed. As it rotates, the cylinder is spun in ink. Excess ink is removed by a wiper known as a "doctor blade" that runs along the length of the cylinder. After the doctor blade has been applied to the cylinder, ink still adheres to the small depressions and etched lines that make up the image area.

The surface to be printed is pressed onto the cylinder and the ink in the etched image area is transferred. In offset gravure processes, the ink and image on the cylinder are first transferred to a rubber roller and then pressed onto the surface to be printed.

The gravure printing process is illustrated in Figure 5–4.

Screen Printing

Screen printing is an extremely versatile printing technique that can be used to print images onto a variety of surfaces. It is often used to print T-shirts and other fabric items. In screen printing, a finely woven mesh screen and a stencil are used as the image carrier. The screen is usually woven out of polyester, though metal screens are sometimes used in special applications.

Stencils, also made from a variety of materials, are used to cover the non-image areas of the screen. A widely used stencil material is a light-sensitive emulsion that is a product of photomechanical stencil-making techniques. The whole surface of the screen is coated with the emulsion. When exposed to an image projected through a transparency, the non-image areas dry and harden while the image areas remain soft and water soluble. When the screen is washed in water, the image area falls off, leaving a stencil that only allows ink through the unprotected image areas.

The screen is stretched and mounted on a metal or wood frame. If a separate stencil has been made (one that isn't attached to the screen), it is placed underneath the screen, between the screen and the surface to be printed. A rubber blade or

A fine mesh, stretched over a frame, forms the screen.

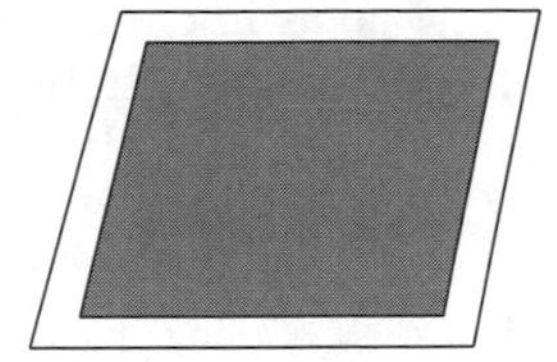

The frame

Non-image areas masked off so they do not print.

Paper is placed beneath the screen.

Ink is spread by a squeegee.

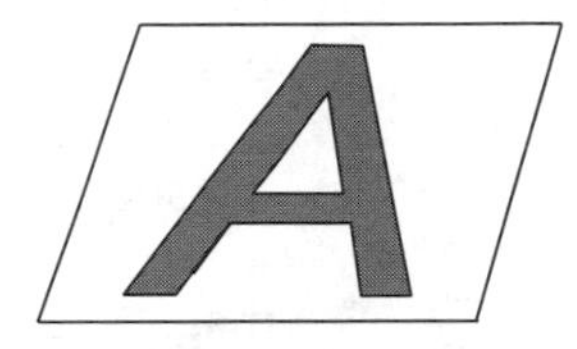

The ink passes through the screen to form the image.

FIGURE 5–5

squeegee is used to apply ink evenly across the top screen's surface. As the squeegee moves across the surface of the screen, ink is forced through the fine openings in the screen fabric and onto the surface that is being printed. Ink does not penetrate the areas that are protected by the stencil. The result is a printed image or pattern.

The screen printing process is illustrated in Figure 5–5.

Electronic Printing

Although electronic printing comprises a very small amount of total printing production in the country, its use has grown in recent years and the next decade could show a dramatic increase in the use of this technique. Electronic printing is important from an environmental standpoint because it is a plateless process and avoids much of the pollution involved in developing, etching, and cleaning printing plates. The two predominant electronic printing processes, **xerography** and **laser printing,** use the same basic principles. The word xerography is derived from the Greek *xeros,* meaning "dry," and *graphe* meaning "writing." The xerographic process dates back to the 1930s and is called "dry writing" because it uses static electricity to deposit a dry, black powder onto a sheet of paper.

Xerography takes advantage of a phenomena known as **photoconductivity.** A material that is photoconductive conducts electricity when it is exposed to light. In xerography, a photoconductive material such as selenium is bonded to the outside of a cylindrical metal drum. If the surface of the drum is given a static charge, anything with a smaller or different charge will adhere to it. If, however, the charged drum is exposed to light, the layer of selenium becomes electrically conductive and loses the static charge by conducting it away to the metal drum.

In xerographic printing, a photoconductive drum is given a static charge in darkness. The image to be printed is then projected onto the drum using a bright light. When the light is projected onto the drum, the areas that are exposed to the light lose their static charge. The areas in darkness retain their charge. The result is a latent image that exists on the drum as a static charge. A conductive, fine black powder is spread on the drum and adheres only to the charged or image carrying areas. This powder is known as ***toner*** and generally consists of fine carbon and ground plastic particles. The toner is the "ink" in the xerographic process.

The toner (which now carries the image) is transferred to a sheet of paper by rolling the paper around the drum. As the paper is rolled on, an electric wire applies a static charge to it. The paper is given a higher static charge than the drum so that the toner will travel from the drum to the paper. Once the toner image is transferred to the paper, the image is fused onto the paper with heat and pressure. This ensures that the image is permanent and the toner won't come off the paper. In some systems, the static image-carrying charge on the drum is transferred directly to the paper before toner is applied. Then toner is applied to the paper instead of the drum and sticks only to the paper's image areas.

The key difference between xerography and laser printing is the way that the image is projected onto the drum. In xereography this is done by reflecting light off a hard-copy original and projecting the reflected light through a system of lenses and mirrors. In laser printing, the image is created electronically on a computer. The digital information is then translated and projected directly on the drum by an LED laser.

In general, electronic processes do not produce the quality and resolution required for many professional printing jobs. These processes are therefore used mostly for lower quality or small batch work. As the resolution of laser printers improves, however, the use of laser printing in the printing industry is becoming more common.

Finishing (Postpress)

The last stage of the printing process involves finishing activities. Once the image has been printed onto that desired surface, that surface may be coated, laminated or waxed for durability. Web fed substrates are cut into sheets and stacked. Stacks of sheets may be cut further depending on the application. Cut sheets may then be folded into pamphlets or bundles. The pamphlets and bundles are then assembled and collated in preparation for binding.

Binding can involve stapling, stitching, and gluing. If you examine the spines of most books (including this one) you will see that the pages have been grouped into bundles that have been stitched together and glued to the cover. By far, the most widely used chemicals in postpress operations are the adhesives used to assemble and bind printed materials.

INGREDIENTS AND MATERIALS BALANCES

A variety of chemical-containing raw materials are used throughout the printing industry. Use of these raw materials results in the production of undesirable or hazardous waste streams. The materials may contain toxins, carcinogens, and volatile organic compounds (VOCs). In a pollution prevention (P_2) approach to managing waste streams, attention must be placed on the raw materials used in a particular process. By substituting alternative, less-polluting materials in the raw material selection stage, printing facilities can reduce pollution at the source. Some examples of substitute materials include alcohol-free fountain solutions, water-based inks, and low-solvent or solvent-free photographic developers. Pollution prevention practices are described in greater detail.

This section will lay out the principal ingredients used in different stages of the printing process, while "Hazardous Waste Stream Identification" will outline some of the waste streams produced by these chemical ingredients.

Materials and Chemicals Used in the Preparation Stage

As discussed in the prior section, preparation or prepress activities include all the steps taken to produce an image carrier for the printing stage. Many materials are used in the copy preparation, photography, and platemaking steps.

Adhesives are used to prepare copy, while photoprocessing chemicals are used to prepare transparencies and photographs of the pages to be printed. Acids, light-sensitive emulsions, and more photoprocessing chemicals are used to prepare the different plates or stencils in the image carriers. The principal materials used in the preparation stage are listed in Table 5–1.

TABLE 5–1
Preparation activities: principal materials and chemicals.

Process/Operation	Materials Used	Chemicals
Cleaning glass plates, lenses, photographic equipment	Film/glass cleaner	Acetone, hexane, perchloroethylene, 1-1-1 trichloroethane, ethanol, isopropanol
Developing negatives, transparencies, and prints	Developer, fixer	Hydroquinone, sodium sulfate, potassium sulfite, ammonium thiosulfate, silver, lead, chromium
Cleaning metal plates	Acids	Phosphoric acid, acetic acid
Making photopolymer plates for letterpress	Plates, plate developer and finisher, plate toner	Photosensitive polymers, benzyl alcohol, diethanolamine, polyvinyl alcohol, ethylene glycol, thiol compounds
Making photopolymer plates for flexography	Flexo platemaking solution	Methacrylate monomer, organic phosphorous compounds, petroleum distillates, anionic surfactants, potassium hydroxide
Developing and finishing lithographic plates	Plates, developer, finisher, image	Aluminum, copper, chromium, benzyl alcohol, preserver diethanolamine, polyvinyl alcohol, ethylene glycol, acetic acid, dextrin, sodium hydroxide, sodium sulfite, N-methylpyrrolidone, phosphoric acid, stoddard solvent
Applying light sensitive resists to gravure plates	Resist solution, photosensitizer, surface cleaner	Gelatin, alum, potassium dichromate, silver halide emulsions, isopropanol

TABLE 5–1
(continued)

Process/Operation	Materials Used	Chemicals
Etching gravure plates	Etching bath	Ferric chloride (for copper plates), ferric sulfate, copper salts, ammonium chloride, sodium chloride, hydrochloric acid (for chromium plates), nitric acid (for zinc plates)
Electroplating copper onto gravure cylinders	Copper plating bath	Copper sulfate, sulfuric acid, copper
Electroplating gravure cylinders for durability	Chromium plating bath	Chromic acid, sulfuric acid, chrome
Making screen printing stencils	Photographic stencil film, photoemulsions	Alkalies, plasticizers, surfactants, PVA, PVAC, methyl ethyl ketone, benzoate esters, citrate esters
Blockout (screen filler) for screen printing stencils	Pigmented polymers, solvents	Methylene chloride, methanol, methyl cellulose acetates, phthalocyanine pigments

From the U.S. EPA, *Printing Industry and Use Cluster Profile*, EPA 744-R-94-003, June 1994; and the U.S. EPA, *Printing and Allied Industries*, EPA 530-SW-90-027g.

Materials and Chemicals Used in the Printing Stage

In the printing or press stage, the image carrier is used to transfer the desired image to the printing substrate. The primary ingredients used in the printing stage are inks, cleaning solvents and substrate materials (paper, plastic, fabric, etc.). The principal materials and associated chemicals used in printing operations are listed in Table 5–2.

TABLE 5–2
Printing activities: principal materials and chemicals.

Process/Operation	Materials Used	Chemicals
Ink pigments	Pigments, dyes	Titanium dioxide, iron blues, molybdated chrome orange, phthalocyanine pigments, and pigments containing barium, chromium, copper, lead, zinc, cyanide, aluminum, cadmium, nickel, and cobalt
Letterpress and lithography inking	Ink, varnish	Petroleum distillates, vegetable oil, resin, rosin, toluene, isopropanol, xylene, and pigments containing barium and copper
Wetting lithographic plates	Fountain solutions	Isopropyl alcohol, 2-butoxy ethanol, gum arabic, ethylene glycol, phosphoric acid

TABLE 5–2
(continued)

Process/Operation	Materials Used	Chemicals
Flexography inking	Flexo ink, varnish	Benzisothiazolinon, ethylenediamine, ammonium hydroxide, isopropanol, toluene
Gravure inking	Gravure ink, varnish	Toluene, hexane, mineral spirits, heptane, methyl ethyl ketone, lactol spirits, petroleum naptha, VM&P naphtha, toluene, xylene, alcohols
Screen print inking	Screen printing ink	Glycol ethers, petroleum distillates, ketones, esters, acrylate monomers, acrylate oligomers, isocyanates, acrylic, vinyl, urethane, styrene, cellulosic, polyamide, resins, silicones, and pigments containing lead, chromium, and cadmium
Letterpress and lithography cleanup	Wash solvent	Ethanol, turpentine, acetone
Gravure cylinders cleanup	Wash solvent	Toluene, ethanol, mineral spirits, acetone, isopropanol
Cleaning/reclaiming screen printing screens	Screen reclamation solvents	Mineral spirits, toluene, xylenes, terpenes, acetone, methyl ethyl ketone, ether acetates, alcohol

From the U.S. EPA, *Printing Industry and Use Cluster Profile*, EPA 744-R-94-003, June 1994; and the U.S. EPA, *Printing and Allied Industries*, EPA 530-SW-90-027g.

Materials and Chemicals Used in the Finishing Stage

Finishing activities include final page preparation (cutting, folding, collating) and binding. Finishing activities are not particularly chemical intensive, although some binding methods rely upon various types of glue. Glues and adhesives may contain solvent-based thinners. The most common chemicals associated with binding glues and adhesives are: paraffin wax, water-soluble latex, toluene, 1-1-1 trichloroethane, isopropanaol, methanol, acetone, ethylene dichloride, and methyl ethyl ketone.[8]

Pollution

Pollution from the printing and publishing industry comes primarily from wastewater discharges, hazardous waste generation, and fugitive air emissions. Hazardous wastes are defined under the Resource Conservation and Recovery Act (RCRA) and may be listed wastes or characteristic wastes. Listed hazardous wastes can be found in Volume 40 of the Code of Federal Regulations, part 261 (40 CFR 261). Characteristic RCRA wastes are nonlisted wastes that meet RCRA criteria for ignitability, corrosivity, reactivity, or toxicity. Regulations under RCRA are promulgated and enforced by the EPA. Pollution is generated in almost all stages of the printing process, from prepress photography to postpress finishing and binding.

Pollution from Preparation Activities

As described in earlier, most of the chemical usage in preparation activities occurs during photoprocessing and image carrier preparation (platemaking). It follows that these are also the activities that generate the most pollution.

Photoprocessing pollution is generated from chemicals used to develop and fix negatives, transparencies, and prints used to prepare copy. Photoprocessing waste includes spent developer, fixer, and rinsewater, all of which are in liquid form. Spent solutions may be stored and disposed of or discharged from the facility as process wastewater with photoprocessing rinsewater. As rinsewater usually runs continuously in photoprocessing assemblies, high volumes of wastewater may be generated. Spent developer is usually caustic (high pH) and spent fixer is usually acidic (low pH). Spent fixer may also contain the heavy metals cadmium, chromium, zinc, and lead. Depending on their corrosiveness (caustic or acidic) or toxicity (heavy metal or silver content), photoprocessing wastes may be characterized as hazardous waste. If the photoprocessing solutions are discharged, pretreatment to neutralize; pretreatment to neutralize the solutions and remove silver may be necessary. Photoprocessing chemicals that contain VOCs may also be sources of air pollution by their release "by their release of VOCs" into the ambient air as they evaporate.

Platemaking pollution consists mostly of hazardous waste and process wastewater that is caustic and may contain heavy metals. Platemaking requires the use of acids, solvents, metal salts, and photopolymer solutions. The acids used to clean, etch, and electroplate different types of printing plates include phosphoric acid, acetic acid, sulfuric acid, and chromic acid. If the acids are used in etching baths, they usually contain metals etched off the plates. These metals include copper, chromium, lead, nickel, and zinc. Because of their acid content and possible metal contamination, spent etching and electroplating solutions are generally characterized as toxic hazardous waste. They are listed under 40 CFR §261.31 (hazardous wastes from nonspecific sources), EPA Hazardous Waste No. F006 (wastewater treatment sludge from electroplating operations). The photopolymers used to generate plastic printing plates create waste streams similar to photoprocessing wastes, though no heavy metals are involved. Although the developing, fixing, and toning chemicals may be different (see Table 5–2), photopolymer wastes will have the same caustic or acidic characteristics as the photoprocessing wastes previously described.

Screen stencils for screen printing are made from photoemulsions that also generate photoprocessing wastes. Solvents are usually present in the stencil films and block out solutions. Spent solvent wastes may be characterized as hazardous wastes due to their ignitability and toxicity. A number of spent solvents are listed hazardous wastes under 40 CFR §261.31, EPA Hazardous Waste Nos. F002–F005 (spent halogenated and non-halogenated solvents). Solvents are also a primary source of VOC air emissions in printing facilities. Rinsewater used to remove soluble stencil films after exposure to light may be caustic or acidic. (The section of this chapter dealing with regulatory impact will discuss characterization, treatment, discharging, and reporting requirements for handling preparation waste.)

Pollution from Printing Activities

A great deal of pollution may be generated in the pressroom. During press operations, high volumes of ink are applied to a variety of substrates to produce printed images and text. During lithographic printing, high volumes of fountain solutions are used to wet lithographic plates between ink applications. Between press runs, printing hardware (including plates, rollers and screens) is cleaned with a variety of solvents and chemicals. Together, these activities result in hazardous waste generation, process wastewater, and fugitive air emissions.

Waste Ink

Waste ink may be generated during ink blending or application to printing hardware and substrates. Waste ink can contain toxic metal pigments, organic compounds, oils, and solvents. Many of the primary chemicals in printing inks are listed in Table 5–3. Many of these constituents could result in the characterization of waste ink as hazardous due to its toxicity. Some of these chemical components will make waste ink a hazardous waste. In some printing applications, solvent-based inks with low flashpoints are used. Waste generated from using these inks may be characterized as hazardous due to its ignitability. The thinner and solvent content of ink can be a source of VOC emissions.

Spent Fountain Solutions

As previously described, lithographic **fountain solutions** are used to repel ink from the non-image areas of lithographic plates. Fountain solutions are mostly alcohol based. The primary chemical components of fountain solutions are listed in Table 5–3. If high concentrations of alcohol are present, spent fountain solutions may be highly flammable. In addition, as fountain solutions are applied to the lithographic printing plates they gradually pick up a great deal of ink. Discarded spent fountain solutions may be characterized as hazardous waste due to their ignitability from alcohol content or their toxicity from the ink constituents they contain. With their high alcohol content, fountain solutions can also be a source of VOC emissions.

Cleanup Waste

Some of the solvents used to remove ink buildup from printing hardware between press runs are listed in Table 5–2. Inks vary between each printing process and a number of different solvents are used for cleanup. As they are used, the solvents will pick up ink constituents. Spent solvent from pressroom cleanup is usually characterized as hazardous waste. The solvents are applied to hardware surfaces with rags, shop towels and disposable wipes. Most of these solvents are listed hazardous wastes under 40 CFR §261.31, EPA Hazardous Waste Nos. F002 - F005 (spent halogenated and non-halogenated solvents). Reusable rags and towels are generally sent to industrial laundries while solvent laden disposable wipes may be characterized as hazardous waste. Pressroom cleanup solvents are often the largest source of VOC air emissions in printing facilities. VOCs are released when solvents evaporate from pressroom hardware and contaminated shop towels. In water-based ink operations, solvents may not be used in cleanup. Instead, water is used to rinse printing hardware and ink-contaminated wastewater is generated. Depending on its level of ink constituents, this wastewater may be considered hazardous when discarded or discharged. Wastewater is also generated in screen printing cleanup when caustic solutions are used to dissolve and remove stencils during screen reclamation. This wastewater may be considered hazardous depending on its corrosiveness.

Pollution from Finishing Activities

In general, finishing activities in the printing industry are not sources of wastewater discharges or hazardous waste. Cutting and assembly operations will generate a solid waste, mostly in the form of excess paper and cover materials. Paper waste should be sent for recycling when possible. Lacquers or coatings applied to covers or glossy pages may be a source of air emissions due to their VOC content. In addition, some binding operations use glues and adhesives that may release VOC air emissions as well. Excess glue waste should be properly disposed.

FEDERAL ENVIRONMENTAL REGULATIONS

The printing and publishing industry is impacted by all of the major federal environmental media statutes (air, water and waste). In addition, provisions under Super-Fund may also apply to printing and publishing establishments. This section will discuss some of the primary implications for printing and publishing facilities under federal environmental regulations. In many cases, state and local environmental regulations may be more stringent and demanding than federal requirements. It is always the responsibility of the industrial facility to determine what its pollution management responsibilities are.

Air

The Clean Air Act of 1990 (CAA) is the most current and comprehensive set of federal environmental regulations designed to protect and improve air quality around the United States. Each section of the CAA is structured around a specific air pollution problem area. Of greatest interest to the printing and publishing industry are the sections on ground level ozone pollution and hazardous air pollutants.

Under the CAA, the EPA has required states to develop operating permit programs for air pollution sources. All ***major source*** facilities will be required to secure permits under the program. Major sources are defined under the CAA as industrial facilities that emit or have the potential to emit 10 tons per year or more of any hazardous air pollutant, 25 tons per year or more of any combination of hazardous air pollutants, or 100 tons per year of any air pollutant.[9] Facilities in **ozone nonattainment areas** may also be considered major sources based on the following definition:

Ground level ozone a gaseous air pollutant, is a strong pulmonary irritant and a principal component of smog. Ground level ozone is formed by the emission of VOCs and nitrogen oxides into the ambient air. A printing facility may be subject to VOC emission regulations if it is in an ozone nonattainment area and its annual VOC emissions meet or exceed EPA criteria for a major source. Information on whether a facility is located in a nonattainment area should be available through state environmental agencies or the EPA. Nonattainment areas are grouped into five categories based on air quality. Whether a facility falls under the major source for nonattain-

TABLE 5–3
Major source VOC thresholds.

Nonattainment Category	Major Source VOC Emissions Threshold (Tons Per Year)
Marginal	100
Moderate	100
Serious	50
Severe	25
Extreme	10

ment areas definition depends on its level of annual VOC emissions. The threshold level of annual VOC emissions that defines a major source is different in each category. These threshold levels are listed in Table 5–3.

To calculate the total VOC emissions for a printing facility for an year, the net VOC emissions attributed to every VOC-containing chemical used that year must be calculated. The net VOC emissions for a chemical is calculated by multiplying the total weight used in an year by the manufacturer-specified VOC content. The manufacturer-specified VOC content is given on the Material Safety Data Sheets (MSDS) that accompany the chemical. Summing up the net VOC emissions from all the VOC-containing chemicals used in a year yields the total VOC emissions from the printing facility that year.

Under CAA requirements each state must develop and administer a State Implementation Plan (SIP) to control air quality in ozone nonattainment areas. Under SIPs, major sources in ozone nonattainment areas are required to install Reasonably Available Control Technology (RACT) to control their VOC emissions. To aid states in setting RACT requirements, the EPA has issued sets of standardized RACT guidelines for different industrial categories. Several of these Control Techniques Guidelines (CTGs) have been issued for different printing industries including flexographic printing, rotogravure, and, most recently, lithographic printing. CTGs for the printing industry include performance testing, monitoring, and operating requirements that affect a facility's overall VOC emissions.

The CAA established strict guidelines for facilities that emit any one of almost 200 hazardous air pollutants (HAPs). Many of these hazardous air pollutants, including benzene, methyl ethyl ketone, toluene, and trichlorethylene, may be emitted by printing facilities. These materials may be present in the solvents and inks used in printing. For sources that fall under certain industrial categories and emit these pollutants, the EPA may establish Maximum Achievable Control Technology (MACT) requirements. MACT standards have been set by the EPA for several printing industries, including publications rotogravure, product package rotogravure, and wide web flexographic printing. The standards apply to major sources, which are defined under the issued MACT standards as facilities emitting ten tons per year of any HAP or twenty-five tons per year of any combination of HAPs.

Water

The most recent amendments to the Federal Water Pollution Control Act (Clean Water Act, or CWA) were in 1987. CWA establishes the basic framework for federal water pollution control and will probably be amended and reauthorized one more time before the end of the decade. Under CWA requirements, facilities are divided into direct and indirect dischargers.

Direct dischargers are facilities that discharge pollutants directly into the waters of the United States. Direct dischargers are regulated under the CWA through the National Pollutant Discharge Elimination Systems (NPDES), which requires permits for all point source discharges into navigable waters. NPDES permit requirements are listed under 40 CFR §122 (National Permit Discharge Elimination System Permit Regulations). If a printing facility decides to discharge process wastewater directly into navigable waters, it must complete a permit application that details the nature of the discharge (location, average flows, treatment, etc.) and report chemical oxygen demand, total organic carbon, total suspended solids, ammonia, temperature and pH.[10] In addition, the facility must test for 126 priority pollutants listed under 40 CFR §122. Commonly used printing solvents and ink wastes on the list include 1-1-1 trichloroethane, methylene chloride, naphthalene, toluene, lead, silver and cadmium. A facility also may be required to list or test for pollutants not listed under 40 CFR §122. NPDES permits listed by the EPA and/or a state agency authorized to do so will set effluent limitations, monitoring, reporting, and recordkeeping requirements.[11]

Indirect dischargers are facilities that discharge process water to a Publicly Owned Treatment Work (POTW). In general, POTWs are designed to treat municipal wastewater through physical and biochemical treatment processes. These treatment processes cannot remove metals from wastewater and may also be compromised by the presence of oily wastewater, toxic chemicals, and hazardous materials. Facilities discharging process water to POTWs must ensure that their effluent meets both general and industry category pretreatment standards.

General pretreatment standards (GPS) set minimum discharge requirements for all industrial discharges to prevent damage to POTWs and reach surface waters. GPSs are listed under 40 CFR §403 and include prohibitions on the following kinds of discharges: fire hazards, acids, obstructive solids, heated effluent, oils and pollutants that result in toxic gases, and vapors or fumes within a POTW.[12] Dischargers to POTWs are also subject to several notification requirements under GPS. No categorical pretreatment standards (CPS) have been set for the printing industry, although categorical pretreatment standards do exist for photoprocessing, ink formulating, and metal-finishing facilities. Printing establishments are exempt from chemical milling and etching standards under the metal-finishing CPS and may be exempt from photoprocessing CPS depending on the volume of film they process. Details on these CPS under the CWA can be found in 40 CFR Parts 433–460. A printing facility can meet GPS and CPS by using physical and biochemical means to pretreat its process wastewater before discharge.

Waste

As previously discussed, printing and publishing facilities may be subject to regulations under the 1984 RCRA amendments. RCRA is the comprehensive federal environmental law governing hazardous and solid waste management.

Hazardous wastes are defined, characterized, and listed under 40 CFR Part 261. Earlier in this chapter we noted that printing and publishing facilities may be sources of both characteristic and listed RCRA hazardous waste. These wastes may include the following: spent photoprocessing chemicals, spent etching baths, spent solvents, and ink wastes. It is the generator's responsibility to inventory and test all waste to determine what kind of hazardous waste, if any, it generates.

If a facility generates any hazardous waste it must obtain an EPA identification number that will be active so long as the facility generates hazardous waste. The facility will be given generator status as either a ***large quantity generator*** (at least 1,000 kg. per month or greater than 1 kg. of acutely hazardous waste per month), ***small quantity generator*** (between 100 and 1,000 kg. per month and up to 1 kg. of acutely hazardous waste per month) or ***conditionally exempt small quantity generator*** (less than 100 kg. per month and/or up to 1 kg. per month of acutely hazardous waste). Hazardous waste generators are subject to a number of recordkeeping, reporting handling, storage, and disposal requirements under RCRA. These requirements may vary for different status generators. In general, hazardous waste generators must identify and classify all hazardous waste as well as maintain paperwork on all hazardous waste generated. Generators also must package and label hazardous waste according to EPA and U.S. Department of Transportation (DOT) regulations and have a licensed transporter bring the hazardous waste to an approved Treatment, Storage and Disposal Facility (TSDF).

Underground storage tanks (USTs) are also regulated under RCRA. RCRA guidelines for USTs are set in place to prevent environmental damage from leakage. RCRA regulations cover both USTs used to store RCRA hazardous wastes and USTs used to store petrochemicals or hazardous materials other than RCRA hazardous wastes. Underground storage tanks are defined under RCRA as tanks used to store petroleum or other chemicals that are a least 10 percent (by volume, including pipes) underground. In most cases, USTs at printing facilities are exempt from RCRA regulations. RCRA exemptions exist for USTs storing heating oil used on premises, septic tanks and other tanks for collecting waste water and storm water, and flow-through process tanks and emergency spill tanks that normally empty and are emptied immediately after spills.[13] Tanks at printing facilities that may not be exempt include USTs used to store ink, solvents, and fountain solutions. Under RCRA regulations specified in 40 CFR §280, these tanks may be subject to a number of requirements including design and installation, general operations, leakage detection and reporting, and leakage response.

SUPERFUND

The SuperFund Amendments and Reauthorization Act (SARA) of 1986 incorporated both the Comprehensive Environmental Response, Compensation and Liability Act (CERCLA) and the Emergency Planning and Community Right-To-Know Act (EPCRA).

CERCLA established a national framework for responsible party liability and federal response in hazardous substance releases and environmental disasters. Some 700 hazardous substances are defined and listed by CERCLA under 40 CFR §302. A number of these are solvents, ink constituents, and platemaking chemicals used by the printing industry, including acetone, lead compounds, chromium compounds, methyl ethyl ketone, perchloroethylene, phosphoric acid, and toluene. CERCLA establishes reportable quantities for each hazardous substance based on its toxicity and threat to the environment. The reportable quantities apply to a 24-hour period. Unless exemptions apply, a facility that releases a hazardous substance in an amount equal or greater than its reportable quantity must notify the National Response Center immediately. Facilities may also be required to report the release of other toxic chemicals listed under 40 CFR 372. Under CERCLA, printers may be identified as "potentially responsible parties" during incident investigation and can be held financially responsible for cleanup costs when hazardous substance releases occur.

EPCRA (also known as SARA Title III) was established partly in response to the 1984 Union Carbide methylisocyanide disaster in Bhopal, India. To avoid the incidence of such manmade hazardous materials disasters in the United States, EPCRA aims to ensure that the public and local governments are aware of chemical hazards within a community and are equipped to respond in case of an emergency. An emergency planning and response framework is mandated under EPCRA requirements that establish state emergency response commissions and local emergency planning committees. These entities must develop written emergency response plans according to uniform national guidelines. Community awareness of chemical hazards is ensured through EPCRA requirements that facilities with *above threshold planning quantities* of extremely hazardous substances or CERCLA hazardous substances on site must report the quantities to state and local emergency response and planning commissions. In addition, any releases of these substances at or above reportable quantities must be reported to the state and local emergency response and planning committees. EPCRA-designated extremely hazardous substances that may be used at printing facilities include ammonia, hydroquinone, sulfuric acid, and toluene diisocyanate.

POLLUTION PREVENTION AND CONTROL

The 1990 Pollution Prevention Act brought to national focus the importance of avoiding the generation of pollution at the source rather than controlling pollution after it has been created. The Pollution Prevention Act stresses that source reduction is the best way to protect the environment from pollution. The core tenet of the pollution prevention (P_2) approach is: *If pollution is not created or generated in the first place then it does not need to be controlled, nor can it be regulated.* This approach is particularly appealing and beneficial to small manufacturing businesses like printing where the expenses for raw materials and pollution control equipment can be burdensome. In 1990, the EPA's science advisory board found that "preventing pollution at the source is usually a far cheaper, more effective way to reduce environmental risk, especially over the long run."[14]

Pollution Prevention Opportunities

The Pollution Prevention Act establishes a four-step waste management hierarchy that ranks waste management strategies from most desirable to least desirable. Whenever possible, prevention or source reduction should be the first strategy employed to minimize waste. Methods of achieving source reduction include "input substitution or modification, product reformulation, process modification, and improved housekeeping."[15] Pollution that cannot be prevented should, in order of preference, be recycled, treated, or disposed of.

The printing industry lends itself easily to the application of pollution prevention. The primary sources of pollution in printing are the raw materials used, particularly photoprocessing chemicals, cleaning solvents, and printing inks. Changing these materials (raw material substitution) or altering the way they are used (process change) are ideal tools for pollution reduction. Some examples of how pollution prevention methods can result in waste reduction at printing facilities are discussed here.

Pollution Problem Spent photoprocessing solutions that are corrosive or tainted with silver are a source of pollution.

Raw materials substitution. The problem of silver contamination can be avoided by seeking alternatives to silver halide photographic films. Many silver-free films have appeared on the market over the years, including vesicular films, photopolymer films, and electrostatic films.

Process changes. The volume of spent solutions generated can be reduced by minimizing the amount of fresh solutions needed to process negatives and film. Adding chemicals such as ammonium thiosulfate to fixing baths through automated systems can extend bath life. Proper storage can reduce the amount of chemicals that are spoiled from exposure to air and reduce air emissions if the chemicals contain VOCs. Using squeegees to wipe excess liquid from film and paper between baths will reduce the amount of photoprocessing chemicals that are lost.

Pollution Problem Cleaning metal plates, etching images onto printing plates, and adding chrome-plated finishes generates toxic and corrosive spent chemicals.

Raw materials substitution. Instead of raw copper plates, lithographers can use presensitized lithographic plates. The use of both plastic and photopolymer printing plates also eliminates the need for corrosive etching and plating baths.

Process change. The plate cleaning step to prepare plates for etching or remove photochemicals after etching can be changed from chemical cleaning to mechanical cleaning. Mechanical cleaning techniques include brushing, abrasive polishing, and dry-ice pellet blasting. Etching and plating bath life can be extended through careful monitoring. Another way to extend bath life and reduce the volume of chemicals entering rinsewater is to properly drain excess fluid off the plates before rinsing. Drained fluid should then be returned to the chemical baths.

Pollution Problem The use and cleanup of printing ink results in high VOC emissions and the generation of hazardous spent solvent waste.

Raw materials substitution. Water-based inks, UV-cured inks or soy-based inks are viable alternatives to solvent-based inks. These inks do not emit VOCs and can be cleaned with water-based detergents instead of solvents. Less flammable and toxic alternative cleaning solvents containing glycol ethers and hydrocarbons have also entered the market.

Process change. Proper ink storage can prevent VOC emissions and spoilage. Cutting down on press cleaning can reduce the use of cleaning solvents. Shortening the time between runs to prevent ink from drying out and cleaning only when new colors are used are effective measures. If presses are to be left unused for a period between runs, commercially available sprays can be used to prevent the ink from drying out. Automatic ink leveling systems reduce ink spoilage and waste. Newer electrostatic screen printing technologies can eliminate the use of ink completely.

Pollution Problem Fountain solutions generate VOC emissions and become ignitable hazardous waste.

Raw materials substitution. Alternative low-VOC fountain solutions using alcohol substitutes should be used.

Pollution Problem Cutting pages to size during finishing operations generates high volumes of paper waste.

Process change. Methods should be employed to maximize the printed area of the page.

Recycling, Treatment, and Control

Although less desirable than P_2 practices, recycling, treatment, and control of waste may be necessary if printing pollution cannot be prevented at the source.

Recycling opportunities exist throughout the printing process. Used photographic film and paper can be sent out for silver recovery. In some instances, spent photoprocessing chemicals can be treated and reused after the silver has been removed through electrolytic recovery. Waste ink can be sent out to be recycled or blended in-house to make black ink. Runoff fountain solutions can sometimes be collected and reused. Excess cleaning solvents can be collected in drip pans and reused either to clean press hardware or to formulate fresh ink. In some applications, solvents may be recovered and reused through distillation processes. Waste paper trimmings from finishing operations should be recycled.

Treatment and control of pollution should be used only as a last resort. A facility should make every effort possible to maximize the amount of pollution prevented and recycled. Pollution that cannot be prevented or recycled must be treated to prevent environmental damage. Ion exchange treatment systems are used to remove dissolved metal ions from wastewater. Ion exchange can be used to treat photoprocessing wastewater with residual silver content that can't be removed through electrolytic recovery. Ion exchange systems can also be used to treat metal-contaminated wastewater from plate-processing activities as well as wastewater contaminated with ink constituents. Spent ion exchange media must be sent for disposal. Metals can also be precipitated out by passing the wastewater through chemical reactors. Chemical precipitation or flocculation systems generate sludge that must be treated and properly disposed. Corrosive wastewater, including photoprocessing and plate processing water, must be treated by limestone chip or chemical neutralization. VOC emissions that can't be prevented should be captured and treated. Air pollution control equipment that may be installed for managing VOC emissions includes incinerators, condensers, and carbon adsorbers. Incinerators or afterburners will burn most of the VOCs in the air that is fed into them. Condensers and carbon adsorbers collect VOCs from air for reprocessing or disposal.

SUMMARY

- The five most widely printing processes used today are letterpress, flexography, lithography, gravure and screen printing. In addition the use of electronic printing is growing in popularity and will become fairly common in the next decade.

- The major stages in the printing process include prepress, press and postpress. Key raw materials used in printing include photoprocessing chemicals used in prepress, inks, solvents and substrates used in press and adhesives used in postpress activities.
- Pollution from the printing and publishing industry comes primarily from wastewater discharges (photoprocessing, platemaking and fountain solution wastewater), hazardous waste generation (contaminated electroplating solutions and spent solvents) and VOC emissions (solvent based inks and cleaners).
- Pollution from printing and publishing activities is regulated under the 1990 Clean Water Act, RCRA, SARA and the 1990 Clean Air Act.
- Pollution prevention (widely known as P_2) advocates the reduction of pollution at the source over recycling, treatment or disposal. The printing industry is well-suited to the application of P_2 techniques as the primary sources of pollution in printing are the raw materials used (photoprocessing chemicals, cleaning solvents and printing inks). Changing these materials (raw materials substitution) or altering the way they are used (process change) are ideal ways to apply pollution prevention.
- Some P_2 techniques used in the printing industry include substituting soy-based inks for solvent-based inks, using silver-free films, using plastic and photopolymer printing plates, improving solvent storage and application and using low-VOC fountain solutions.

QUESTIONS FOR REVIEW

1. List and describe some of the major activities in the following printing stages: prepress, press, postpress.
2. Explain the difference between lithography and relief printing.
3. What are the key raw materials used in the printing process? Examine some items around your home that have been printed. Try to find at least one printed item from each of the following *substrate* materials: paper, plastic, metal, cloth.
4. Identify the major sources of air, water, and waste pollution generated by the printing industry.
5. Suppose you are the manager of a small, family-owned printing facility. Recently, you've done several things to cut down on pollution at your plant. For each activity listed below, decide whether the activity represents prevention, recycling, treatment, or disposal. Explain your reasoning, then rank the activities in order from most preferred to least preferred. For each activity that you identify as prevention, describe whether it represents *raw material substitution* or a *process change.*
   - Increasing the capacity of your plant's metals treatment system to clean up more ink-tainted wastewater.
   - Collecting all wastepaper scraps from cutting operations once a month and sending them to a nearby paper plant where they'll be reused to make paper.
   - Buying a desktop publishing system and laser printer to print small-batch orders.
   - Collecting all waste glues, excess varnishes, and unusable lacquer once a month to be landfilled by a contractor.
   - Switching from oil-based inks to soy-based inks.
   - Installing squeegees to thoroughly wipe ink off plates before the plates are cleaned.
6. What are some ways that you, as a facility manager, can reduce pollution at the source in a printing plant?

NOTES

1. U.S. Department of Commerce. 1994. *U.S. Industrial Outlook 1994—Printing and Publishing.* Washington D.C.
2. U.S. Department of Commerce. August 1995. *1993 Annual Survey of Manufacturers, Statistics for Industry Groups and Industries.* M93(AS)-1. Washington, D.C.
3. U.S. Department of Commerce. June 1995. *1992 Census of Manufacturing.* MC92-1-27A, MC92-1-27B, MC92-1-27C.
4. U.S. Department of Commerce. 1987. *Standard Industrial Classification Manual.* PB87-100012. Government Printing Office.
5. Adapted from the following: U.S. Environmental Protection Agency. June 1994. *Printing Industry and Use Cluster Profile.* EPA 744-R-94-003; and U.S. Environmental Protection Agency. *Printing and Allied Industries.* EPA 530-SW-027g.
6. U.S. Environmental Protection Agency. *Printing Industry and Use Cluster Profile,* EPA 744-R-94-003, June 1994 and U.S. Environmental Protection Agency, *Printing and Allied Industries,* EPA 530-SW-90-027g.
7. U.S. Environmental Protection Agency. F*ederal Environmental Regulations Potentially Affecting the Commercial Printing Industry,* EPA 744B-94-001, March 1994.
8. U.S. Environmental Protection Agency. *Federal Environmental Regulations Potentially Affecting the Commercial Printing Industry,* EPA 744B-94-001, March 1994.
9. Ibid.
10. Ibid.
11. Ibid.
12. U.S. Environmental Protection Agency. *Reducing Risk: Setting Priorities and Strategies for Environmental Protection,* EPA-SAB-EC-90-021, September 1990.
13. U.S. Environmental Protection Agency. *Pollution Prevention Fact Sheet,* Office of Pollution Prevention, August 1991.

READINGS

3M/Printing Industries of America. September 1992. *Environmental Management Program Manual.* 3M/PIA.

Eyraud, P., and D.J. Watts. May 1992. *Waste Reduction Activities and Options at a Printer of Forms and Supplies for the Legal Profession.* EPA Environmental Research Brief, EPA/600/S-92/003.

Printing Industries of America. 1995. *Print Market Atlas (1995).* PIA Office of The Chief Economist.

U.S. Environmental Protection Agency. August 1990. *Guides to Pollution Prevention: The Commercial Printing Industry.* EPA/625/7–90/008.

U.S. Environmental Protection Agency. July 1993. *Project Summary: Ink and Cleaner Waste Reduction Evaluation for Flexographic Printers.* EPA/600/SR–93/086.

U.S. Environmental Protection Agency. October 1983. *Summary of Available Information on the Levels and Control of Toxic Pollutants Discharges in the Printing and Publishing Point Source Category.* EPA 440/1–83/400.

6

Laboratory Chemical and Waste Management

Mignon J. Clarke

Upon completion of this chapter, you will be able to meet the following objectives:

- Identify the basic categories of waste produced from a laboratory.
- Understand the process for determining if a waste is a hazardous waste.
- Identify the possible reactions that can occur when incompatible materials or chemicals are mixed.
- Identify the three most common treatment and disposal technologies for hazardous waste and the problems associated with each.
- Describe the concept of lab-packing materials.
- Identify applicable environmental regulations.
- Understand the concept of engineering controls, administrative controls, and personal protective equipment.

INTRODUCTION

Laboratories, for the purposes of this chapter, are defined as facilities that use chemicals to perform a variety of functions, including chemical analysis, research, and education. The activities conducted within them are extremely varied and frequently change. However, the use of chemicals is the common thread among all laboratory facilities. Facilities included in this definition include: research and development laboratories, such as government labs, industrial labs, and commercial labs; commercial testing laboratories including labs that analyze hazardous waste samples; academic laboratories, such as those found in universities, high schools, and educational or scientific organizations; and medical laboratories, including dental and hospital labs.

Types of Laboratory Waste

Laboratories are capable of producing a variety of wastes, including the basic categories of wastewater, air emissions, radioactive, biological, solid, and hazardous waste. Each of these categories is regulated by various federal and state laws. Waste categories can be further divided into waste streams. Waste streams, for purposes of this chapter, are defined as wastes generated by the same process or wastes that have common characteristics. The remainder of this chapter will concentrate on the management of nonhazardous and hazardous solid waste from laboratory facilities.

LABORATORY WASTE MANAGEMENT

The best designed and managed facilities can accumulate large and varied inventories of chemicals. Often these inventories are so large that they become difficult to manage and, over time, are forgotten. Laboratories typically respond to this situation by declaring much of the unused chemicals as waste.

Laboratories have a unique set of waste management issues that are inherent to their operations. Unlike industrial generators who typically have a few large-volume waste streams, laboratories typically have small volumes of many different waste streams. In addition to managing wastes that are generated from unused chemicals, laboratories must also manage waste generated from sample residues, spent solvents, and other analytical processes. However, the overwhelming majority of wastes coming from laboratories are hazardous and nonhazardous solid wastes generated from the disposal of unused chemicals.

Hazardous Waste

Solid waste is regarded as hazardous if through improper handling it is capable of causing injury or death, or of polluting the environment. Labs that generate solid waste must determine if their waste meets the criteria for hazardous waste (see Figure 6–1). This may be accomplished through the use of process knowledge or waste analysis as specified by environmental protection requirements discussed later in this chapter.

Waste Stream Identification and Classification

The process of identifying hazardous waste is a complicated and sometimes burdensome task. The responsibility of correctly identifying and characterizing solid waste is the sole responsibility of the generator. This process is regulated by a set of federal regulations know as the Resource Conservation and Recovery Act (RCRA).

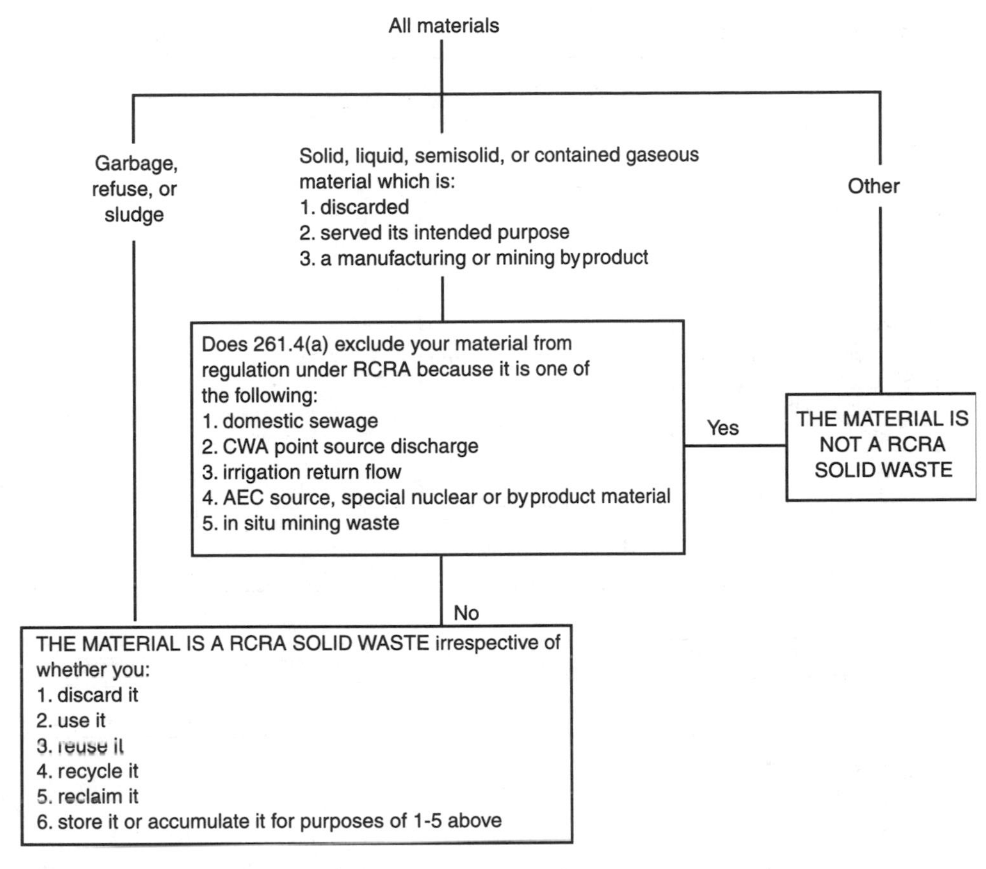

FIGURE 6–1
Definition of a solid waste.

Generally speaking, the hazardous waste determination process is a three step process (Figure 6–2). The generator must first determine if his solid waste is specifically excluded from federal or state regulations. Secondly, if the waste is not excluded, the generator must determine if the solid waste is considered a listed waste. A **listed waste** is one that appears in Part D Subpart 261 of Title 40 of the Code of Federal Regulations. These wastes may be waste commercial chemical products, specific wastes, and wastes from specific processes. Thirdly, the generator must determine if the solid waste is hazardous by characteristic. It is considered to be a characteristic hazardous waste if the solid waste exhibits one or more of the following characteristics: ignitability, corrosivity, reactivity, or toxic by the Toxic Characteristic Leachating Procedure (TCLP). Analytical testing or knowledge of the solid waste is necessary to make this determination. Analytical test methods for determining hazardous waste characteristics are specified in Part 261 Subpart C of Title 40 of the Code of Federal Regulations.

RCRA provides exemptions to the hazardous waste management requirements for laboratories that collect solid waste samples for the sole purpose of testing to de-

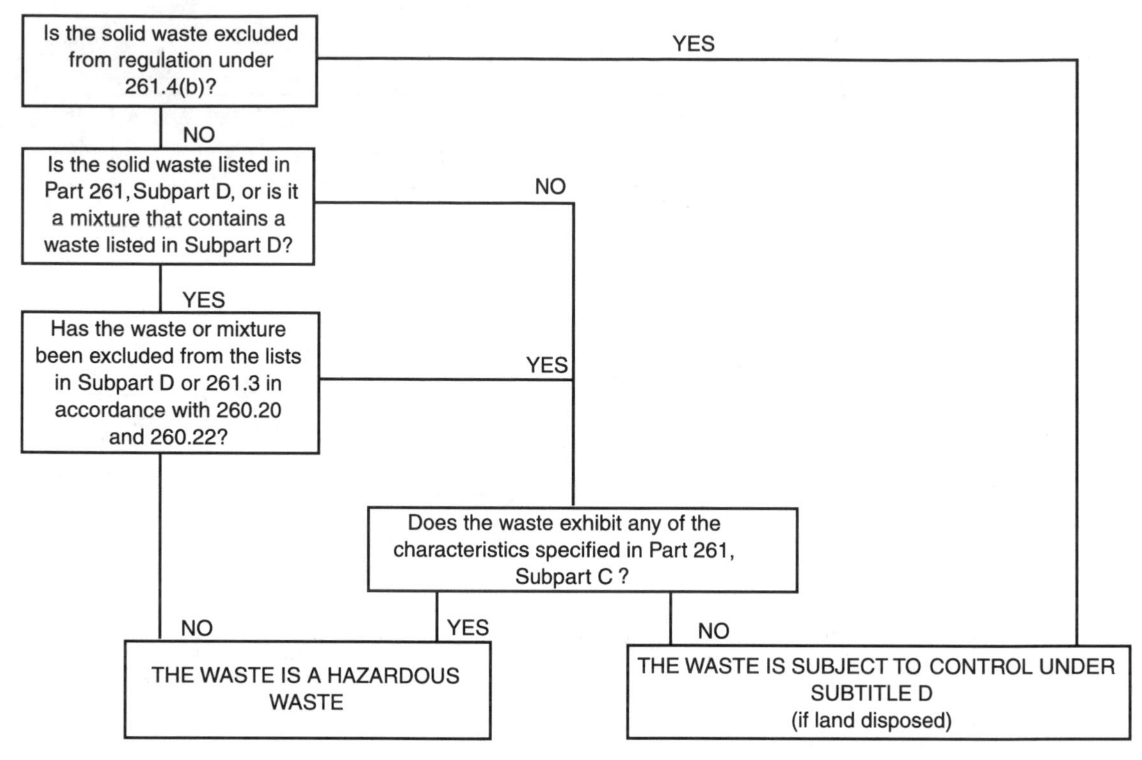

FIGURE 6–2
Definition of a hazardous waste.

termine their characteristics or composition, or to conduct treatability studies. Qualifications for these exemptions are described in Part 261 Subpart A of Title 40 of the Code of Federal Regulations. Laboratories should consult with their legal and environmental departments to determine the applicability of the exemptions.

Laboratories produce many different types of potentially hazardous waste. Some of these wastes are listed in Table 6–1.

TABLE 6–1
Commonly generated laboratory hazardous waste.

Type of Waste	Use in Lab
Spent solvents	Used in cleaning, extraction, or other analytical processes
Unused reagents	Typically chemicals that are no longer needed, do not meet specifications, are contaminated, have exceeded their storage life, or are otherwise unusable in the lab
Reaction products	Wastes of known or unknown composition that are typically produced by research and academic labs
Testing samples	Samples that are not totally consumed by analytical procedures
Contaminated materials	Glassware, protective clothing, and other miscellaneous solid waste

TABLE 6–2
Generator status.

Type of Generator	Hazardous Waste Generated Quantity	Storage Time (Without a RCRA Permit)
CESQG	<100 kg., 1 kg. acute 1,000 kg. max per month	—
SQG	100 kg.–1,000 kg. 1 kg. acute 6,000 kg. max per month	180 days; 270 days if shipped more than 200 miles
LQG	>1,000 kg. per month	Less than 90 days

In addition to determining which wastes are regulated, generators must also determine their generator status. There are three types of generator status recognized by the EPA: conditionally exempt small-quantity generators (CESQG), small-quantity generators (SQG), and large-quantity generators (LQG). The classifications are based on the amount of hazardous waste generated in one calendar month. The generator status also determines the amount of hazardous waste that can be stored at the facility, storage time, and storage conditions. Laboratory facilities can qualify for any of the three categories (Table 6–2). It is important to note that requirements are based on the facility managing hazardous waste without obtaining a permit to treat, store, or dispose of the hazardous waste.

Hazardous Waste Storage

In our definition of hazardous waste it was noted that hazardous waste is regarded as hazardous if it is capable of causing injury or death, or of polluting the environment. In keeping with this definition, let's consider how a hazardous waste is generated in a laboratory facility. A large percentage of the hazardous waste comes from waste chemicals. Some of these chemicals are ignitable, corrosive, reactive, and toxic. Therefore, the same precautions that apply to the storage of chemicals apply to the storage of hazardous waste generated from chemicals. One must keep in mind that this concept is only applicable when the composition of the waste is known. Many waste streams are mixtures of chemicals whose exact composition is seldom known.

Potential Hazards

The most common potential hazards related to the storage of hazardous waste chemicals are discussed next.

Mixing incompatible wastes in the same container. Accurately assessing laboratory waste stream composition can be a complicated task. Many laboratory facilities have multiple generators. As a result, laboratory waste streams have the potential to be composed of many different types of chemicals. This makes it difficult to properly assess potential incompatibilities. To avoid guessing what chemicals are in a particular waste stream, an accurate account of what chemicals are generated and their quantities should be maintained. Some common incompatible reactions include:

- Heat generation, explosions and fires
- Polymerization of reactants
- Generation of toxic fumes, gasses, or vapors

Spills. Spills of chemical hazardous wastes sometimes occur from metal containers corroding or the waste of materials being stored in containers that are incompatible with them. Laboratories usually have an abundant supply of empty glass bottles and metal containers that are commonly used to store small amounts of waste. However, glass containers have the potential to break when dropped and metal tends to corrode over time. Spills can also occur during the addition or removal of waste from containers. Under RCRA regulations cleanup materials and debris from a hazardous waste spill may be considered hazardous by what is known as the **mixture rule.** The mixture rule means, in simplest terms, if your waste is hazardous because it is a listed waste, anything that is mixed with it may also carry the hazardous waste designation.

Fire and safety hazards. Housekeeping practices can significantly impact safety performance in hazardous waste storage areas. As stated previously, material collected during the cleanup of a spill has the potential to be characterized as hazardous waste. In addition, an unattended spill can create a contamination hazard as well as an occupational exposure hazard. Other housekeeping concerns may include the condition of storage containers, labeling of the hazardous waste container, and inspection of waste containers. The RCRA regulations address these housekeeping concerns for the storage of hazardous wastes.

Reducing hazards associated with the storage of laboratory waste can be managed by sound laboratory management practices. However, it is extremely important to store laboratory hazardous waste in a manner that is consistent with federal and state regulations. The consequences for noncompliance with these laws can result in civil and criminal actions. Before storing hazardous waste in your laboratory facility, consult with your environmental and legal departments.

LABORATORY WASTE DISPOSAL AND TREATMENT

Until recently, storage and disposal of laboratory waste consisted of pouring it down the laboratory drain or throwing it in the dumpster. Years ago it was not uncommon to see laboratory building designs that included drainage lines identified as "Toxic Waste" or "Hazardous Waste." A common engineering practice was to include a limestone effluent trap to neutralize any acids that might be present.

In 1976, Congress passed the RCRA, a law directing the EPA to implement a program that would protect human health and the environment from improper hazardous waste management. RCRA created a new dimension in managing laboratory waste. Many laboratories generating hazardous waste were now required to dispose of their waste at a RCRA-permitted treatment, storage, and disposal facility (TSDF). To use such facilities, laboratories may find it necessary to obtain an EPA identification number. This number is a twelve-digit number issued by the EPA or, in some cases, a state agency and is used to identify generators, transporters, and TSDFs when shipping and disposing of hazardous waste. This number may be obtained by submitting a "Notification of Hazardous Waste Activity" form.

On-site Treatment of Laboratory Hazardous Waste

The EPA regulates the treatment of hazardous waste through RCRA and in most cases requires the generator to obtain a permit in order to treat, store, and dispose of hazardous wastes. Materials generated as a direct result of laboratory experimentation in some cases may be treated as part of the experiment to reduce the amount and toxicity of the waste. However, if the laboratory collects waste that meets the definition of hazardous waste for later treatment, it may be required to obtain a RCRA permit. The generator should consult with its environmental and legal departments before the waste is generated to ensure compliance with applicable regulations. In addition, personnel must be properly trained and protected to prevent exposure or injury of personnel and property.

Once the decision is made as to whether the laboratory will obtain a permit or treat hazardous wastes in the course of each experiment, the laboratory must assess the major chemical waste streams associated with its operations. After this task is complete, the laboratory must select methods and procedures to effectively and safely treat the waste. Some of the more common methods include acid and base neutralization, distillation, metal precipitation, and reactive treatment.

Off-site Treatment of Laboratory Hazardous Waste

If a laboratory decides not to treat hazardous waste on-site it will have to select a facility to dispose of its wastes. Selecting an EPA-permitted TSDF should involve research into its background and qualifications. Check the reputation of the firm, the number of years it has been in business, the status of the operating permits, results of recent EPA and OSHA inspections, and whether it is involved in any legal or court actions. Trade associations and the local Chamber of Commerce are valuable resources. The selection process may also involve a visit to the facility to observe its operating conditions. The laboratory must also select a reputable transporter to haul its waste to the facility. The selection process for the transporter should be similar to the selection process for the TSDF.

The three most common treatment and disposal technologies applied to laboratory hazardous wastes are **incineration, chemical treatment,** and **solidification** through encapsulation. Incineration is the thermal treatment of hazardous waste using high temperatures to change chemical, physical, or biological characteristics or composition. It is typically the most expensive of the alternatives. One of the most common problems associated with the incineration of waste is incomplete combustion. Chemical treatment involves the use of chemicals to alter waste properties by destroying the hazardous component or preparing the still hazardous material for further processing. Chemical treatment may include neutralization, oxidation, reduction, hydrolysis, and precipitation. Common problems associated with this method include limited solubility of some metals, reaction-inhibiting impurities in the waste, and the potential to generate hazardous byproducts. The third treatment option is solidification. Solidification of hazardous waste through encapsulation can be described as surrounding the hazardous compound with a nonreactive nonhazardous media. (Portland cement is an example of a solidifying media.) The end product of this treatment is a stable, presumably nonleaching, compound suitable for disposal. One of the problems associated with this technology is that if the hazardous compound was not chemically altered, its initial hazardous properties still exist in the capsulated form. Therefore, long-term effectiveness is dependent on the stability of the encapsulating media. Landfilling a laboratory's hazardous waste is also an alternative for its disposal. However, laboratory waste may be subject to EPA land disposal restrictions. Laboratory wastes identified in any of the EPA restricted lists may not be land-disposed unless the restricted wastes are treated according to specific standards. The treatment and disposal of laboratory waste may be a combination of any of the previously discussed technologies.

Lab Packing

Lab packing is the most common preparation method for shipping laboratory chemical waste offsite. Lab packs are multiple containers of compatible wastes placed together in a 55-gallon drum and packed with absorbent material for cushioning. The lab pack is a recognized packaging unit of the U.S. Department of Transportation (DOT) that allows different materials from the same hazard class to be packaged together in specified containers for transport, treatment, and disposal. This method is fairly simple; however, it is time consuming and requires trained personnel familiar with the composition of the waste.

Waste laboratory chemicals are first assigned to compatible groups. DOT recognizes such groups as flammable liquid, flammable solid, oxidizer, corrosive, Poison B, and "Otherwise Regulated Materials." These groups are used to identify the most applicable hazard associated with the material being shipped. Once this is done, the shipper has to determine the proper shipping name for the material. This information can be obtained from the Hazardous Materials Table in Title 49 of the Code of Federal Regulations Part 172, Subpart 101.

In addition to assigning proper shipping designations, DOT imposes other requirements when shipping the lab pack to an off-site location. These requirements include:

- Wastes from only one DOT hazard class may be packed in each shipping container.
- Wastes packaged in the lab pack must be compatible with each other and with the container itself.
- Individual containers placed in the lab pack that contain liquids must be surrounded by a compatible material capable of absorbing the total liquid contents. This material must also serve as a shock absorber.
- Lab packs must be properly marked, labeled, and packaged.

The next step in the lab-packing process involves assessing individual chemical compatibility. Unknown substances are of particular concern when assessing chemical compatibility. Chemical analysis and process knowledge may be used to properly identify unknowns. Chemical incompatibilities could result in one or more of the following:

- Heat generation, explosions, and fires
- Polymerization of reactants

- Generation of toxic fumes, gases, or vapors
- Formation of greater toxicity compounds
- Formation of pressure or shock sensitive compounds

Once the segregation, identification, and classification of the laboratory waste is completed, the next step in the process is to create a lab pack inventory sheet. This will allow the generator to effectively track the waste throughout the disposal process. Information included on the inventory sheet should include the following:

- Chemical name
- Hazard class
- Size and quantity of the container
- Number of containers
- Container identification number(s)
- EPA waste code(s)
- Reportable quantities (RQ)

The DOT and the EPA require the use of the appropriate state manifest or the Uniform Hazardous Waste Manifest whenever lab packs containing hazardous waste are transported off site for disposal. The purpose of the manifest is to create a written record that identifies all parties involved in the transport and disposal of hazardous waste.

OVERVIEW OF ENVIRONMENTAL REGULATIONS

Regulatory requirements can be very cumbersome and sometimes confusing. Current environmental regulations control the release of pollutants to the air, water, and land. Some of the environmental regulations that may effect the laboratory are discussed below.

Clean Air Act (CAA)

This law addresses air pollutants released from facilities into the atmosphere. The act requires each state to develop its own plans for complying with federal air guidelines; therefore, requirements may vary from state to state. Laboratory facilities that include incineration of animal carcasses and other wastes (including hydrocarbons, radioactive materials, biological hazards, or particulates) that may be released into the air may be required to obtain an air quality control permit.

Clean Water Act (CWA)

The Clean Water Act affects those laboratories that discharge wastewater effluents into sewers or bodies of water such as lakes, rivers, and streams. The CWA is a federal program that empowers states to issue water discharge permits. The permit associated with this law is called a National Pollutant Discharge Elimination System (NPDES) permit. This permit may set discharge limits, require monitoring of certain pollutants, or state reporting requirements. The act also regulates discharges from facilities to sewer systems that are connected to treatment works, or a publicly owned treatment works (POTW) that has NPDES permits. In this case, the POTW is responsible for meeting NPDES permit limits. To achieve this goal, the POTW may impose discharge limits to all facilities it serves.

Toxic Substances Control Act (TSCA)

This act has limited applicability for most laboratories. It regulates the manufacture and use of chemicals. TSCA authorizes the EPA to obtain data from industry regarding the production, use, and health effects of chemical substances and mixtures, and regulate the manufacture, processing, and distribution in commerce, as well as use and disposal of a chemical substance or mixture. The passage of this law was primarily in response to the large number of polychlorinated phenols (PCBs) discovered in the environment.

Resource Conservation and Recovery Act (RCRA)

RCRA governs the disposal of hazardous waste. It can be considered as a "cradle-to-grave" approach to managing hazardous waste. It makes it possible for a generator to be made liable for waste even after it has been disposed of.

Comprehensive Environmental Response, Compensation and Liability Act (CERCLA)

CERCLA, or SuperFund, regulates past disposal sites and the release of regulated hazardous wastes to the environment. The law includes a protocol that must be followed in the event of a release of regulated material. 40 CFR 302 explains the reporting steps and the minimum reportable release quantities for each regulated material. The relevance of this law to the laboratory is the potential for the release of a regulated material (waste or nonwaste) from the laboratory facility.

LABORATORY POLLUTION PREVENTION

Pollution prevention (P_2) includes a wide variety of strategies and concepts. Most environmentalists would agree that the overall emphasis of P_2 centers on the elimination or reduction of waste. Pollution prevention as it applies to laboratory activities must incorporate both source reduction and end-treatment strategies.

This section will briefly discuss the benefits of pollution prevention in the laboratory, pollution prevention strategies, and how to implement a successful laboratory pollution prevention program.

Benefits of a Laboratory Pollution Prevention Program

Economic Benefits

Economic advantages in the areas of resource conservation and disposal cost are an obvious benefit from an effective laboratory P_2 program that encourages frugality and reuse of existing resources, and as a result reduces the frequency of purchases. Reducing resource requirements is a form of source reduction.

The cost of waste disposal escalates yearly because of increasingly stringent and restrictive treatment and disposal requirements. These costs are not likely to decrease in the near future. In addition to initial disposal costs, laboratories must also be conscious of future liabilities associated with the management and disposal of waste. What may be considered legal and technically correct today may not be in the future.

Reducing the amount of waste generated by the laboratory will often result in a reduction of compliance activities, such as recordkeeping, self-audits and inspections, and training.

Personnel Morale

A successful P_2 program includes active participation and input from all persons involved in laboratory operations, a feature that promotes good morale among employees. As with any business, when employees believe that their opinions are taken seriously a positive morale develops that usually increases productivity and reduces safety issues. Laboratories that keep the lines of communication open between management and technical staff find that each group has noteworthy contributions in regard to the daily operation of the laboratory. The human factor is an integral part of a successful prevention program.

Public Image

Positive public image is considered to be a benefit received from a strong P_2 program. Public perception of a company's environmental performance can affect the company's ability to operate in a community. In this day and age communities are becoming more active in the environmental arena. An informed public is not willing to allow companies perceived as having a poor environmental track record to operate in their community. Poor public image translates into lost business and revenues for the company.

Once a positive relationship between the community and laboratory facility has been established, solicitation of public approval will be eased for such things as renewing permits and expanding operations. Community outreach is more than a neighborly gesture: It is good business.

Pollution Prevention Strategies

Chemical Procurement and Usage

A common approach to purchasing materials is to buy at the lowest per-unit cost. Unfortunately, this often means the materials must be bought in large quantities. In many instances, only a small amount of the chemical is used while the excess is kept in stock for future use. As a result, the excess chemicals remain in storage beyond the life of the initial project creating various safety and housekeeping problems. Poor housekeeping may contribute to unnecessary worker exposure, fires and explosions, and other occupational and environmental hazards. Laboratories often lack the necessary space to provide optimum storage conditions for large quantities of chemicals. This also hampers efforts to keep chemical storage areas clean and free of debris, and minimize the potential for spills.

To avoid buildup of surplus chemicals, laboratories should only purchase quantities that will be used in the foreseeable future. One way to ensure immediate use is through planning. Each project manager or researcher should give careful consideration as to how much material will actually be needed to complete his or her project. In some cases, it is difficult to predict the exact quantities needed. Project priorities, chemical availability, and requisition processing time are variables that are not easily controlled by the experimenter. These circumstances apply more specifically to specialty chemical purchases. However, the same principle can be applied in part to community stock chemical purchases. Quantities of community or commonly stocked chemicals should be minimized without completely depleting the main source at any given time. Purchasing frequency will depend on the demand for the chemical, recommended storage conditions, and risk potential.

In addition to predicting future chemical usage, researchers must try to lessen the quantities of chemicals used in analytical procedures and experiments. This reduction may be achieved by reducing the scale of the experiment or sample sizes. Smaller samples will lessen the amount of extraction solvents used in analytical procedures. In a similar fashion, reduced scale experiments generate less product and waste. Developing activities that encourage reduction of chemical usage will help to achieve P_2 goals.

The procurement process of a laboratory P_2 program should also include measures to reduce the toxicity of the chemicals used and stored in the laboratory. The chemical's toxicity not only affects the manner in which it may be used, but how it may be disposed, too. Reducing the toxicity of a chemical may require the selection of an alternative chemical. Chemical substitutes can be found for a number of common laboratory activities. A list of potential chemical substitutes should be developed and made available to laboratory personnel. In the past, activities such as the cleaning of equipment and glassware had been conducted with solvents and degreasers. These solvents have since been replaced by commercial products that work as well as the solvents but are safer and less toxic. Occupational exposure is of great concern to laboratory personnel. Effectively reducing the risk of exposure by eliminating the source is by far the best solution. Analytical procedures should be assessed to determine if substitution is a viable option.

Toxicity as it relates to the disposal of a chemical can be easily assessed in terms of economics. Highly toxic chemicals cost more to dispose of. Purchasing less-toxic chemicals results in reduced disposal costs.

Chemical Inventory Management

A comprehensive approach to chemical inventory management requires substantial information and a modern file or data management system. At any given time, laboratory personnel should be able to locate every chemical used in the laboratory. In small laboratories, locating a chemical may not be a problem; however, in facilities with multiple laboratories and stockrooms, location can be a difficult task. Many resources are available to laboratory facilities to help them establish an inventory network system.

The bar code inventory system has proven effective at some facilities, particularly when a centralized storage and data management system is used. This system allows tracking of a container as it moves throughout the organization. Systems can be adapted to suit the needs of the laboratory. Large facilities may elect to use a central mainframe computer. Small facilities, by contrast, may choose a personal computer or card file system. Laboratory personnel at each facility must decide which system is most appropriate for them.

Information in the database should include but not be limited to the name and location of the chemical, location of the Material Safety Data Sheet, quantities on hand, and special hazards associated with the chemical. The inventory system must also be able to identify chemicals that may degrade over time into potentially explosive or highly hazardous compounds.

There are several advantages to tracking chemicals in a laboratory facility. Tracking chemicals will assist procurement personnel in identifying usage patterns for community stock chemicals, and chemical purchases can be adjusted to meet laboratory consumption. Secondly, the system can identify chemicals that are not being used in a timely manner. Chemicals that no longer have an identified use can be removed from the regular inventory and moved to the excess chemical inventory. A third advantage of tracking chemical inventories is that the information contained in the database may be used to satisfy federal, state, and local reporting requirements. Remember, in most cases databases may be customized to meet specific facility needs.

Once the method of managing the chemical inventory is chosen, the laboratory must delegate the responsibility of maintaining and developing protocol and procedures for its use. The procedures should include provisions for the inspection of all chemical storage areas.

Managing Surplus Chemicals

Surplus chemicals are inherent to all laboratory operations. Unfortunately, over time these surplus chemicals in many cases become waste. Many of these chemicals are still useful if an alternative user can be located. Locating a user for containers that have been previously opened or where age is a factor may be difficult. However, simple assay and purification programs can be used to ensure the chemical's purity.

Laboratories may select a passive or an active chemical exchange program to redistribute their surplus chemicals. A **passive exchange program** provides a listing of available chemicals to potential users. Interested parties then direct inquires to the designated group responsible for the inventory, or pass on the information to the person in possession of the chemical. By contrast, an **active exchange program** involves a paid intermediary or broker who unites interested parties. Some brokers take physical possession of the chemicals. Such an exchange program requires additional time and effort. However, this system can prove to be very effective because of the greater variety and quantity of chemicals made available to potential users. Care must be exercised if the active program is preferred by an organization, because future liability is always a concern. The most effective exchange programs are often in-house operations.

Other potential off-site users may include universities, secondary school systems, or laboratories affiliated with your company. If a laboratory has access to other laboratories within the company, an effort should be made to exchange surplus chemicals with one another. One of the additional benefits of off-site exchange is that it increases visibility in the community.

Implementing the Laboratory Pollution Prevention Program

Assessing laboratory chemical management practices is one of the most important steps in implementing the P_2 program. Information gathered during the assessment will assist in determining the extent of the program's implementation. The following are some of the areas that should be addressed during the assessment.

Chemical Inventory Assessment

A chemical inventory assessment should be conducted to gather information about the various types of chemicals stored in the laboratory facility. Information should include location, quantity of chemical, chemical or trade name, expiration date, and if there are unique hazards associated with the chemical. The date on which the chemical was received is also helpful.

The next step is to note current storage conditions and to identify obvious housekeeping problems. Information should be gathered noting if the chemical is stored with compatible materials, if the chemical is placed in the proper storage area (flammable, corrosive, etc.), and if the chemical is stored within the recommended temperature range. Examples of housekeeping issues include evidence of a previous spill or gross container contamination.

The last step of the chemical inventory assessment is to identify hazards associated with the storage of laboratory chemicals. Examples of items to include on the assessment are the storage of chemicals that have the potential to degrade or decompose into dangerous compounds, and the storage of large quantities of flammable or ignitable materials. Improper storage of these chemicals has the potential to produce catastrophic results.

Chemical Management Procedure Review

All written procedures concerning the management of laboratory chemicals should be reviewed to determine if they adequately address identified hazards, regulatory compliance, and industry best-management standards. The review should verify that each procedure incorporates laboratory P_2 concepts and philosophies. Examples of procedures that should be reviewed include, but are not limited to, procurement, storage and handling of chemicals, and waste disposal.

The review process may also involve interviewing appropriate laboratory personnel. The interviews will provide information regarding the utilization of the written procedures. This information can then be used to determine if revisions to the procedures or additional training are necessary.

LABORATORY SAFETY AND HEALTH

Safety and health considerations should be part of all laboratory activities. As we learn more about the hazards in our workplaces, the job of managing risks to provide a safe, healthful workplace becomes increasingly challenging. To complicate the matter, there are regulatory requirements imposed by the Occupational Safety and Health Administration (OSHA) and the EPA that must be met by the lab. OSHA has detailed standards addressing work practices, personal protective equipment, exposure limits, and facility construction requirements to protect workers.

Safety and Health Programs

The laboratory is a workplace that offers a variety of safety and health problems. The key to solving these problems depends on the development, implementation, and maintenance of a comprehensive, proactive safety and health program. An effective program will require the participation of all laboratory personnel and the utilization of laboratory resources. One of the main points of consideration is assigning accountability and responsibility. The program must identify key positions and assign individuals to them. Secondly, laboratory personnel must understand their individual responsibilities in maintaining safe work environments. All laboratory personnel should be held accountable for their actions.

The safety and health program must include provisions for employee training. Personnel who are potentially exposed to hazards in the laboratory should be provided with written materials on the nature of the hazard. Laboratory workers must be able to identify hazards and risks involved in the work they perform. Individuals trained in handling hazardous materials are better equipped to minimize the risk of exposure to themselves as well as their coworkers. The ultimate responsibility for ensuring a safe work environment rests with the laboratory employee. Mechanisms should be provided to empower employees to mitigate or correct any unsafe acts or conditions.

Laboratory management is also responsible and accountable for safety. Management should encourage thorough training of laboratory employees. The method of training, when not otherwise specified, should include formal classroom training, informal training via routine safety meetings, and on-the-job training. The training should always be conducted by a qualified safety person and be properly documented.

The safety and health program should include, but not be limited to, the following issues:

- Respiratory protection
- Hazard communication
- Fire safety training
- Emergency response and evacuation
- Handling of radiological materials
- Interpretation of Material Safety Data Sheets
- First aid and CPR
- Engineering controls
- General laboratory safety
- Chemical hygiene
- Personal protective equipment

Regulatory Overview of Laboratory Occupational Safety and Health Regulations

OSHA has a variety of regulations addressing the protection of workers. Title 29 of the Code of Federal Regulations, Part 1910, deals with general industry requirements. Table 6–3 lists occupational regulations that may be applicable to laboratory operations.

Hazard Communication Standard

The hazard communication standard, sometimes referred to as the worker "Right-To Know" law, ensures that employees are properly informed of the chemical dangers associated with any hazardous materials they handle in the workplace. The law establishes employer responsibility by requiring a written Hazard Communication Plan and a program to explain how the plan will be carried out. The program for

TABLE 6–3
OSHA regulations.

Regulation	Description
29 CFR 1910.132	General requirements for personal protective equipment
29 CFR 1910.1000	General
29 CFR 1910.1001–1045	Z tables
29 CFR 1910.133(a)	Eye and face
29 CFR 1910.95	Noise exposure
29 CFR 1910.134	Respiratory protection
29 CFR 1910.135	Head
29 CFR 1910.136	Foot
29 CFR 1910.137	Electrical protective devices
29 CFR 1910.120	Hazardous waste operations and emergency response
29 CFR 1910.1200	Hazard communication
29 CFR 1910.1450	Laboratory standard

employees should address container labeling requirements, the use of Material Safety Data Sheets, employee training, and a list of the hazardous chemicals in the workplace. The program must also include methods the employer will use to inform workers of workplace hazards.

Laboratory Standard

This standard requires laboratories using hazardous chemicals to take additional measures to inform workers of chemical hazards they may be exposed to under normal conditions or in a foreseeable emergency. The law affects all employers engaged in the laboratory use of toxic substances.

A written program must be developed that includes a chemical hygiene officer and criteria for determining and implementing control measures to reduce employee exposure to hazardous chemicals. This plan is referred to as the Chemical Hygiene Plan. The contents of the plan must include:

- Standard Operating Procedures (SOPs) to follow when laboratory work involves the use of hazardous chemicals
- Criteria determining the need for, and the nature of, exposure control strategies to reduce personnel exposures
- A requirement that controls measures including lab hoods and other local exhaust ventilation be properly selected, designed, installed and maintained, along with procedures to ensure satisfaction of the requirement
- Information and training procedures
- A provision for medical consultation and evaluation
- Circumstances under which a particular laboratory operation will require approval prior to implementation
- Identification of personnel responsible for implementing and maintaining the hygiene plan
- Additional protective measures for work performed with carcinogenic materials and other particularly hazardous chemicals

There are many benefits from maintaining an accurate and up-to-date chemical hygiene plan. The plan requires each laboratory to carefully define what hazards exist in the facility, the programs necessary to control the hazards, and what means can be used for assessing its effectiveness and degree of implementation. In addition, the hygiene plan reflects the laboratory's commitment to maintain a safe work environment and provides a consistent and uniform method of standardizing work practices within and between laboratories.

Personal Protective Equipment Program

Activities that involve the use of hazardous materials may require the use of personal protective equipment (PPE) to prevent harmful chemicals from being absorbed through the skin, and from being inhaled, ingested, or injected into the body. Laboratories should develop and implement a written program to address the use of PPE.

The two basic objectives of any PPE program should be to protect the wearer from safety and health hazards, and to prevent injury to the wearer from incorrect use and or malfunction of the PPE. To accomplish this, the program should include hazard identification, medical monitoring, environmental surveillance selection, use, maintenance, and decontamination of PPE. The program must also include policy statements, procedures, guidelines, and a method for review and evaluation. It should be available to all employees.

The first step in the PPE selection process involves an assessment of the type of work and the area in which it will be conducted. A thorough understanding of the activities will help to identify any physical, chemical, or biological hazards that may be encountered. Laboratories may find the Material Safety Data Sheet particularly helpful in determining potential chemical hazards. In some cases it will be necessary to conduct atmospheric testing to determine exposure levels. The following information must be known before PPE can be properly selected:

- Contaminate(s)
- Concentration(s) or potential concentrations
- Harmful concentration levels (PELs)
- Immediately dangerous to life and health (IDLH) conditions
- Potential skin absorption and irritation sources
- Potential eye irritation sources
- Explosive sensitivity and flammability ranges
- Biological and physical hazards
- Engineering controls
- Climatic conditions
- Duration and type of work

Once the hazards have been identified and concentrations are known, the selection of PPE may begin.

PPE is fabricated from a wide variety of synthetic and natural materials and is commercially available from literally hundreds of sources in the United States. The right PPE for the job will involve knowing what materials will and will not do. This job is best left to the safety professional. However, there are some general rules that should always be followed:

- If IDLH conditions are present, a Self Contained Breathing Apparatus (SCBA) is required.
- If the IDLH condition is an absorption hazard, Level A protection must be worn. (Level A provides the greatest amount of respiratory and skin protection.)
- There is no such thing as "impermeable" plastic or rubber clothing.
- Chemical resistance may vary from manufacturer to manufacturer.

Before PPE is prescribed to mitigate hazardous conditions, engineering and administrative controls should be used. Engineering controls form the first line of defense against exposure to toxic or hazardous agents in the laboratory. This may consist primarily of ventilation equipment. The effectiveness and efficiency of the system will depend on regular preventative maintenance and routine measurements. The second line of defense for protecting workers is the use of administrative controls. These controls include training, labeling and posting, and standard operating procedures.

The selection process for PPE must also take into account the hazards created by the PPE itself. Individuals using PPE may experience any one or a combination of the following: reduced visibility, reduced dexterity, reduced ability to communicate, and increased physical and emotional stress. Therefore, decisions regarding the level of protection must also consider the increased safety hazard involved in donning any article of PPE.

Laboratory Personal Protective Equipment

Chemical-resistant clothing is the most common form of PPE found in laboratories. This class of clothing, for the purpose of this section, refers to coveralls, aprons, shoes, and gloves.

Table 6–4 outlines the type of clothing recommended to protect laboratory workers.

TABLE 6–4
Laboratory personal protective clothing.

Body Part Protected	Type of Clothing	Type of Protection	Use Consideration
Body	Aprons, sleeve protectors	Splash protection for arms, legs, chest	Useful for sampling, labeling, analysis activities. Should be used only when the possibility of whole body contamination is low
Eyes and face	Face shield, safety glasses, goggles, sweat band	Protects against chemical splashes; impact-resistant eyewear protects against projectiles	Useful for sampling, labeling, analysis activities

TABLE 6–4
(continued)

Body Part Protected	Type of Clothing	Type of Protection	Use Consideration
Ears	Ear plugs, ear muffs	Protects against physiological damage and psychological disturbance	Must comply with OSHA regulations. Care should be exercised when working directly with chemicals; could introduce contami-nantes to the ear
Hands	Gloves	Protects hands from chemical contact	Wear sleeve over glove cuffs to prevent splashed liquid from entering glove
Feet	Safety boots and shoes, chemical-resistant boots and shoes	Protects feet from chemical contact, puncturing, crushing, compression	Must meet OSHA requirements

SUMMARY

- Laboratories are defined as facilities that use chemicals to perform a variety of functions including chemical analysis, research, and education. The activities conducted within them are extremely varied, and frequently changed. However, the use of chemicals is the common thread among all laboratory facilities.
- Laboratories are capable of producing a variety of wastes. The basic categories include wastewater, air emissions, and radioactive, biological, solid and hazardous waste.
- Unlike industrial generators who typically have a few large-volume waste streams, laboratories typically have small volumes of many different waste streams. The overwhelming majority of waste generated by laboratories is hazardous and non-hazardous solid wastes generated from the disposal of unused chemicals.
- In addition to managing wastes that are generated from unused chemicals, laboratories must also manage waste generated from sample residues, spent solvents, and other analytical processes.
- If a laboratory decides not to treat hazardous waste on-site, it will have to select a facility to dispose of them.
- The three most common treatment and disposal technologies applied to laboratory hazardous wastes are incineration, chemical treatment, and solidification through encapsulation.

- There are several regulations that may effect laboratory health, safety, and environmental operations, including the Toxic Substances Control Act, RCRA and OSHA.
- A sound laboratory P_2 program will benefit the laboratory by providing economic advantages, impacting personnel morale, and enhancing public image.

QUESTIONS FOR REVIEW

1. Describe the basic categories of waste that can be generated from a laboratory. Also, describe the five commonly generated potentially hazardous wastes.
2. What are the three most common potential hazards related to the storage of hazardous waste chemicals? Why is it important to know what chemicals are in a particular waste stream?
3. What are the three most common treatment and disposal technologies applied to laboratory waste? Briefly describe each.
4. Describe three ways laboratory personnel may reduce the amounts of chemicals used in the laboratory.
5. What is the most common form of personal protective equipment used in the laboratory?

ACTIVITIES

1. What type of engineering controls, administrative controls, and personal protective equipment (PPE) can be used to protect laboratory workers using nitric acid to prepare samples for testing? Use the following information about nitric acid to assist you.
   - Nitric Acid, HNO_3
   - Miscible in water
   - Appearance and odor: Colorless, yellow, or red liquid with suffocating, acrid odor
   - Routes of exposure: inhalation, skin absorption, ingestion
   - Effects: Nitric acid vapor or mist is an irritant of the eyes, nose, throat, and skin. Can cause severe skin and eye burns and severe breathing difficulties.
2. Discuss the environmental laws that may apply to a laboratory that generates a variety of waste streams including ignitable solvents and toxic heavy metals. The laboratory also has a water treatment plant and an incinerator that is used to burn animal carcasses.
3. Determine how a small laboratory could track chemicals from "cradle to grave." Discuss what resources and information will be needed to effectively complete this task.

REFERENCES

American Chemical Society. 1993. *Less Is Better: Laboratory Chemcial Management for Waste Reduction. 2nd Ed.* Department of Government Relations.

American Chemical Society. 1990. *The Waste Management Manual: For Laboratory Personnel.* Department of Government Relations and Science Policy.

Ashbrook, P.C. 1991. Safe Laboratories: *Principles and Practices for Design and Remodeling.* Lewis Publishers, Inc.

CRC Handbook of Laboratory Safety. 2nd Ed. CRC Press. 1971

Carson, H.T. 1992. *Hazardous Materials Management.* Institute of Hazardous Materials Management.

Kaufman, J.A. 1990. *Waste Disposal in Academic Institutions.* Lewis Publishers. Inc.

National Safety Council. 1975. *Handbook of Occupational Safety and Health.*

Stricoff, R.S. 1990. *Laboratory Health and Safety: A Guide for the Preparation of a Chemical Hygiene Plan.* Wiley-Interscience Publication.

Wentz, C.A. 1989. *Hazardous Waste Management.* McGraw-Hill.

7

An Introduction to Nuclear Waste Issues

Thomas E. Byrne

Upon completion of this chapter, you will be able to meet the following objectives:

- Give an overview of the nuclear fuel cycle.
- Describe in detail the mining, milling, chemical conversion, enrichment, processing, and reprocessing of nuclear fuels.
- Describe the considerations of waste management.
- Define low-level waste and high-level waste.
- Discuss regulations related to radioactive waste.
- Describe the long-term disposal plans for radiological waste.

INTRODUCTION

The nuclear industry has the step-by-step utilization of resources as do other industries. In the nuclear industry, the movement of fuel is called the **nuclear fuel cycle.** The nuclear fuel cycle deals with the extraction, conversion, separation, purification, use, treatment, storage, and disposal of nuclear fuel and radioactive waste (see Figures 7–1 and 7–2).

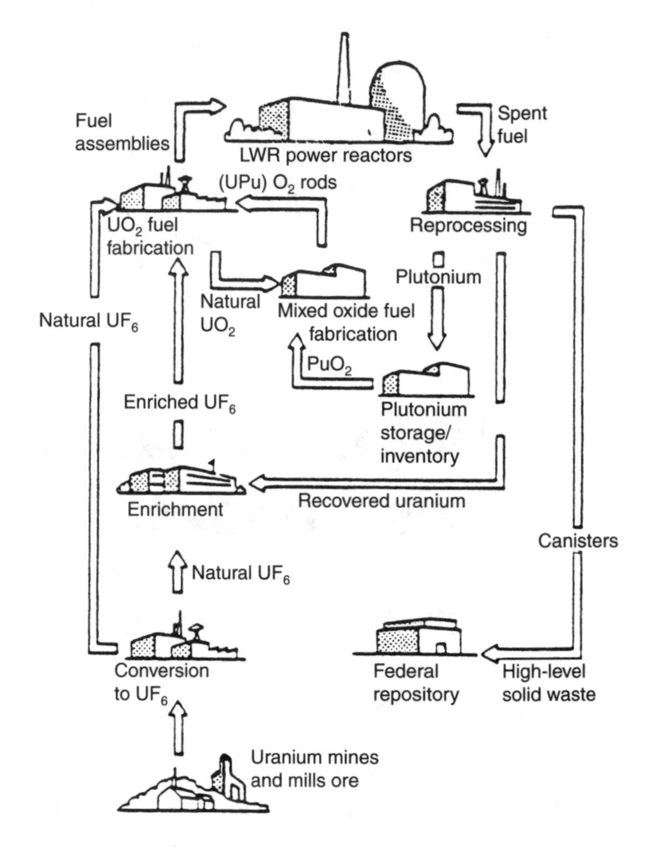

FIGURE 7–1
Chemical and metallurgical operations in the nuclear fuel cycle for LWRs with plutonium recycle.

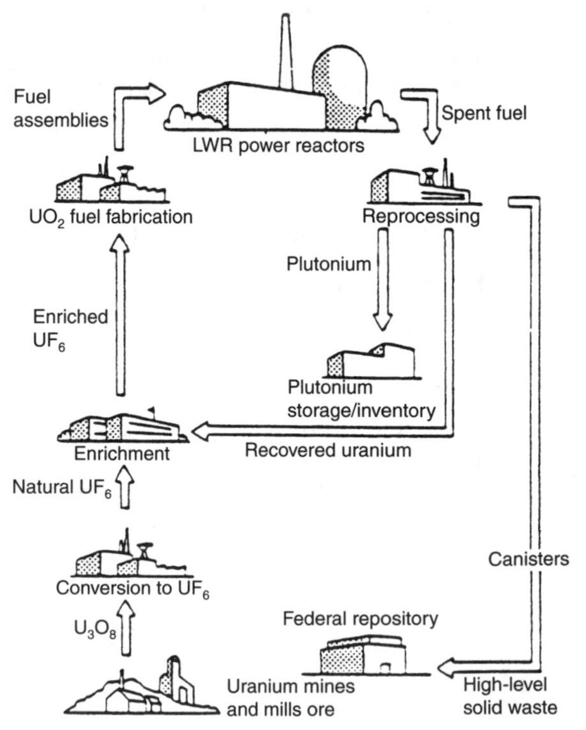

FIGURE 7–2
Chemical and metallurgical operations in the nuclear fuel cycle for LWRs without plutonium recycle.

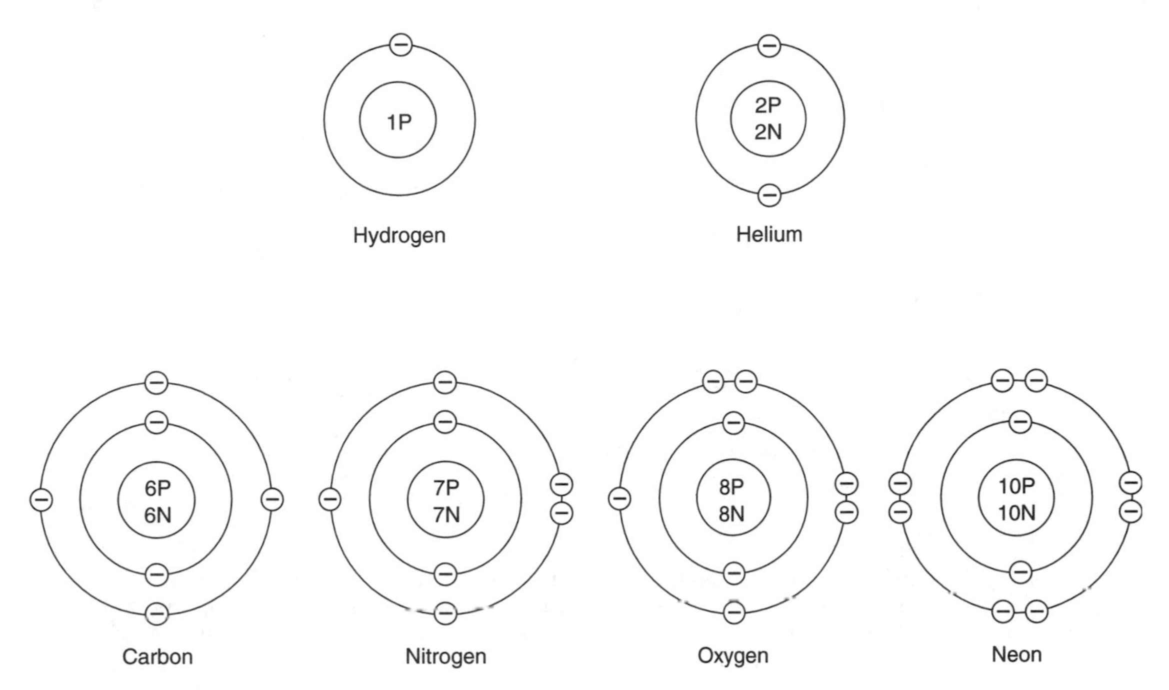

FIGURE 7–3
Electrons travel around the nucleus in paths called "orbits" or "shells." The first orbit (K shell) or energy level can hold either one, or a maximum of two, electrons. The second orbit (L shell) or energy level can hold from one to eight electrons.

The smallest form of an element is an **atom.** An atom is composed of a nucleus, containing neutrons with neutral charge and protons with positive charge, and negatively charged electrons in orbit around the nucleus (Figure 7–3).

There are ninety-two naturally occurring elements. The atoms of these elements have a different atomic number ranging from 1 to 92. The atomic number of an atom is the number of protons within its nucleus. An atom may also have a different number of neutrons in the nucleus that will cause a change in its mass (also called the atomic mass). This difference in neutrons is responsible for the isotopes of the elements (Figure 7–4).

Some isotopes of atoms are not stable and decay or break down with the release of radiation. The types of radiation observed in this decay are referred to as **alpha, beta,** or **gamma decay.** In alpha decay, a helium nucleus is released with a relatively short range of travel. In beta decay, a particle the size of an electron is produced, with either negative or positive charge and an intermediate range of travel. With gamma decay, an X-ray like particle of high to very high energy is produced, with a potentially great range of travel. The interaction of radiation with human cells depends on the energy of the radiation, the deposition of radiation (external or internal), the length of time of exposure, and the cell types involved (for example, lens of the eye or bone marrow). (For additional information concerning radioactive decay, radiation dose, and biological effects, refer to Volume 1 of this series, *Introduction to Environmental Technology*, Chapter 10, "Ionizing Radiation and the Environmental Worker."

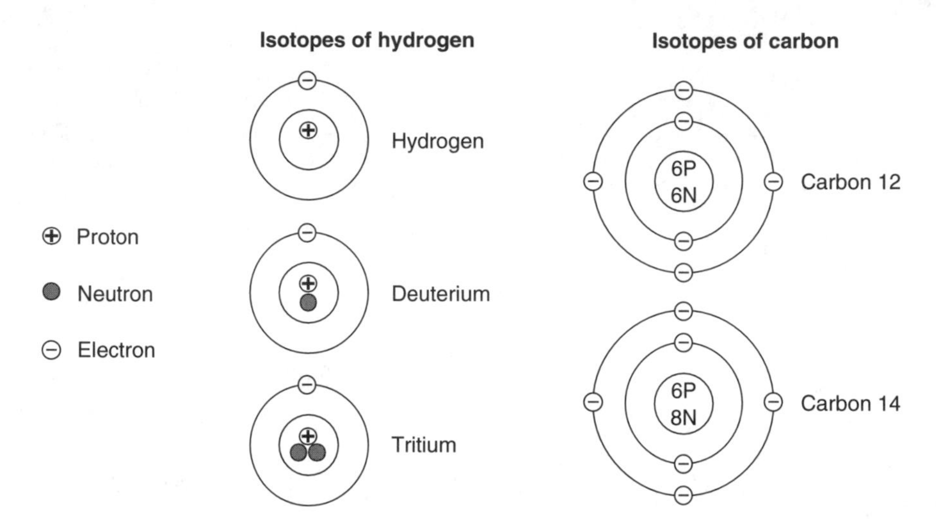

FIGURE 7–4
Change in the atomic mass number that one would observe with different isotopes. Notice the increase in the number of neutrons in deuterium and tritium. Also, note that the number of neutrons has increased in radioactive carbon-14 above the number of neutrons in stable carbon-12.

THE NUCLEAR FUEL CYCLE

Mining

Uranium is present in the crust of the earth's surface in concentrations of about 2 parts per million. In some areas of the earth, uranium is more abundant and mining takes place. Uranium metal is not usually found free in rocks but rather is complexed with other elements. Two forms are most common in U. S. soils: uranium oxide (or pitchblende) and uranium hydrous silicate (or coffinite). Most of the uranium ore mined in the United States is found in the form of pitchblende and coffinite deposits. Uranium ore is mined by typical methods of open pit mining and underground mining. In Colorado, Wyoming, and the area of the Black Hills, uranium oxide ($U_3 0_8$) may be found in 0.3 percent of the rock deposits.

Problems exist with the mining of uranium ores. Radioactive particles are contained in the dust created in the mining process. In the natural breakdown and decay of uranium to lead or bismuth, many intermediate radioactive particles (daughter products) are formed. If taken into the body, these radioactive particles may cause damage to biological tissues and human organs. If a worker inhales uranium dust, his body will receive an internal dose of radiation from the radioactive decay of uranium and the radioactive daughter products. To avoid these internal doses, radiation workers must wear respirators and underground mines must be well-ventilated.

Milling

After mining, the raw ore is crushed, then ground into smaller pieces and leached (by pouring acidic chemicals through it) to dissolve uranium minerals from the rocks. In these processes, more radioactive daughter products escape and may be inhaled;

thus, respirators may be necessary to prevent internal deposition of uranium dust or radioactive daughter products. The separation of the uranium from the mill tailings occurs next and this is followed by purification by ion exchange or solvent extraction techniques. Both processes serve to concentrate the uranium metal with the uranium attached to resins in the ion exchange technique or precipitated out in the solvent extraction technique. The concentrated form of uranium oxide is called "yellow cake" and may be 70-85 percent concentrated in this phase.

Chemical Conversion

To remove more impurities and obtain a gaseous form of uranium, additional processing takes place. After roasting, fluid-bed reduction with ammonia and additions of elemental fluorine occurs, resulting in a chemical conversion to uranium hexafluoride ($U\ F_6$). This chemical conversion is necessary to separate the different isotopes of uranium. Isotopes are varieties of the same chemical element with the same number of protons, but a different number of neutrons and different atomic mass. We are interested in separating U-235 from U-238 for use in nuclear reactors.

Enrichment

For commercial or military applications, uranium must be enriched from its natural state, in which only 0.7 percent is the U-235 isotope. This isotope must be concentrated to levels as high as 90 percent U-235 for some applications. Several methods can enrich the uranium isotope, including gaseous diffusion, which has been employed in this country for large-scale enrichment. The sites of these industrial plants are Oak Ridge, Tennessee; Paducah, Kentucky; and Portsmouth, Ohio. The basis for the separation is the difference in mass of $^{235}U\ F_6$ and $^{238}U\ F_6$ with the lighter $^{235}U\ F_6$ traveling faster and, thus, allowing for a separation. A second method of isotope separation is achieved by the use of the uranium in gaseous form in a centrifuge and is processed at very high speed. In this method, the heavier molecules of $^{238}U\ F_6$ are slung to the outer walls of the centrifuge while the lighter uranium isotope remains in the center. Another enrichment technology is atomic vapor laser isotope separation (AVLIS), in which melted and vaporized uranium is exposed to laser light with U-235 becoming positively charged. In an electric field, the uranium with a positive charge is separated from the uncharged uranium isotope.

Product Processing

Following the enrichment process, the uranium is converted into an oxide UO_2. The oxide form is most frequently used in nuclear reactors for power production. The uranium oxide is formed into pellets and the pellets are inserted into rods that are sealed. The fuel rods are arranged into bundles that make up the reactor core.

Reprocessing

Nuclear fuel does not burn to form ashes and some of the uranium is still in a usable form in used or "spent" fuel. The process of recovering the desirable uranium or other radioactive elements is called **reprocessing.** In this process, the fuel rods are crushed or sheared to expose the uranium in the rod's center. Nitric acid is used to leach out the uranium and plutonium by solvent extraction. A chemical extraction called the "Purex Process" is used to separate the uranium from the plutonium. The recovered fuel is shipped to an enrichment plant for further use. The radioactive waste has a very high level of radioactivity and must be treated with the worker's safety in mind. Reprocessing also allows radioactive gases to escape the fuel rods during the crushing and shearing. These gases, if inhaled, will cause internal doses of radiation to the workers. Proper ventilation and respirator protection are necessary to provide the necessary level of protection.

WASTE MANAGEMENT

Each step in the nuclear fuel cycle generates radioactive wastes. These wastes exist in solid, liquid, and gaseous forms. Mill tailings are a low-level radioactive waste (LLW); however, they occupy a large area. Many techniques are used to reduce the volume of LLW to lessen the number of waste shipments and extend the operating life of existing waste burial sites (Figure 7–5).

Reprocessed reactor fuel is a small volume of waste, but a high-level radioactive waste (HLW). Other HLWs can include research reactor waste and nuclear medical waste. Low and intermediate aqueous wastes are treated by ion exchange, precipitation, filtration, and evaporation to convert them to solid form. Solid radioactive waste may be sealed in cement or glass to prevent it from dissolving in water and being carried away to contaminate land or water supplies. High-level wastes are usually stored underground at the fuel reprocessing plant.

Long-term disposal options for HLW include incorporation into glass, ceramics, or composite materials, which are stored in steel drums for long-term underground storage. Radioactive gases are trapped or are preferentially absorbed to charcoal

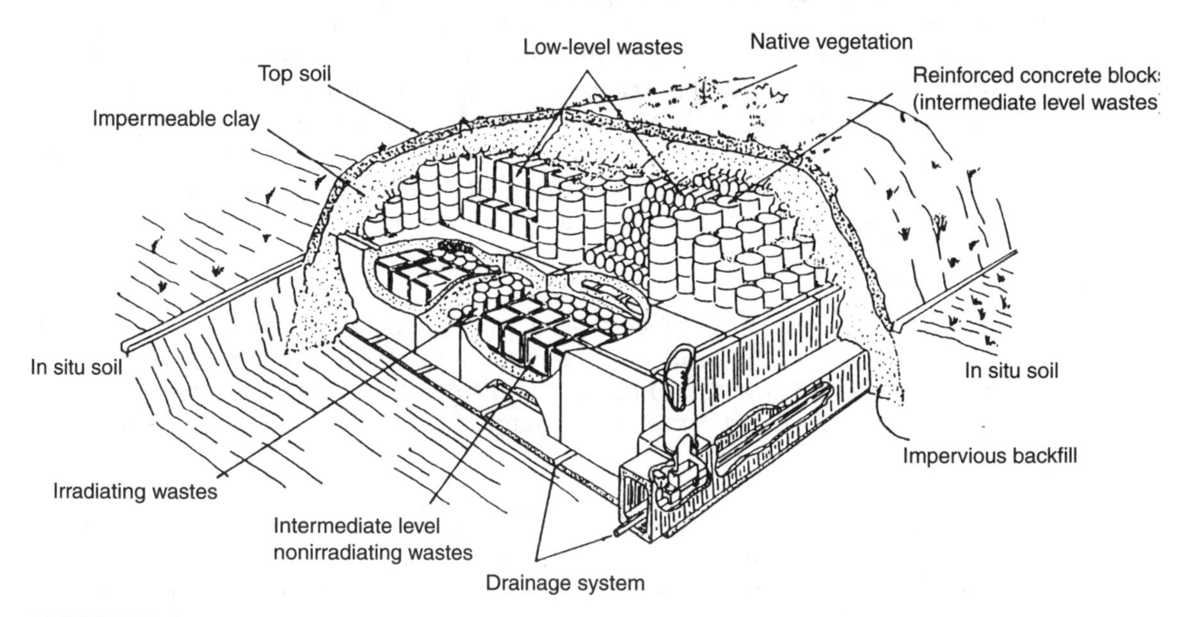

FIGURE 7–5
An earth-mounded concrete bunker depicting the approximate locations of wastes, which are separated according to level of activity.

(krypton and iodine) or silica gel (tritium) and are converted to a solid radioactive waste form, which is then disposed of.

REGULATIONS

Due to the dangers that radioactive materials may cause if not properly handled, packaged, and transported, several regulations dealing with radioactive waste have been signed into law. The following section is a brief listing of regulations related to radioactive waste. An understanding of the scope of the regulations may be of benefit in practical waste remediation situations.

Nuclear Regulatory Commission

The Clean Air Act Amendments, signed by the president on August 7, 1977, gave the EPA and the Nuclear Regulatory Agency authority to set air quality standards for radioactive substances and emissions.

The Uranium Mill Tailings Radiation Control Act of 1978, signed by the president on November 8, 1978, authorized the Nuclear Regulatory Agency to regulate the handling and disposal of wastes from the processing of uranium.

The Low-Level Radioactive Waste Policy Act of 1980, signed by the president on December 23, 1980, established a federal program for the interim storage of spent nuclear fuel away from reactors.

The Nuclear Waste Policy Act of 1982, signed by the president on January 7, 1983, set a deadline for establishing a permanent underground repository for high-level nuclear waste.

The Low-Level Radioactive Waste Policy Amendments Act of 1985, signed by the president on January 15, 1986, imposed strict deadlines for states to set up disposal facilities for low-level radioactive waste.

The Energy Policy Act of 1992 (Title III), signed by the president on October 24,1992, deals with the Nuclear Regulatory Commission's responsibilities regarding the storage and disposal of high-level radioactive waste.

Environmental Protection Agency

The Toxic Substances Control Act, signed by the president on October 11, 1976, required the federal government to regulate the manufacture, processing, distribution, use, or disposal of chemical substances and mixtures that may present an unreasonable risk to the public health.

The Resource Conservation and Recovery Act of 1976 (RCRA), signed by the president on October 21,1976, authorizes the EPA to regulate the treatment, storage, transportation, and disposal of hazardous wastes.

The Comprehensive Environmental Responses, Compensation and Liability Act of 1980 (CERCLA), signed by the president on December 11, 1980, requires the identification and cleanup of inactive hazardous waste sites by responsible parties. CERCLA was amended by the SuperFund Amendments and Reauthorization Act of 1986, which was signed by the president on October 17, 1986.

The Safe Drinking Water Act (Title XIV), an amendment by Congress to the Public Health Service Act, assures that the public is provided with safe drinking water. The EPA issued radiological provisions on July 9, 1976.

Department of Energy

The Uranium Mill Tailings Remedial Action Amendments Act of 1988, signed by the president on November 5, 1988, allows the Interior Department to transfer Bureau of Land Management lands to the Department of Energy for surveillance and maintenance of slightly radioactive mill tailings.

LONG-TERM DISPOSAL OF RADIOLOGICAL WASTE

In 1983, Congress charged the Department of Energy with selecting a site where HLW could be stored underground. This waste consists of spent fuel from reactors and weapons materials. A major component of this waste is plutonium with a radiological half-life of 24,000 years (radiological **half-life** is the time it takes for one half of a radioactive isotope to decay). The safety standards of the repository must meet or exceed standards set by the EPA and the Nuclear Regulatory Agency. This requires that the waste be prevented from contaminating the environment for 10,000 years. Several locations have been considered for this disposal site: Yucca Mountain (a Nevada test site), a Texas site, and a Washington State site. Work has been in progress on the Yucca Mountain site for several years, with a grid of tunnels in the center of a mountain range. Workers will, in the future, place canisters filled with radioactive waste in this system of tunnels (Figure 7–6).

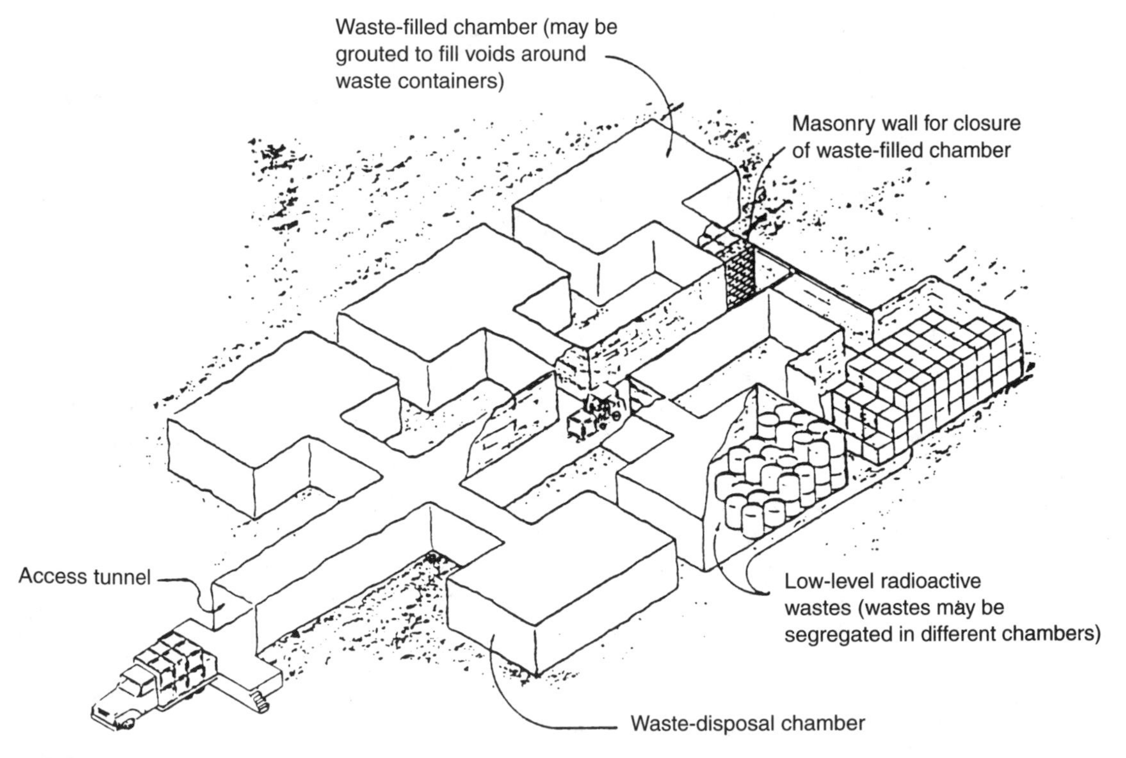

FIGURE 7–6
Disposal in deep mined cavities in bedded limestone or salt. Wastes may be segregated by chamber if required.

In approximately 50 years, the tunnel system will be full and the mountain will be permanently sealed. The heat from the high-level waste will cause the canisters to break down and/or corrode in 300 to 1,000 years. The mountain range will then serve to shield the environment and future populations from the radioactive waste. The watertable lies 2,500 feet below the mountain and is considered safe from radioactive contamination. Geologists continue to research the stability of the mountain range for earthquakes or other events that would penetrate the repository. Political discussion as to when or if Yucca Mountain will be used for long- term storage of radiological waste continues at the time of publication of this chapter.

SUMMARY

- The nuclear fuel cycle deals with the extraction, conversion, separation, purification, use, treatment, storage, and disposal of nuclear fuel and radioactive waste.
- In several of these procedures, workers must protect themselves with respirators and other protective items.
- Waste management involves the techniques for dealing with solid, liquid, and gaseous radiological waste. The reduction of the volume of low-level waste and the safe disposal of high-level waste are key issues in waste management.
- Regulations concerning radiological materials address how these substances are to be handled, packaged, and transported. The Nuclear Regulatory Commission, EPA, and Department of Energy are charged with responsibility of oversight as listed in the chapter.
- The long-term disposal of radiological waste is under the responsibility of the Department of Energy, with construction at Yucca Mountain proceeding at time of publication of this chapter.

QUESTIONS FOR REVIEW

1. What is an isotope?
2. What are the differences between three types of radioactive emissions—alpha, beta, and gamma?
3. What are the most common uranium oxides in the Unites States and where are they found?
4. What dangers must workers be protected from in mining and milling of radioactive ores?
5. In the enrichment process, what steps can be used to separate the isotopes?
6. What is the "Purex Process" and what does it allow to be separated?
7. What is low-level waste and high-level waste?
8. What radiological regulations are the responsibility of the Nuclear Regulatory Commission?
9. What radiological regulations are the responsibility of the EPA?
10. What are a few of the considerations for the long-term disposal of radiological waste?

ACTIVITIES

1. In 1983 Congress charged the Department of Energy with the selection of a site where waste could be stored underground. The DOE has narrowed the search to the sites listed in the chapter. What do you think are the environmental considerations for selection? Review newspaper and magazine articles related to underground storage of radiological waste. This issue is still developing.
2. In the United States, 103 nuclear reactors (at time of publication) are in service. What will be done with the radiological waste from these reactors if underground storage is not used? How would you suggest the nuclear industry proceed with waste management?

8

The Pharmaceutical and Cosmetic Industries

S. Merris Sinha

Upon completion of this chapter, you will be able to meet the following objectives:

- Understand the diverse nature of pharmaceutical and cosmetic processes, and their varied operations for manufacturing, purifying, and packaging drug and cosmetic products.
- Learn about the typical ingredients used to manufacture pharmaceutical and cosmetic products.
- Review the system for maintaining and improving process, product, and service quality performance.
- Discover how the diverse manufacturing processes, reactions, and hazardous materials produce a wide range of chemical and biological wastes.
- Learn about the critical regulatory requirements for the pharmaceutical and cosmetic industries since the wide variety of operations and products developed by the pharmaceutical and cosmetic industries subject them to almost all environmental, health, and safety regulations applicable to manufacturing facilities.
- Become familiar with the treatment technologies the industry is using to manage wastes after they are generated.
- Identify the pollution prevention efforts that companies are using for cost containment in the pharmaceutical and cosmetic industry.

INTRODUCTION

This chapter deals with the current environmental, health, and safety (EHS) conditions surrounding the manufacturing processes of the pharmaceutical and cosmetic (P&C) industries. Of course, the processes vary from company to company, not to mention that the regulations will vary from state to state. However, it is vital for burgeoning environmental professionals to have an accurate knowledge of general manufacturing processes, waste streams, available technology, and regulatory issues affecting the P&C industries. This chapter gives a brief overview of these topics. It is impractical to list here all the laws and regulations that may apply to the P&C industries, or to give exact interpretations for individual manufacturing situations; therefore, when in doubt about the impact of a particular process or regulation, it is essential that you consult with appropriate agencies, and legal inhouse experts.

Information in this chapter is aimed primarily at EHS professionals who have some familiarity with the P&C industry. If you have never tackled the terminology found in the P&C industry, you may want to visit a medical library and review a medical dictionary and books such as the **U.S. Pharmacopoeia, National Formulary, Food Chemical Index,** and the **USAN** and the **USP** Dictionary of Drug Names.

PHARMACEUTICAL MANUFACTURING PROCESSES

The pharmaceutical manufacturing industry encompasses the manufacture, purification, and packaging of chemical materials to be used as medication or have a therapeutic value for humans or animals (U.S. EPA, 1981). The industry is characterized by a broad range of products that include natural substances from plants or animals, metal organics, and wholly inorganic materials. The industry is also characterized by diversity of processes and plant sizes, as well as waste quantity and quality (U.S. EPA, 1985). Products of the industry are basically split into four categories based on the Standard Industrial Classification (SIC) system: medicinal chemicals and botanical products (SIC 2833); pharmaceutical preparations (SIC 2834); invitro and invivo diagnostic substances (SIC 2835); and biological products, except diagnostic substances (SIC 2836).

Process of Developing New Drugs and Products

To create the variety of pharmaceutical products, the industry uses many complex batch processes and technologies. Although it is impossible to discuss all of the processes that apply to pharmaceutical manufacturing, there are essentially five common methods used: research and development; chemical synthesis; fermentation; extraction; and formulation (U.S. EPA, 1986). Many of the processes encompassed in these five methods are proprietary, but the general processes are identified here (U.S. EPA, October 1991). Figure 8–1 describes the broad range of manufacturing processes used in the pharmaceutical industries.

Research and Develpment

Research and development (R&D) encompasses several fields of study, including chemical research, microbial research, microbiological research, and pharmacological research. It should be no surprise that the development of a new drug requires

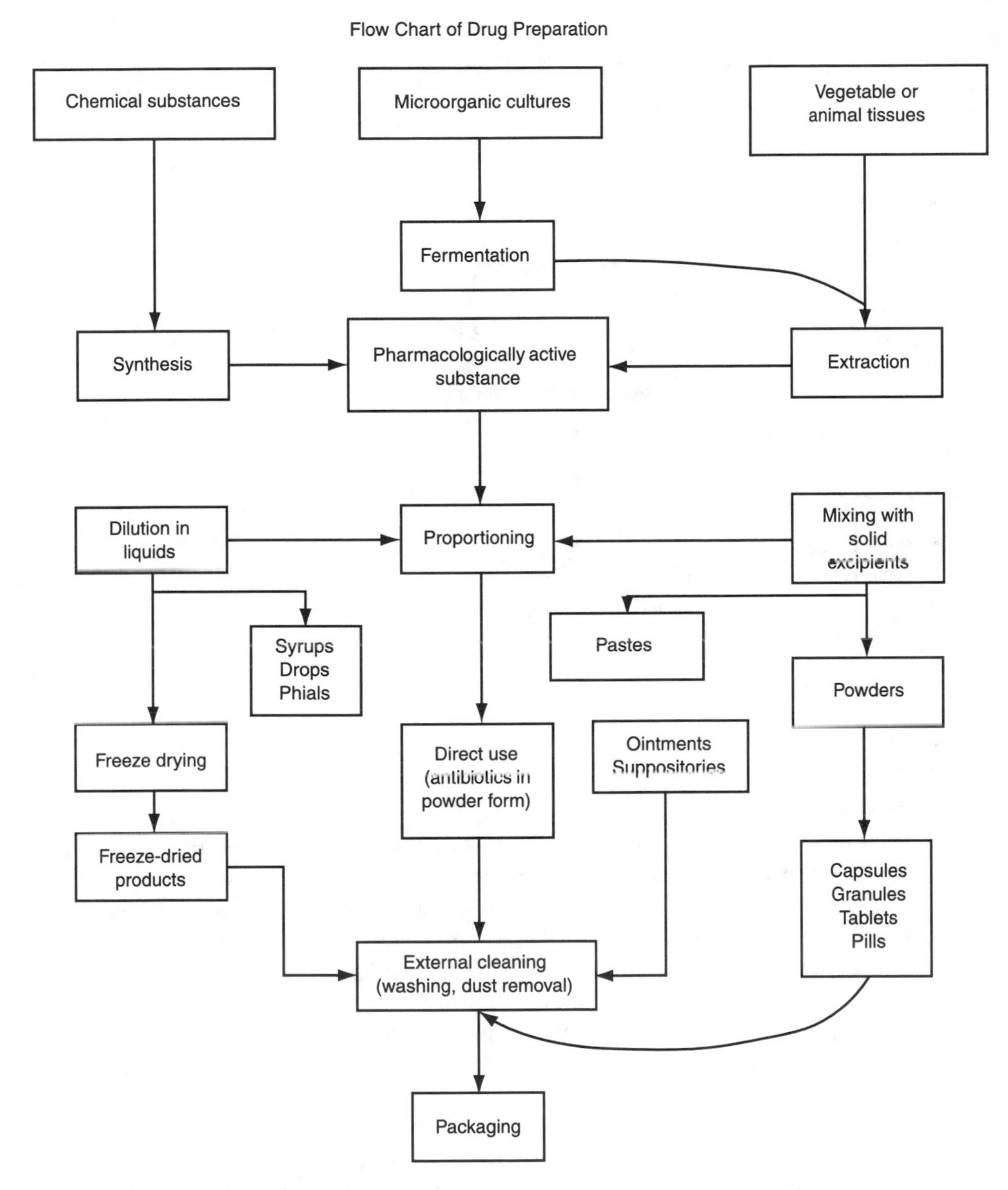

FIGURE 8–1
General pharmaceutical manufacturing process.

the cooperative efforts of many trained personnel specializing in the following areas: medicinal, organic, and analytical chemistry; microbiology; biochemistry; physiology; pharmacology; and toxicology, chemical engineering, and pathology. Figure 8–2 depicts an ideal drug discovery process.

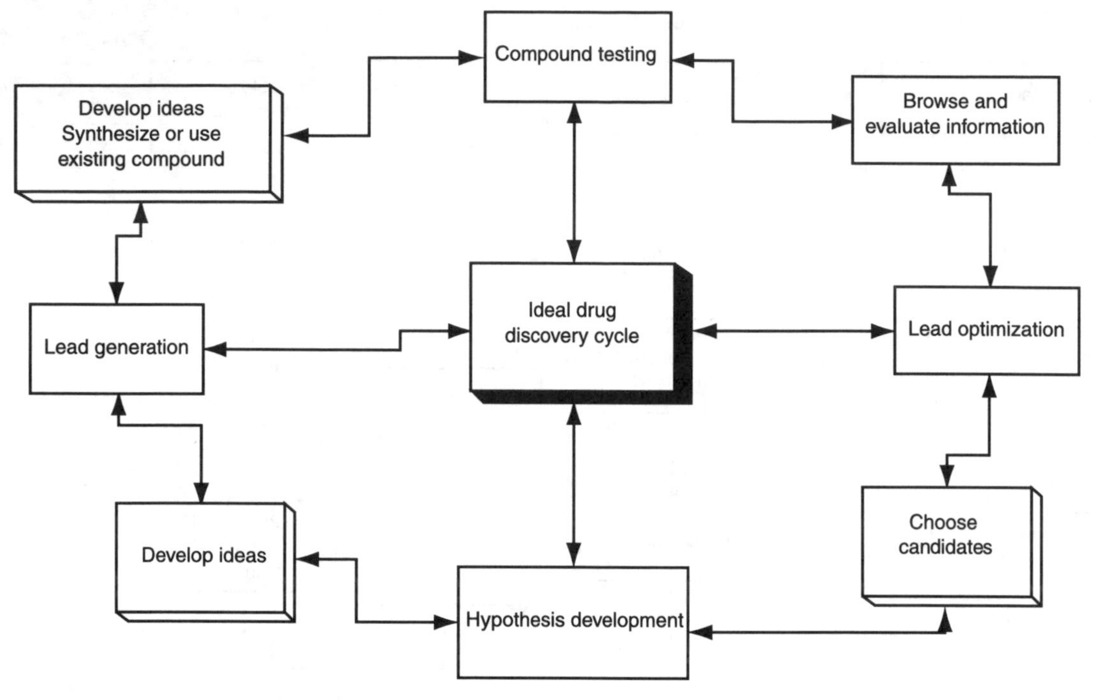

FIGURE 8–2
Ideal drug discovery process.

Chemical Synthesis

Pharmaceutical manufacturers produce most drugs by chemical synthesis. In a typical manufacturing plant, operations include using one or more batch reactor vessels in a series of reactions, separation, and purification steps. In these vessels, manufacturers employ many types of chemical reactions, recovery processes, and chemicals to produce a variety of drug products. Chemicals used in chemical synthesis operations range widely and include organic and inorganic reactants and catalysts. In addition, manufacturers may use many solvents listed as priority pollutants for product recovery, purification, and as reaction media. Each of these manufacturing processes must conform to its own rigid product specifications.

For the highest through-put products, engineers arrange reaction vessels and ancillary equipment into separate, dedicated process units. Sometimes, depending on the market for the product, companies may manufacture some products in a single product "campaign," which may last a few weeks or months. During such a campaign, operators or computerized controllers will add the required reagents and monitor process functions (i.e., flow rate, pH, temperature) according to Good Manufacturing Practice (GMP) protocols. Finally, at the end of the campaign the process equipment is thoroughly cleaned. Furthermore, personnel tightly control campaign schedules to ensure timely product delivery and availability of raw materials and process equipment.

Fermentation

The production and separation of medicinal chemicals such as steroids, vitamins, and antibiotics from microorganisms occur using batch fermentation processes

(U.S. EPA, October 1991). Generally, fermentation processes consist of two major steps: inoculum, seed preparation, and fermentation, followed by crude recovery and purification. Sterile inoculum preparation starts in the lab using a carefully maintained population of a microbial strain. Next, a few cells from this culture are matured into a dense suspension through a series of test tubes, agar slant, and shatter flasks. For further propagation and maximum cell growth, the operators transfer the cells to a seed tank that operates like a full scale fermenter.

Operators then charge a sterilized fermenter with material from the seed tank through a series of sterilized lines and valves and the fermentation commences. During the fermentation cycle, sterile air usually agitates and aerates the contents, with instruments carefully monitoring dissolved oxygen content, pH, temperature, and several other parameters.

Once the cells mature, operators filter the fermentation broth to remove the solid residues resulting from the fermentation process. Next, the filtrate is processed to recover the desired product using methods such as solvent extraction, precipitation, and ion exchange or adsorption chromatography (Bailey and Ollis, 1977). In solvent extraction, the aqueous filtrate is combined with an organic solvent (typically methylene chloride or butyl acetate) to transfer the product into the solvent phase. Final operations entail recovery of the product by further processes of extraction, precipitation, or crystallization.

Formulation

The last phase, formulation, is the preparation of dosage forms such as tablets, capsules, liquids, parenterals, creams, and ointments. Table 8–1 outlines the standard dosage forms and their constituents. A few of the commonly manufactured dosage forms are discussed here.

Manufacturers produce tablets in three varieties: plain compressed, coated, and molded. The tablet dosage form depends on the desired release characteristic of the active ingredients, which can be slow, fast, or sustained. To control their release characteristics, the tablets are sprayed or tumbled with a coating material. Initially, the operators produce tablets by blending the active ingredient with fillers, such as starch, sugar, and binders. Manufacturers may also use the technique of "wet granulation," which involves blending the powdered active ingredient and filler, then wetting them with a binder solution. Next, the processes consist of forming coarse granules, then drying and mixing them with a lubricant such as magnesium stearate. If products are too unstable for wet granulation or too fragile for direct compression, "slugging" is necessary. Slugging involves using heavy-duty tablet presses to compress large tablets, which are then ground and screened to a desired mesh size and recompressed into finished tablets.

Tablet compression follows wet granulation, direct compression, and slugging. Compression operations create the tablets by pressing this tablet mixture using a die, which holds a measured amount of the material, while a punch compresses the tablet. If multilayered tablets are needed, manufacturers employ presses with several feed hoppers, partially compressing each layer until the final compression of the last layer.

Capsule manufacturers use the procedure of dipping pins into a solution of gelatin that is maintained at a specific temperature. Next, the deposited gelatin film is separated from the pins, then dried, and trimmed. The next procedure involves taking the separated gelatin sections and filling them with the product and, finally, joining the capsules. If soft capsules are needed, the process involves placing two continuous gelatin films between rotary die plates. As the plates are brought together and sealed to form two halves of the capsule, the drug is injected into the capsule.

To formulate solutions, syrups, elixirs, suspensions, and tinctures, manufacturers mix the solutes with a selected solvent in a glass-lined or stainless steel vessel. Then they filter the solutions and pump the liquid into storage tanks for quality control inspection prior to packaging. Suspensions and emulsions are frequently prepared using colloid mills and homogenizers. Preparation of many of these liquid dosage forms entails adding preservatives to prevent mold and bacteria growth. For some products, like parenteral dosage forms, operations include sterilization by steam autoclave or the use of bacteria-retaining filters and sterile containers.

To prepare ointments and creams, manufacturers normally melt a base, which is usually the petroleum-derivative petrolatum, or an oil-in-water or water-in-oil mixture. Then they blend the mixture with the drug, cool the product, and pass the mixture through a colloid or roller mill.

Cosmetic Industry

Cosmetic manufacturing also involves batch processes that can be similar to pharmaceutical manufacturing. In fact, many pharmaceutical companies also produce cosmetics. Cosmetic manufacturers also formulate emulsions, lotions, and creams that are very similar to pharmaceutical dosage forms. To make things even more complicated, certain cosmetic products are also classified as pharmaceutical products. For example, cosmetic preparations that function as skin treatments are covered by SIC Code 2844, which also covers pharmaceutical products (U.S. EPA, 1982). This is because products that are cosmetics but are also intended to treat or prevent disease, or otherwise affect the structure or functions of the human body are also considered to be drugs (FDA, 1992).

The Cosmetic, Toiletries, and Fragrance Association (CTFA) maintains that the processing of the cosmetic industry consists largely of mixing and packaging operations (Rose Sheet, 9/25/95, p. 5). However, more and more cosmetics are being formulated as over-the-counter drugs (OTC) and prescription products (Bollinger, 1989). A major consideration in these types of cosmetic formulations is that the products must not only be aesthetically elegant, but may also need to stabilize active drug ingredients. In such cases, efficacy of an active drug ingredient needs to be documented and/or submitted to the U.S. Food and Drug Administration (FDA) for approval. Active ingredients in these types of cosmetic products could include agents that are anti-inflammatory, antibiotic, antifungal, antibacterial, antiparasitic, antipruritic, antiseptic, keratolytic, antiactinic, antiviral, antiseborrheic, and antipsoriatic (Bollinger, 1989).

Despite the overlap of cosmetic and drug products, manufacture of a typical cosmetic product is usually much simpler than a drug product. A cosmetics product manufacturer only needs to make safety determinations, but does not need government approval or review before a product is marketed. But the cosmetic product and the individual ingredients must be safe (21 USC §361). Furthermore, the manufacturer must choose from a list of color additives already approved by the FDA for use in the cosmetics (21 USC §376, 21 CFR §73, §74, §82).

Typically, most cosmetics are either oil-in-water or water-in-oil emulsions that are stabilized by using soaps, synthetic detergents, waxes, alcohols, and various solvents.

Coloring agents added to many cosmetic products are either naturally occurring or compounds belonging to the coal-tar group of dyes. Perfumes, an essential component of many cosmetics, come from plant oils, animal secretions, or chemical substances. Additionally, preservatives are used to prevent growth of microorganisms in cosmetics products. Commonly used preservatives include: organic acids, alcohols, aldehydes, essential oils, phenolic compounds, esters, and antioxidants (Parmeggiani, 1983). Manufacture of creams and emulsions involves heating oils, waxes, emulsifiers, and other oil-solubles in a steam-jacketed kettle. The water-soluble components are dissolved in the aqueous phase and heated in another steam-jacketed kettle. The warmed inner phase is slowly added to the outer phase, stirred constantly and homogenized to assure efficient emulsification (Kirk-Othmer, 1982).

Loose face powders are commonly manufactured by screening all materials into a spiral mixer, blending, and then adding inorganic colorants and perfume. The finished powder, after passing quality control for shade, is stored. If pressed powders are manufactured, the loose powder is agitated before pressing, prepressed by applying moderate pressure to the material placed in pans, and then the cakes are placed on to a hydraulic press (Kirk-Othmer, 1983).

COMPOSITION OF PHARMACEUTICAL AND COSMETIC PRODUCTS

As mentioned earlier, the terms "drugs" and "cosmetic" are not mutually exclusive (21 USC §509, §359). For instance, in many cases the composition of cosmetic products is very similar to pharmaceutical products. A cosmetic product intended to beautify the skin may claim to accomplish it through a physiological action and, therefore, may be also considered a drug (Eldermann, 1988). The raw materials used in the process can also be similar, as it is in many cases in the manufacturing process. However, regulatory requirements, the time it takes to get the product to the market, and marketing efforts for drugs and cosmetics can be quite different.

Table 8–1 outlines dosage forms of pharmaceutical products and composition.

TABLE 8–1
Pharmaceutical dosage forms and constituents.

Dosage Form	Constituents and Properties
Liquid Solutions	
Aromatic waters	Volatile solids and oils, water
Liquors or solutions	Water, chemicals
Syrups	Sweetener, solvent, medicinal agent
Elixirs	Sweetened hydroalcoholic solution, may be medicated
Spirits, essences	Alcohol, water, volatile substances
Tinctures	Natural drugs, extracted with appropriate solvents
Colloidions	Pyroxylin in ether, medicinal agent, castor oil, camphor

TABLE 8–1
(continued)

Dosage Form	Constituents and Properties
Liniments	Oily or alcoholic solutions, suspensions
Mucilages	Colloidal polymer solutions
Parenteral solution	Sterile, pyrogen-free, isotonic, pH close to that of blood; oily or aqueous suspension
Ophthalmic	Sterile, isotonic, pH close to that of tears; viscosity builder
Nasal	Aqueous, isotonic, pH close to that of nasal fluid; sprays or drops
Otic	Glycerol based
Mouthwash, gargles	Aqueous, antiseptic
Inhalations	Administered with mechanical devices
Enemas, douches	Aqueous solution or suspension, may include medicinal agent
Liquid Dispersions	
Suspensions	Powder suspended in water, alcohol, glycol, or an oil; viscosity builder, wetting agents, preservatives
Emulsions, lotions	Oil-in-water or water-in-oil
Gels, jellies, magma	Viscous, colloidal dispersions
Semisolid and Plastic Dispersions	
Ointments	Hydrocarbon (oil), absorptive water-washable or water-soluble bases; emulsifying agents; glycol, medicating agent
Pastes and cerates	Ointments with high-dispersed solids or waxes, respectively
Suppositories	Theobroma oil, glycerinated gelatin, or polyethylene glycol base plus medicinal agent
Solids	
Bulk powder	Comminuted or blended, dissolved, or mixed with water
Effervescent powder	Comminuted or blended, dissolved in or mixed with water
Dusting powder	Contains adsorbents
Insufflation	Insufflator propels medicated powder into body cavity
Lyophilized powders	Reconstitution by pharmacist of unstable products
Capsules	Small-dose bulk powder enclosed in gelatin shell; active ingredient plus diluent
Masses or molded solid pills	Adhesive or binding agents facilitate compounding; prepared by massing and piping

TABLE 8–1
(continued)

Dosage Form	Constituents and Properties
Trouches, lozenges, pastilles	Prepared by piping and cutting or disk candy technology; compounded with glycero-gelatin
Tablet triturates	Small molded tablets intended for quick complete dissolution (e.g., nitroglycerin)
Granules	Particle size larger than powder
Compressed tablets	Dissolved or mixed with water; great variety of shapes and formulations
Pellets	For prolonged action
Coated tablets	Coating protective; slow release

Source: Adapted from U.S. EPA, October 1991.

Literature and patent reviews performed annually in *Cosmetic and Toiletries* magazine give an update on cosmetic raw materials and new uses appearing in the patents and scientific literature. Since the cosmetic industry uses a nomenclature very different from any other industry, to understand it and the ingredients, the reader is referred to 21 CFR §701.30 for a list of ingredients named by the FDA; **International Cosmetic Ingredient Dictionary;** U.S. pharmacopoeia; National Formulary; Food Chemical Index; and USAN and the USP Dictionary of drug names (Schuller and Romanowski, 1995). Table 8–2 gives examples of materials used to create some common cosmetic products.

TABLE 8–2
Cosmetic products and raw materials.

Product Type	Potential Raw Materials
Antiflammatory lotion	S-lactoylglutathions, glycerin, 1, 2-butylene glycol, ethyl alcohol, polyoxyethylene lauryl ether, methylparaben, citric acid, sodium citrate, fragrance, and color
Antiperspirants	1, 3-butylene glycol, hexylene glycol, steareth-100, dibenzyl monosorbitol acetal, stearic acid, cyclomethicone, ethyl alcohol, aluminum chlorhydrex, triclosan, propylene glycol, hydroxypropylcellulose, N-(2-hydroxyethyl) acetamide, cetyl alcohol, mineral oils
Powders	Talc, sodium polyacrylate
Bath preparations	Chitosan succinate
Dandruff preparations	Ethyl alcohol, polyoxyethylene oleyl alcohol ether, glycerin, perfume, sodium malate
Cleansing creams	Smectite clay, octamethyl cyclotetrasiloxane, mineral oil, perfume, methylparaben, glycerin

TABLE 8–2
(continued)

Product Type	Potential Raw Materials
Face masks	Carboxyl-modified polyvinyl alcohol, calcium chloride, methylparaben, glycerin, perfume, polysorbate 20, sodium alginate, borax
Lip preparation	Mica powder, bengars, Red No. 202 and 226, titanium dioxide, antioxidant, perfume, cetyl 2-ethylhexanoate, octyldodecyl myristate, microcrystalline wax, paraffin, petrolatum, allantoin, Category No. 1 skin protectant
Makeup emulsion	Bengars, yellow iron oxide, black iron oxide, titanium dioxide, mica, talc, microcrystalline wax, petrolatum, mineral oil, cholesterol, squalene, candelilla wax, carnauba wax, octyldodecyl myristate, perfluo ropentylte-tradecane, perfluorohexyl lauryl ether, perfluorodecaline, benzoyl peroxide
Nail products	Nitrocellulose, alkyd resin, acrylic resin, tributyl citrate acetate, camphor, organic bentonite derivative, isopropanol, ethyl acetate, butyl acetate, toluene, butyl alcohol, pigments, nylon-12 powder, glyoxal, polyester resin, benzophenone-1, D&C violet, polyamide resin mixture, isopropyl alcohol, squalene, urea, fragrance, dibenzylidene sorbitol
Toothpaste	Sodium fluoride-stannous fluoride
Mouthwash	Ethyl alcohol, castor oil, flavor, sorbitol, glycerin, polygalactosamine A-HCL, benzethonium chloride, Category No. 1 antimicrobial
Tanning products	OTC Category No. 1 ultraviolet absorbers
Shampoos	Sodium lauryl dihydroxypropyl phosphate, triethanolamine lauryl phosphate, lauryldimethylamine oxide, cationic cellulosic, glycerol, sorbitol, propylene glycol, dodecyldimethylammonium caprolactum bromide, lauric acid diethanolamide, zinc pyrithione, colloidal sulfur, coal tar

Source: Adapted from Fox, 1989, and Estrin, 1984 (p. 556).

QUALITY ASSURANCE AND QUALITY CONTROL

Quality assurance and quality control (QA/QC) is traditionally an inspection and testing system for maintaining and improving process, product, and service quality performance. If you are an EHS professional practicing within the P&C industry, you need to be knowledgeable of the company's good laboratory practices (GLPs), Good Manufacturing Practices (GMPs), and indirectly, with Good Clinical Practices (GCPs.) Because of the rigid specifications and safety issues, GMPs, validation, and quality-assurance programs are required by the FDA. **Validation** is defined as the scientific study of a process to prove that the process is under specified control, and to determine process variables and acceptable limits.

Recently, many companies have shifted from traditional thinking and integrated the QA/QC process into their total quality system (Nally & Kieffer, 1993). As the P&C companies respond to rising customer expectations, costs, and increased regulatory pressures, many are strategically restructuring their organizations so that the products are validated and improved continuously.

HAZARDOUS WASTE STREAMS

The diverse nature of pharmaceutical and cosmetic processes, varied operations, reactions, and hazardous materials produces a wide range of chemical and biological wastes. Many hazardous waste streams could potentially be generated by mixing nonhazardous or infectious wastes with hazardous wastes. For the purposes of this section, it is assumed that most, if not all, of the cosmetic process wastes are similar in nature to waste products from the pharmaceutical processes. However, it should be noted that the amount and characteristics of cosmetic wastes may be quite different, depending on how the products were manufactured. Waste streams from the general P&C processes are discussed here.

Research efforts may generate a significant amount of waste materials, including biological and medical wastes. Some of these wastes include halogenated and nonhalogenated solvents, photographic chemicals, radionuclides, corrosives, and oxidizers (Zanowiak, 1982). Other common types of wastes can also include biomass, natural products, and a myriad of reagents (Venkataramani, 1994).

Chemical synthesis wastes usually involve an organic mother liquid that contains unconverted reactants, reaction byproducts, and residual products in the organic solvent base. Synthesis operations may also generate acids, bases, cyanides, and metals. In addition, aqueous waste stream results from solvents, filtrates, concentrates, equipment cleaning, wet scrubbers, and spills. Wastewater from synthesis processes may have high biological oxygen demand, chemical oxygen demand, total suspended solid levels, and pHs from 1 to 11 (U.S. EPA, 1983). Because of the waste stream concentration or toxicity, pretreatment may be required before discharging to sewers. Equipment cleaning accounts for a large volume of solvent and nonsolvent waste generated (Venkataramani, 1994). Recovery of solvents on-site may generate still bottom tars. The use of volatile solvents can also result in air emissions from dryers, reactors, distillation units, storage and transfer, filters, extractors, centrifuges, and crystallizers (U.S. EPA, 1981).

Wastes from natural product extraction include spent raw materials such as leaves and roots, water-soluble solvents, solvent vapors, and wastewaters. Extraction wastewaters typically have low biological oxygen demand, chemical oxygen demand, and total suspended solid levels and pH in the range of 6 to 8 (U.S. EPA, 1983). Ammonia, acids, and bases are commonly used to control pH. These operations generate spent biomass, organic solid, and liquid wastes rich in solvents, natural products, and salts (Venkataramani, 1994).

The fermentation process generates large volumes of wastes such as spent aqueous fermentation medium and solid cell debris. This aqueous medium is very impure, and may contain unconsumed raw materials such as corn steep liquor, fish meal, and molasses. Filtration processes result in large quantities of solids in the form of spent filter cake, which includes solid remains of the cells, filter aids, and some residual product. After product recovery, discharging of spent filtrate as wastewater occurs, along with wastewater from equipment cleaning operations and fermenter vent gas scrubbing. Wastewaters from fermentation operations typically have high biological oxygen demand, chemical oxygen demand, and total suspended solid levels, with a pH range of 4 to 8 (U.S. EPA, 1983).

Depending on the type of drug and dosage form being manufactured, formulation and dosage-form manufacturing may produce solvent wastes, wastewater from cleaning and equipment sterilization, packaging waste, and off-spec products (Venkataramani, 1994). Formulation processes tend to produce fewer wastes than the other manufacturing processes.

It is important to remember that some large pharmaceutical firms can provide products that can furnish up to 90 percent of a hospital. These companies also manufacture medical devices, hospital gowns, laboratory equipment, test kits, medical-imaging equipment, circuit boards for medical equipment, and so on. Many of these types of full-service pharmaceutical companies may also use continuous manufacturing processes (rather than just batch chemical processes) typically found in die-cutting, metal, and plastic extrusion industries. These operations also produce metal and plastic wastes, and packaging process wastes. A complete list of wastes from these additional types of operations is beyond the scope of this chapter. However, Table 8–3 describes existing and potential waste streams for the model P&C industry. Some important nonhazardous wastes are also included in this table.

TABLE 8–3
Typical pharmaceutical process wastes.

Waste Description	Process Origin	Composition
Process liquors	Organic synthesis	Contaminated solvents
Spent fermentation broth	Fermentation processes	Contaminated waters
Spent natural product, raw materials	Natural product, extraction processes	Leaves, tissues
Spent aqueous solutions	Solvent extraction processes	Contaminated water
Leftover raw material, containers	Unloading of materials into process equipment	Bags, drums (fiber, plastic, metal), plastic bottles
Scrubber water from pollution control equipment	Dust or hazardous vapor generating processes	Contaminated water
Volatile organic compounds	Chemical storage tanks, drums	Solvents

TABLE 8–3
Typical pharmaceutical process wastes.

Waste Description	Process Origin	Composition
Off-spec or out-dated products	Manufacturing operations	Miscellaneous chemicals
Spills	Manufacturing and lab operations	Miscellaneous chemicals, mercury
Wastewater	Equipment cleaning, extraction residues	Contaminated water, pheno-based
Spent solvents	Solvent extraction or wash practices	Contaminated solvents
Used production materials	Manufacturing operations	Filters, tubing, diatomaceous earth
Used chemical reagents	R & D operations	Miscellaneous chemicals, solvents, acid/alkaline wastes, radioisotopes, formaldehyde, photographic chemicals, silver
Spent ethylene oxide	Sterilization operations	Ethylene oxide
Miscellaneous wastes	Maintenance operations	Waste lube oils, vacuum pump oils, cleaning solvents, paint stripping wastes, leftover paints and accessories, spent fluorescent lamps, trash
Used packaging material	Packaging operations	Plastic, wood, cardboard, foam products
Infectious/medical wastes	R & D, manufacturing operations, off-spec products	Vials, biomass, blood products, human animal specimens
Incinerator exhaust	On-site incinerators	Metals, oxides
Natural gas combustion products	Steam boilers	Carbon compounds, oxides of nitrogen and sulfur, boiler blowdown, cooling tower sludges, and sediments

Source: U.S. EPA, October 1991, 1983, and adapted from Venkataramani, 1994.

REGULATORY CONSIDERATIONS AND TRENDS

Since the wide variety of operations and products developed by the P&C industries subject them to almost all EHS regulations applicable to manufacturing facilities, the following discussion provides only a brief summary of critical regulatory requirements for the P&C industries. Emphasis is placed on the governmental agencies responsible for implementing the federal regulations pertinent to the P&C industries. Specific state and local agency regulations that may be more stringent than their federal counterparts are beyond the scope of this chapter; however, you should remember that local governmental regulations must be attended to by the P&C industries. For example, Proposition 65 in California requires label warnings of carcinogenicity and reproductive toxicity associated with ingredients classified as toxins by the state. Likewise, P&C firms are more impacted by state and local agency interactions and inspections.

Federal Food and Drug Administration (FDA)

The FDA is charged with protecting the health of people against impure and unsafe foods, drugs, cosmetics, medical devices, and so on. All cosmetics, drug soaps, drugs, and combinations thereof come under the FDA's jurisdiction. The Food, Drug & Cosmetic Act (FD&CA) and all pursuant regulations applicable to a given category of product must be satisfied, including formulation, manufacturing conditions, documentation, and labeling. The National Center for Devices and Radiological Health (NCDRH) and the National Center for Drugs and Biologics (NCDB) are also of relevance to the pharmaceutical industries. The Office of Medical Devices of the NCDRH develops FDA policy on safety, efficacy, classification, and labeling of medical devices. The NCDB regulates biological products shipped in interstate and foreign commerce, and tests and establishes standards for biologicals (Rose, 1984).

If a cosmetic product is also a drug, then generally its safety and efficacy must be approved by the FDA prior to marketing. Approval is sought through a new drug application (NDA.) An NDA may be approved for either prescription or OTC sale of a product. Another way to market a drug product is to have it comply with one of the monographs in the FDA's OTC Drug Review. If a product's ingredients and its claims are part of an ongoing OTC review, the product can be marketed. Otherwise, if a drug product cannot meet the requirements of the OTC monograph, it must undergo an NDA process.

The FD&CA [USC 21, §201(i)] defines "cosmetics" as articles intended to be rubbed, poured, sprinkled, or sprayed on, introduced into, or otherwise applied to the human body or any part thereof, for cleansing, beautifying, promoting attractiveness, or altering the appearance, and articles intended for use as a component of any such articles. The definition does not include soap, however, but the burden to demonstrate "exempted soap" status rests with the soap manufacturer and distributor. Any soaps that claim a "medicinal" or "medicated" effect may also need to comply with the FDA's drug compliance requirements (Wood, 1989).

The FDA is also responsible for enforcement of the provisions of the Fair Packaging & Labeling Act (FPLA) that specifies requirements for labeling of products sold at retail. Mandatory labeling requirements include ingredient listing, net weight and name, and address of manufacturer, packer, or distributor (USC 21, §362; 21 CFR §701.3). The labeling on cosmetics must be truthful and not misleading (Brady,

1989). If a claim of a physiological effect is made for a cosmetic, the FDA may regulate the product as a misbranded cosmetic and it may have to go through an NDA. The U.S. Federal Trade Commission (FTC) also requires that labeling or advertising claims must be reviewed by it or the FDA prior to marketing. The FTC also issues guides to designate what trade and advertising practices are unacceptable under the Robinson-Patman Act and other statutes (Jess, 1995).

New Drug Approval Process

Manufacturing of new drug products consists of a complex procedure of approval from the FDA. The development of a new drug usually begins with the discovery of a new chemical entity. The R&D staff test the action of the new chemical under laboratory conditions, invitro and/or invivo. If the results indicate that the chemical agent has benefits over existing drugs, then preliminary animal testing is initiated. Preliminary animal studies investigate if the drug is safe for human use, and use acute toxicity testing and chronic testing for carcinogenicity and teratogenicity. If no serious safety problems are discovered, the manufacturer may submit an NDA. The FDA has thirty days to consider the submission. If there is no objection from the FDA, Phase I studies begin.

Phase I studies are usually carried out with fewer than 100 people. If these studies determine that the drug is safe, the manufacturer may then start a limited Phase II human studies to evaluate the dosage needed. Phase III studies extend the observation made in Phase II to a larger group of patients who are treated in supervised studies. These studies confirm the clinical value of the new drug by comparing it with existing drugs, track effects of the drug over time, and provide statistics on adverse reactions (Clayman, 1991).

With the completion of the three phases of studies, the manufacturer submits an NDA to the FDA. This submission includes all the safety data (human and animal) and the efficacy data (21 USC §505; 21 CFR §314.1 et seq.). In certain cases, the FDA may require Phase IV studies. This entire process can take five to ten years; typically, only a small proportion of new drugs are eventually approved for marketing (DiMasi, 1991). It is no wonder that the new drug approval process may change significantly in the coming years. The high pressure of cost containment and high R&D costs have induced pharmaceutical firms and the FDA to make changes in the drug development process (DiMasi, 1994). To make drug development easier, the FDA is also publishing a number of guidelines to harmonize its regulations with other nations' (Jass, 1995).

The regulatory requirements do not end, but are only beginning once a drug is approved by the FDA. The drug manufacturer still has postmarketing regulatory requirements after receiving FDA approval. For example, the manufacturers must continue to comply with the FDA's GMP regulations (21 USC §351(a)(2)(B); 21 CFR §211.11, et seq.). If the product is marketed as a prescription drug, the manufacturing data must be made available to the FDA during inspections, which can occur every two years (Brady, 1989).

Environmental Protection Agency

The EPA has direct responsibility for safe release and disposal of industrial chemicals and hazardous waste, be it discharged into waters of the United States, released into the air, or disposed on land. One of the best sources for staying current with the ever-increasing and changing environmental regulations is the **Semiannual Regulatory Agenda.** The EPA publishes the **Semiannual Regulatory Agenda** in the **Federal Register,** which is organized by acts, proposed rule, and final rule stages. Under each heading, the EPA lists rules it expects to propose and finalize in the future. (Please note that by the time this book is published, many of the proposed rules and finalized rules may be modified.)

Federal Insecticide, Fungicide, and Rodenticide Act

The Federal Insecticide, Fungicide, and Rodenticide Act (FIFRA) authorizes the EPA to register and regulate all pesticides, including all preservatives. For example, any soap product that makes claims such as a "disinfectant" or as a "flea killing" pet product comes under FIFRA's domain.

Toxic Substances Control Act

The Toxic Substances Control Act's (TSCA) intent is to prevent unreasonable risk of injury to health or the environment associated with the manufacture, processing, distribution in commerce, use, or disposal of chemical substances. The following three major programs are initiated under TSCA:

1. Anyone who intends to manufacture a new chemical substance must notify the agency at least ninety days prior to manufacture, under the premanufacture notification (PMN) program.
2. If a chemical is believed to pose an unreasonable risk to health or the environment. The EPA can require testing of a new or existing chemical if there is insufficient data.
3. The EPA can control existing chemicals that are found to pose an unreasonable risk.

Manufacturing, processing, distribution and disposal of polychlorinated biphenyl (PCBs) and certain uses and disposal of chlorofluorocarbons, asbestos, and dioxin are also limited under the TSCA (Homburger, 1982).

Clean Air Act

The 1990 Amendments to the Clean Air Act (CAA) issued several new regulations and schedules of compliance that significantly impact P&C manufacturing operations. Most of these stringent regulations have been enacted to meet mandates to limit the ozone levels in urban areas. For instance, the EPA recently issued "Report to Congress: Study of VOC Emissions from Consumer and Commercial Products," which describes the basis of future federal rules limiting volatile organic compounds (VOCs) in consumer products. The agency expects to issue the regulation of VOC products in 1997 for twenty-four categories of consumer products, including hair sprays, hair mousses and gels, nail-polish removers, shaving creams, antiperspirants, and deodorants. Many states, such as New Jersey and California, are already moving forward to limit VOCs in personal care products (Jass, 1995). To meet these regulations, P&C firms are looking into new ingredients, new formulation approaches, and new pressurized and dispensing systems. The greatest product changes to date are occurring in hair sprays (Price, 1995). Furthermore, the EPA's Office of Air Quality Planning and Standards is currently developing regulations to control air emissions from pharmaceutical manufacturing processes, including standards for storage tanks, process vents, equipment leaks, and wastewater treatment systems.

Clean Water Act

Recently, the EPA proposed effluent limitations guidelines and standards for the pharmaceutical point-source category under the Clean Water Act (CWA). These effluent limitation guidelines attempt to require in-plant emissions limitations to control air emissions by preventing them from being emitted during wastewater collection and treatment (PhRMA, 1995). These proposed rules would also regulate the release of pollutants created by manufacturers of "cosmetic preparation ... that function as a skin treatment" (Rose Sheet, 1995).

Resource Conservation and Recovery Act

Over the last decade, waste regulations have become more complex and demanding on hazardous waste generators. The Resource Conservation and Recovery Act (RCRA) Subtitle C regulates the control of hazardous waste disposal from cradle to grave. By definition, RCRA regulates disposal of wastes that are reactive, corrosive, toxic, or ignitable. If a waste displays hazardous characteristics, the generator must meet requirements for waste packaging, storage, labeling, transport, and disposal. In addition to federal regulations, many more stringent state regulations monitor nonhazardous and industrial waste from P&C industries. For instance, in New York the following N700 series of wastes are regulated: cosmetic rinse water, cosmetic sludge from cleanup operations, dust-collector residues, extenders, gelatins, gum arabic, lanolin, off-specification products, and prescription drugs (Stimson, 1993). Additionally, many states have a tracking program for medical wastes similar to the hazardous waste program.

Under RCRA's "mixture and derived-from" rules, any listed hazardous waste that is combined with a nonhazardous waste become a hazardous waste; nonhazardous wastes obtained from EPA listed process are also considered hazardous wastes. The EPA is proposing to change this regulation in the near future under the Hazardous Waste Identification Rule (HWIR), which will amend how listed the "mixture and derived from" rule affects low-risk wastes. This may exempt from regulation many such wastes. Depending on the final rules, changes may significantly impact the management of many waste streams generated by the P&C industries.

Nuclear Regulatory Commission

The Nuclear Regulatory Commission (NRC) is an independent government agency formed to regulate and inspect the uses of radioisotopes, along with nuclear power plants, reactors, reactor fuel, and products. Licenses are granted to persons and to organizations that own and use radioactive materials. The NRC regulates disposal of low-level radioactive wastes (10 CFR Part 20). These regulations specify that animal carcasses and liquid scintillation fluids containing less than 0.50 microcuries/gram of tritium or carbon-14 are "biomedically exempt," and may be discarded without special procedures. The NRC allows wastes that contain radiological material with a half-life of sixty-five days or less to be stored until a minimum of ten half-lives have passed. If wastes are measured less than or equal to background levels of radiation, the waste can be disposed of as nonradioactive waste (U.S. EPA, December 1991).

Occupational Safety and Health Administration (OSHA)

The P&C industries and the R&D laboratories are subject to the relevant standards found within *OSHA General Industry Standards and Interpretations* (29 CFR 1910). The General Industry Standards address a broad range of safety and health requirements, including lighting, noise control, electrical and fire safety, emergency planning, medical examinations, and exposure to toxic chemicals. Although an explanation of all the applicable OSHA regulations to the P&C industries is impossible here, a summary of the regulations that place an additional burden on them follows.

Occupational Exposure to Hazardous Chemicals in Laboratories (29 CFR 1910.1450) emphasizes the use of work practices and effective worker protection appropriate to the unique nature of the laboratory. The laboratory standards requires compliance with OSHA's permissible exposure levels, and with the employer's chemical hygiene plan. The standard requires that the chemical hygiene plan includes appropriate work practices, standard operating procedures, methods of control, measures of appropriate maintenance, use of protective equipment, medical examinations, and special precautions for work with hazardous substances (OSHA, 1994).

Bloodborne Pathogens (29 CFR 1910.1030) is a law requiring that an infection control plan be prepared by every person who handles, stores, uses, processes, or disposes of infectious medical wastes. The plan must include requirements for personal protective equipment, housekeeping, training, and a procedure for reporting exposures. The term **biological hazard** or **biohazard** is taken to mean any viable infectious agent (etiologic agent) that presents a risk or a potential risk to the well being of humans. All infectious/medical material must also be handled according to Universal Precautions (OSHA Instruction CPL 2-2.44A).

Department of Transportation

The Department of Transportation (DOT) regulates transportation modes as well as the transportation of hazardous materials in the United States. Materials covered by hazardous waste and hazardous material regulations, as well as infectious waste, are also regulated by the DOT during transportation. The DOT oversees many aspects of transporting hazardous materials, including packaging, labeling, marking, placarding, vehicle requirements, and emergency response.

Regulatory Trends

The green movement is impacting packaging and processing of many products. Furthermore, corporate policies of many companies are touting social and environmental responsibility and good business practices. Because of the shift to green marketing by many companies, the FTC, and the FDA are prohibiting misbranding of products as environmental friendly (Section 5 of FTC Act, "environmental advertising and marketing practices"). Many states are also adopting "rates and dates" legislation to reuse packaging. In addition to the FDA's monitoring of indirect food additives into food products because of leaching from packaging, coalitions are forming to limit intentional the addition of heavy metals into packaging and packing components (CTFA, 1994). Regulations may be forthcoming that mandate recycling consumer products packaging to stimulate markets of recycled materials.

Other than the issues already mentioned here, P&C companies are being guided by international efforts (e.g., European Community Directives and ISO Standards) to provide evidence that their products do not present a significant risk to the environment. Recent European directives are also concerned with the fate of a product after manufacture and the release of a genetically modified organism. These directives require an assessment of risks of new products due to storage, use, and disposal of new active ingredients (Moran, 1995).

TREATMENT CONTROL TECHNOLOGIES

From the earlier discussion of P&C manufacturing processes, it may be apparent that pharmaceutical and cosmetic manufacturing can be costly and complex. As already mentioned, there may be ten to fourteen stages in the manufacture of a single new chemical entity, each involving many chemicals and generating several waste products. Moreover, the P&C industries are heavily impacted by QA/QC duties. Concerns about quality of the products may drive the industry to use end-of-the-pipe treatment technologies rather than in-process systems (Moran, 1995).

Table 8–4 lists treatment control technologies that have been employed by the pharmaceutical industry to treat process wastewaters, the most prevalent waste type, to meet CWA (U.S. EPA, 1983).

TABLE 8–4
Summary of end-of-pipe treatment processes for pharmaceutical wastewaters (database: 308).*

Technology	Number of Plants
Equalization	62
Neutralization	80
Primary treatment	61
Coarse settleable solids removal	41
Primary sedimentation	37
Primary chemical flocculation/clarification	12
Dissolved air flotation	3
Biological treatment	76
Activated sludge	52
Pure oxygen	1
Powdered activated carbon	2
Trickling filter	9
Aerated lagoon	23
Waste stabilization pond	9
Rotating biological contactor	1
Other biological treatment	2
Physical/chemical treatment	17
Thermal oxidation	3
Evaporation	6

TABLE 8–4
(continued)

Technology	Number of Plants
Additional treatment	40
Polishing ponds	10
Filtration	17
Multimedia	7
Activated carbon	4
Sand	5
Other polishing	17
Secondary chemical flocculation/clarification	5
Secondary neutralization	5
Chlorination	11

*Note: Some plants use more than one treatment process.
Source: U.S. EPA, 1983

The P&C industries create many different types of waste streams; therefore, it is impossible to list all the control technologies they use. The P&C industries use a wide variety of treatment processes: physical, chemical, biological, thermal, and stabilization. Land-ban regulations are affecting the way wastes are being managed and, therefore, new technologies are being created continually. A brief summary of the treatment processes available to the P&C industries is outlined here.

Physical Treatment Processes

Physical treatment processes are those using physical characteristics to effect a separation or concentration in a waste stream. The processes are organized into four groupings: gravity separation, phase change, dissolution, and size/adsorptive/ionic characteristics. Table 8–5 presents the treatment technologies under each of these groupings (U.S. EPA, 1987).

TABLE 8–5
Physical treatment processes and technologies.

Gravity Separation	Dissolution	Phase Change	Size/Adsorptive/Ionic Characteristics
Sedimentation	Soil washing/flushing	Evaporation	Filtration
Centrifugation	Chelation	Air stripping	Carbon adsorption
Flocculation	Liquid/liquid extraction	Steam stripping	Ion exchange
Oil/water separation	Supercritical solvent extraction	Distillation	Electrodialysis
Dissolved air flotation			
Heavy media separation			

Source: Adapted from U.S. EPA, 1987

Chemical Treatment Processes

The chemical treatment processes are the most commonly used waste treatment practices. These include (U.S. EPA, 1987):

- PH adjustment for neutralization or precipitation
- Hydrolysis and photolysis
- Oxidation and reduction
- Hydrogen peroxide oxidation
- Ozonation
- Alkaline chlorination
- Hypochlorite chlorination
- Electrolyte oxidation
- Chemical dechlorination

Biological Processes

Biological degradation is fast becoming a viable approach to waste management. Originally, we used biological degradation of hazardous organic substances in the treatment of municipal wastewater, especially processes based on aerobic bacteria or anaerobic bacteria. Aerobic biological treatment is a treatment technology applicable to wastewaters containing biodegradable organic constituents and some nonmetallic inorganic constituents including sulfides and cyanides. Anaerobic digestion is best suited to wastes with moderate to high pH nonhalogenated hydrocarbons, moderate to low organic loadings, and low to zero biological oxygen demands (U.S. EPA, January 1991). We are now carrying out insitu treatment of contaminated soils using indigenous microbes, selectively adapted microbes, or genetically altered microorganisms. Various companies are developing and evaluating processes based on fungi and other biological communities, but so far they have not fully demonstrated these processes. Biological processes include (U.S. EPA, 1987):

- Aerobic biological treatment
- Activated sludge
- Rotating biological contactors
- Bioreclamation
- Anaerobic digestion
- White-rot fungus

Thermal Destruction Processes

Thermal destruction processes include several energy recovery processes, traditional incineration processes, and several innovative thermal processes. These processes include (U.S. EPA, 1987):

- Liquid injection incineration
- Rotary kiln incineration
- Fluidized bed incineration
- Pyrolysis
- Wet air oxidation
- Industrial boilers
- Industrial kilns (cement kiln, aggregate, clay)

- Blast furnaces (iron and steel)
- Infrared incineration
- Circulating bed combuster
- Supercritical water oxidation
- Advanced electric reactor
- Molten salt destruction
- Molten glass
- Plasma torch

You can incinerate any waste at some cost. Technical limits exist for specific incineration technologies, but there are no technical limits on incineration for any waste type.

Fixation and Stabilization Processes

Fixation and stabilization involves immobilization of the toxic and hazardous constituents in the waste. You can immobilize constituents by either changing the constituents into insoluble forms, binding them in an immobile, insoluble matrix, and/or binding them in a matrix that minimizes the material surface exposed to solvent exposure. Often the immobilized product has structural strength sufficient to help protect it from future fracturing and leaching. Most of these processes are proprietary. They include (U.S. EPA, 1987):

- Lime-based pozzolan processes
- Portland cement pozzolan process
- Sorption
- Vitrification
- Asphalt-based (thermoplastic) microencapsulation
- Polymerization

POLLUTION PREVENTION: BEST DEMONSTRATED PRACTICES

Companies are moving away from compliance-driven activities, to Pollution Prevention (P_2) activities. This trend is augmented by increasing pressures from government agencies, international competition, and consumer interest groups. P_2 efforts are fast becoming a strategic necessity for cost containment in the P&C industry. P_2 efforts not only include practices that minimize waste production from the use of solvents and bulk raw chemicals in drug and cosmetic manufacturing, but also practices that involve suppliers to assist in decreasing waste generation.

The Business Roundtable (BR) started a benchmarking project to determine the common as well as unique elements of successful facility-level P_2 programs (BR, November 1993). The project's key findings from the best-in-class companies who had strong P_2 programs suggest the following actions:

- Create strong support and focal points (or champions) for P_2 programs.
- Implement P_2 programs that work in the site-specific corporate and plant cultures.
- The corporate role should be to develop and establish P_2 goals with each facility staff and to assist in technology transfer across the company.
- The program should ensure the freedom to choose P_2 approaches and methods.

- Progress on goals and initiatives should be reported on a monthly or quarterly basis.
- The P_2 project must be cost-effective to compete with normal capital processes.
- It should not rely only on source reduction techniques, but use the entire range of waste management options.

Introduction of an approved drug into the worldwide therapeutic environment takes time; therefore, opportunities do exist to design and implement proactive safety and P_2 practices that require minimal capital (Forman & Venkataramani, 1994). A few leading pharmaceutical companies are incorporating P_2 practices *early* in the process of the drug development cycle before it becomes locked-in. For example, Merck has dramatically reduced waste loads through the following actions (Venkataramani, 1990):

- Process modification to reduce the use of methylene chloride by 82 percent
- Solvent-distillation and internal recovery of acetone-water mixture to reduce usage of acetone by 80 percent
- Recovery and reuse of isoamyl alcohol in process
- Recovery of isopropyl acetate from its cholesterol-lowering drug
- Recovery of ethyl acetate off-site and sent for nonprocess reuse by others

Table 8–6 lists the primary waste streams associated with pharmaceutical operations, along with suggested waste minimization options. The ideal P_2 assessment procedure is outlined in Figure 8–3.

While early waste-minimization efforts generally focused on hazardous waste reduction, the EPA now encourages attention to all wastes generated using a multimedia approach (U.S. EPA, October 1991). Source reduction is always the most desirable option with recycling, reuse, or reclamation of part or all of a waste stream being next. Both source reduction options and recycling options suited to pharmaceutical and cosmetic manufacturing are discussed in this section. In addition to the specific recommendations provided here, rapidly advancing technologies make it important that you continually educate yourself about improvements that prevent waste and pollution. Information sources include trade associations and journals, chemical and equipment suppliers, equipment expositions, conferences, government bulletin board systems, Internet resources (e.g., http://www.fedworld.gov), and industry newsletters. By keeping abreast of changes and implementing applicable technology improvements, you can help your company take advantage of the dual benefits of reduced waste generation and a more cost-efficient operation.

Source Reduction

You can achieve source reduction of hazardous wastes through changes in products, raw materials, process technologies, or procedural and organizational practices. Pharmaceutical and cosmetic manufacturing is a diverse and very competitive industry. Because of the highly specific and often confidential nature of each company's specific operations, only a very general description of material substitution and process modification is given here.

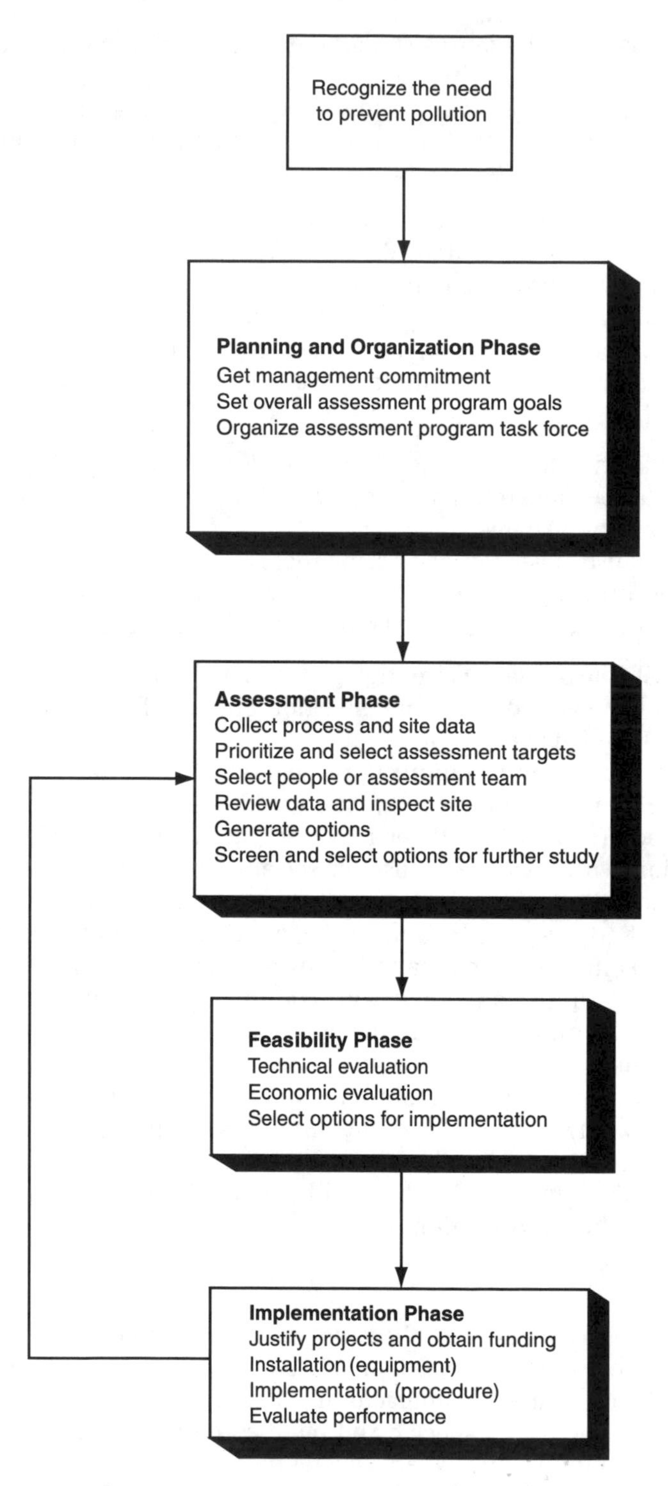

FIGURE 8–3
The ideal P_2 assessment process

Material Substitution

Material substitution is a change in one or more of the raw materials used in production to reduce the volume or toxicity of waste generated. For the pharmaceutical and cosmetic industry, however, product reformulation is likely to be very difficult due to the testing required to ensure that the reformulation has the same therapeutic or cosmetic effect as well as stability and purity profile as the original drug or cosmetic. Furthermore, a considerable amount of time is required for FDA approval of the reformulated drug. An additional concern is the effect the reformulation might have on the product's aesthetic qualities. Changes in characteristics such as taste, color, or dosage form could result in customer rejection of the product. However, many companies have used material substitution successfully in pharmaceutical tablet coating operations to reduce hazardous waste generation.

Other material substitutions that may be suited to pharmaceutical manufacturing include the use of aqueous-based cleaning solutions instead of solvent-based solutions, and the replacement of chlorinated solvents with nonchlorinated solvents. Because of the reformulation difficulties encountered in the production phase, it is best to introduce waste minimization at the research and development phase. For these reasons, as an integral part of R&D activities you should carefully examine all materials that can be used in manufacturing or formulating a pharmaceutical with the aim of reducing residual toxicity.

Process Modification

Besides investigating material substitution options, you can look for source-reduction opportunities that can be accomplished through modification or modernization of the existing process. In most cases the product/process yield determines the product/waste ratio. Reasons for high byproduct yield include inadequate feed-rate control, mix-up, or temperature control. By controlling reaction parameters, you can improve the reactor efficiency and reduce the byproduct formation. Increased automation can also reduce operator errors. For example, automated systems for material handling and transfer, such as conveyor belts for bagged materials, can help reduce spillage. Crystallization, sedimentation, polymerization, and corrosion can cause fouling deposits on interior equipment surfaces. These deposits reduce process-operating efficiencies and increase waste generation. Proper agitator design and optimizing operating temperatures can inhibit fouling deposits.

Another process modification option is to redesign chemical transfer systems to reduce physical material losses. Other design considerations for waste minimization include modifying tank and vessel dimensions to improve drainage, installing internal recycle systems for cooling waters and solvents, selecting new or improved catalysts, switching from batch to continuous processes for solvent recovery, and optimizing process parameters to increase operating efficiency. While process modification can result in significant waste reduction, there may be major obstacles to this approach. Extensive process changes can be expensive; production must be stopped for new equipment installation; and new processes must be tested and validated to ensure that the resulting product is acceptable. In addition, to the extent that processes and process equipment are specified in an approved drug application, FDA approval is likely to be required prior to instituting any changes.

Good Operating Practices

Good Operating Practices (GOPs) are procedures and policies that result in a reduction of wastes. The following is a list of some key overall operating strategies:

- Keep individual waste streams segregated.
- Keep hazardous waste segregated from nonhazardous and infectious wastes.
- Keep recyclable wastes segregated from nonrecyclable wastes.
- Minimize dilution of hazardous wastes.
- Assure that the identity of all chemical and wastes is clearly marked on all containers.

To improve management and control practices you can do the following:

- Centralize purchasing and dispensing of drugs and other hazardous chemicals.
- Monitor drug and chemical flow within the facility from receipt of raw materials to disposal as hazardous wastes. Use computer systems and computer-readable bar-coded labels for incoming chemicals, such as those used in supermarkets.
- Apportion waste management costs to the department that generate the wastes.
- Improve inventory control by requiring users of chemicals with limited shelf-life to use up old stock before ordering or using new stock; ordering hazardous chemicals only when needed and in minimal quantities to avoid outdated inventory; and asking suppliers to take back expired chemicals.
- Implement a facility-wide P_2 program.

Management initiatives can encourage new ideas from knowledgeable employees, which can result in the reduction or recycling of waste. The following is a list of some management initiatives:

- Provide employee training in hazardous materials management and waste minimization. Train employees on handling hazardous materials and P_2. Employees should be aware of waste disposal costs and liabilities, and they should understand the causes of waste generation and potential process upsets.
- Closer supervision of plant personnel and operations can increase production efficiency and reduce waste generation by reducing material costs, spins, and production of off-spec products. Coordination within the overall plant operation can, in turn, increase the opportunity for early detection of mistakes.
- Effective production and maintenance scheduling can help reduce waste generation. Proper scheduling ensures raw materials are used before expiration and products are recovered and processed efficiently, while maintenance scheduling makes sure that work is done on equipment at a time least likely to result in product losses. Minimization of equipment cleaning requirements should be one of the objectives of production scheduling.

Spillage or leakage of hazardous chemicals generates hazardous wastes in the form of liquid waste from washing the spilled toxic chemicals or solid waste from cleanup using absorbent materials. Spill and leak prevention is critical to waste minimization, and a properly trained and spill control team is needed to prevent or contain spills. Methods of reducing or preventing spills include conducting hazard assessment studies, using proper storage tanks and process vessels, equipping all liquid containers with overflow alarms, and testing alarms periodically. Also, you should take steps to maintain the physical integrity of containers, set up administrative controls, and install sufficient secondary containment. Other preventive measures include having a good valve layout, having interlock devices to stop the flow to leaking sections, not allowing operators to bypass interlocks or alter set points, and isolating equipment or process lines that are not in service. Finally, consider documenting all spills and their related dollar value in relation to overall operating efficiency.

Preventive maintenance programs can reduce the incidence of equipment breakdown through routinely cleaning, making minor adjustments, lubricating, testing, measuring, and replacing minor parts. Typically, equipment data cards, master preventative maintenance schedules, deferred preventive maintenance reports, equipment history cards, and equipment breakdown reports are used as recordkeeping documents. Corrective maintenance repairs the unexpected failures as they occur and collects data for use in determining maintenance demands. It is important that you prepare maintenance and operating data sheets for each piece of equipment.

Recovery and Recycle

Recovery and recycling includes direct reuse of waste material, recovering used materials for a separate use, and removing impurities from waste to obtain relatively pure substances. The goal is to recover materials for reuse in a different application. The strict quality-control requirements of the pharmaceutical and cosmetic industry often restrict reuse opportunities, though some do exist. Usually recycling is a much easier option, which you can perform either on or off site.

Solvent Waste Recycling

Solvents are used for equipment cleaning, reaction media, extraction media, and coating media. Processes to recover solvents from concentrated waste streams include distillation, evaporation, liquid-liquid extraction, sedimentation, decantation, centrifugation, and filtration. Some commonly used solvents are (U.S. EPA, 1983):

- Acetone
- Cyclohexane
- Methylene chloride
- Ethyl acetate
- Butyl acetate
- Methanol
- Ethanol
- Isopropanol
- Butanol
- Pyridine
- Methyl ethyl ketone
- Tetrahydrofuran

An important first step in determining the feasibility of on-site distillation and recovery of waste solvents is to separate waste streams according to specific chemical components (U.S. EPA, 1990). This may allow you to use simple batch distillation equipment, which is less expensive than fractional distillation equipment. Many companies have developed individual solvent recycling units suitable for small facilities. In the event that one of the distillation methods is not feasible for your company, you should consider off-site distillation or waste exchange. You can also use solvent wastes with sufficiently low chlorine content as a fuel supplement in cement kilns and some industrial boilers.

The following steps can improve solvent waste recyclability:

- Segregate solvent wastes as follows: chlorinated from nonchlorinated solvent wastes; aliphatic from aromatic solvent wastes; chlorofluorocarbons from methylene chloride; and water wastes from flammables.
- Minimize the solids concentration in solvent wastes.
- Label all solvent wastes and keep records on their compositions and methods of generation.

Waste Exchanges

An alternative to recycling is waste exchange, which involves the transfer of a waste to another company for use as is or for reuse after treatment (U.S. EPA, October 1991). Waste exchanges are private or government-subsidized organizations that help to identify the supply and demand of various wastes. Three types of waste exchanges are available to you: information exchanges, material exchanges, and waste brokers. **Information exchanges** are clearing houses for information on supply and demand, and typically publish a newsletter or catalog. **Material exchanges** take temporary possession of a waste for transfer to a third party, in contrast to **waste brokers,** who do not take possession of the waste, but charge a fee to locate buyers or sellers.

Waste exchanges frequently recycle metals and solvents because of their high recovery value. The most commonly recycled wastes include acids, alkalis, salts and other inorganic chemicals, organic chemicals, and metal sludges (U.S. EPA 1983).

TABLE 8–6
Compendium of pollution prevention methods for the P & C industry.

Waste Category	P_2 and Waste Minimization Methods
Pharmaceutical drugs and off-spec products	Drugs that cannot be used in the U. S. when their potency decreases to less than 95 percent are usable in many other countries where drugs are in scarce supply. Donate products that are still usable. Rework off-spec material. Use automated processing systems. Validate cleaning and reuse. Frequent attention paid to batch records and QA/QC checks can help reduce reject batches. Test raw materials prior to use to decrease rejects.
Equipment cleaning wastes	Maximize the number of campaigns to reduce cleaning frequency. Use the final rinse as the premise on your next cleaning cycle. Use wiper blades and squeegees, and rework remainders into products.

TABLE 8–6
(continued)

Waste Category	P_2 and Waste Minimization Methods
	Use low-volume, high-frequency cleaning (e.g., spray heads). For manual systems, hoses should be on only when the equipment is being cleaned. For automated systems, cleaning cycles should be optimized and validated for minimal use of water. Have parts molded clean and certified by the supplier. Use physical instead of chemical cleaning methods.
Solvents	Substitute less-hazardous cleaning agents or aqueous solvents. Reduce the quantity of solvent used. Reduce analyte volume requirements. Use premixed kits for tests involving solvent fixation. Use calibrated solvent dispensers for routine tests. Segregate solvent wastes. Recover/reuse/regenerate spent solvents. Consider reformulation to eliminate chlorinated solvents, acetone, and methanol usage.
Radionuclides	Use less-hazardous isotopes whenever possible. Segregate and properly label radioactive wastes, and store short-lived radioactive wastes in isolation on-site until their decay permits disposal in trash.
Mercury	Substitute electronic sensing devices for mercury-containing devices. Provide mercury spill cleanup kits and train personnel. Recycle uncontaminated mercury wastes using proper safety controls. Investigate the use of sodium borohydride reduction followed by ultrafiltration to remove colloidal metallic mercury wastes.
Chemicals	Substitute less-toxic compounds and cleaning agents. Reduce volumes used in experiments. Use mechanical handling aids for drums to reduce spills. Establish an internal recycling program.

TABLE 8–6
(continued)

Waste Category	P_2 and Waste Minimization Methods
	Require all new materials to be tested in small quantities before being purchased in bulk. Encourage suppliers to be responsible partners in your P_2 program. Replace oil-based paints with water-based paints in maintenance departments. Use nonchemical pest control methods. Test expired lab chemicals before discarding to see if the chemicals are still potent. Prepare a standard operating procedure manual for R & D waste management. Review and revise the purchasing and inventory of tracking procedures of the R & D section to reduce waste generated from expired chemicals.
Containers	Return empties to supplier. Thoroughly empty and triple rinse with minimal water. Use containers with recyclable liners. Segregate solid wastes. Collect and reuse plastic from in-house molding.
Formaldehyde	Minimize the strength of formaldehyde solutions. Minimize wastes from cleaning. Capture your waste formaldehyde. Investigate reuse in laboratories.
Air emissions	Control bulk storage air emissions (e.g., internal floating roofs). Use dedicated dust collectors and rework dust back into the product. Optimize fossil fuel combustion. Use dedicated vent condensers and return condensate to source, where possible. Find alternatives for chlorofluorocarbons.
Spill and area washdown	Use dedicated vacuum systems. Use dry cleaning methods. Use recycled water. Maintain equipment properly to avoid leaks.
Filtration process wastes	Investigate replacing your rotary vacuum filter with an ultrafiltration process. Eliminate requirements for diatomaceous earth filter aid.

TABLE 8–6
(continued)

Waste Category	P_2 and Waste Minimization Methods
	Examine byproduct uses of filter cake material, such as its resale as a fertilizer or soil additive. Investigate any subsequent treatment required. Prevent spillage by installing v-shaped guides beneath rotary vacuum filters to direct the scraped filter cake onto the center of the conveyor belt. Implement procedures to keep filter aid spillage to a minimum. Investigate methods for reducing water content and odor levels in filter cake wastes. Recover and reuse precious metal catalysts from carbon and filter aids.
Corrosive wastes	Provide your spent acid to a battery manufacturer. Use small amounts of acid to wipe indelible ink off glassware. Investigate changing glassware labeling as a means of reducing sulfuric acid usage. Use existing NaOH waste to neutralize waste HCL.
Used oil	Recover and recycle.
Waste fabrics	Investigate alternative uses for medical gowns, sheets, diapers, and covers.

Source: Adapted from U.S. EPA, December 1991, October 1991, and June 1990; Venkataramani, 1992.

SUMMARY

- Pharmaceutical and cosmetic manufacturing is a complex industry. Each manufacturer has its own unique set of circumstances. However, you can learn the general concerns of the P&C industries. For example, the technology used for manufacturing drugs and cosmetics, as well as the composition of these products greatly influence the diversity of
  a. Manufacturing processes and the resulting waste streams
  b. Regulations affecting the manufacturers
  c. Treatment control technologies used to manage the wastes
  d. Pollution prevention techniques that can be used to contain costs of waste production

- For burgeoning environmental professionals dedicated to serving the P&C industries, it is important to understand the EHS concerns particular to these industries. To serve these industries well, you need two types of information.
  a. *Industry factors:* You need to be familiar with the terminology and technology found in the P&C industry.
  b. *Industry objectives:* You need to understand what these industries are trying to achieve; that is, the nature of their products, and how to control the costs of waste resulting from production without deleteriously impacting the quality of the product.

ACKNOWLEDGMENT

The author appreciates the assistance provided by Dr. Ed Venkat at MERCK Research Laboratories; Ron Meissen at Baxter Health Care Corporation; CTFA; PhRMA; the librarians at the Tufts Center for the Study of Drug Development; Estee Lauder; the EPA and OSHA libraries; and MDL, Inc.

QUESTIONS FOR REVIEW

1. How is a drug different from a cosmetic product?
2. What is the general process for manufacturing a drug product?
3. True or false: Cosmetic manufacturers never manufacture drugs.
4. Why are pollution prevention efforts difficult to implement in the pharmaceutical industry?
5. What regulatory trends could have a huge impact on the P&C industries?
6. What are the most common waste types from the P&C manufacturing processes?
7. True or false: Many companies are strategically restructuring their organizations so that products are validated and improved continuously.
8. Why would you want to incorporate pollution prevention practices early in the process of the drug development cycle?
9. Locate a medical dictionary and, using Table 8–1, define each of the pharmaceutical dosage forms listed in the table.
10. Active ingredients in many types of cosmetic products could include agents that are anti-inflammatory, antibiotic, antifungal, antibacterial, antiparasitic, antipruritic, antiseptic, keratolytic, antiactinic, antiviral, antiseborrheic, and antipsoriatic. Using a medical dictionary, define these terms.

ACTIVITIES

1. Locate some drug patent reviews and analyze the ingredients used by the manufacturer. What types of wastes could potentially be produced in the manufacturing process of these drugs?
2. Pamela works for a small pharmaceutical company. She has to write a report that identifies P_2 efforts for her manufacturing operations. Pamela needs to collect all pertinent process information, assimilate the contents, and list the opportunities for pollution prevention. What recommendations would you make to Pamela for collecting, analyzing, and implementing the P_2 efforts?

REFERENCES

Bailey, J.E. and D.F. Ollis. 1977. *Biochemical Engineering Fundamentals.* New York: McGraw-Hill.

Bollinger, J.N. October, 1989. "Role of the Cosmetic Formulator in the Development of Pharmacologically Active Products." *Cosmetics & Toiletries magazine.* Vol. 104. pp. 43–49.

Brady, R.P. June, 1989. "Cosmetics as Drugs: The Trends and Realities." *Cosmetics & Toiletries magazine.* Vol. 104. pp. 35–40.

The Business Roundtable. November, 1993. "Facility Level Pollution Prevention Benchmarking Study." Courtesy of Ronald E. Meissen, Baxter Healthcare Corporation.

Clayman, C.B., Ed. 1991. "Know Your Drugs and Medications." American Medical Association. New York: Reader's Digest. p. 19.

CTFA. 1994. *CTFA Environmental Manual.*

DiMasi, J.A., et al. 1991. "Cost of Innovation in the Pharmaceutical Industry." *Journal of Health Economics.* Vol. 10. pp. 107–142.

DiMasi, J.A., et al. June, 1994. "New Drug Development in the United States from 1963 to 1992. *Journal of Health Economics.* Vol. 55. No. 6. pp. 609–622.

Eldermann, H.J. July, 1988. "Cosmetic Labeling Issues—and Answers." *Cosmetics & Toiletries magazine.* Vol. 103. pp. 33–46.

Estrin, N.F., Ed. 1984. *The Cosmetic Industry: Scientific and Regulatory Foundations.* New York: Marcel Dekker.

FDA. October, 1992. *FDA's Cosmetic Handbook.* Center for Food Safety and Applied Nutrition.

Forman, A.L., and E.S. Venkataramani, et al. June, 1994. "Environmentally Proactive Pharmaceutical Process Development." MERCK & Co. (unpublished article).

Fox, C. August, 1989. "Cosmetic Raw Materials: Literature and Patent Review." *Cosmetics & Toiletries magazine.* Vol. 104. pp. 55–77.

Homburger, F., Ed. 1992. *Safety Evaluation and Regulation of Chemicals.* London: Karger. pp. 1–15.

Jass, H.E. May, 1995. "Regulatory Review." *Cosmetics & Toiletries magazine.* Vol. 110. pp. 21–22.

Jass, H.E. June, 1995. "Regulatory Review." *Cosmetics & Toiletries magazine.* Vol. 110. pp. 27–28.

Kirk-Othmer. 1982. "Cosmetics." *Encyclopedia of Chemical Technology, Vol. 7.* New York: John Wiley.

Moran, J. October, 1995. "Assessing Pharma Environmental Risks." *Scrip Magazine* #39.

Nally, H., and R. Kieffer. October, 1993. "The Future of Validation: From QA/QC to TQ." *Pharmaceutical Technology.* pp. 106–114.

Parmeggiani, L., Ed. 1983. "Cosmetics." *Encyclopedia of Occupational Health and Safety, Vol. 1.* Geneva: International Labour Office.

PhRMA. August, 1995. "Comments on the Proposed Effluent Limitations Guidelines, Pretreatment Standards and New Source Performance Standards: Pharmaceutical Manufacturing Category."

OSHA. 1994. "Exposure to Hazardous Chemicals in Laboratories." OSHA 3119.

Price, S.N. June, 1995. "Keeping VOCs Under Control." *Cosmetics & Toiletries magazine.* Vol. 110. pp. 42–49.

The Rose Sheet. September 25, 1995. F-D-C Reports. p. 5.

Rose, S.L. 1984. *Clinical Laboratory Safety.* Philadelphia, PA: Lippincott. pp. 1–35.

Schueller, R., and P. May Romanowski. 1995. "Cosmetic Ingredient Nomenclature: Making the Transition from the IUPAC System." *Cosmetics & Toiletries magazine.* Vol. 110. pp. 29–31.

Venkataramani, E.S., et al. August, 1990. "Waste Minimization in a Leading Ethical Pharmaceutical Company." *Environmental Progress.* Vol. 9, No. 3. pp. A10–A11.

Venkataramani, E.S. 1994. "Pollution Prevention in the Pharmaceutical Industry." *Industrial Pollution Prevention Handbook.* H.M. Freeman, Ed. New York: McGraw-Hill. pp. 865–883.

Venkataramani, E.S., et al. 1992. *Create Drugs, Not Waste: Case Histories of One Company's Successes.* Chemtech. pp. 674–679.

Stimson, R.P. June, 1993. "Pharmaceutical Waste: Liability, Drugs, Brokers, Ethics and Politics, Part I." *Pharmaceutical Technology.* pp. 84–87.

Wood, T.E. December, 1989. "Regulatory Considerations for Soap Products in the USA." *Cosmetics & Toiletries magazine.* Vol. 104. pp. 75–79.

U.S. EPA. January 1981. "Enforceability Aspects for RACT for the Chemical Synthesis Pharmaceutical Industry." EPA-340/1-80-016.

U.S. EPA. November 1982. "Development Document for Proposed Effluent Limitations Guidelines, New Source Performance Standards and Pretreatment Standards for the Pharmaceutical Manufacturing Point Source Category." Effluent Guidelines Division, Washington, D.C.

U.S. EPA. September 1983. "Development Document for Effluent Limitation Guidelines and Standards for the Pharmaceutical Manufacturing Point Source Category." EPA440/1-83/084.

U.S. EPA. July 1985. "Enhanced COD Removal from Pharmaceutical Wastewater Using Powdered Activated Carbon Addition to an Activated Sludge System." EPA/600/D-85/153.

U.S. EPA. December 1986. "Development Document for Final Best Conventional Effluent Limitations Guidelines for the Pharmaceutical Point Source Category." EPA440/1-86/084.

U.S. EPA. September 1987. "A Compendium of Technologies Used in the Treatment of Hazardous Wastes." U.S. EPA/625/8-87/014.

U.S. EPA. October 1991. "Guides to Pollution Prevention: The Pharmaceutical Industry." EPA/625/7-91/017.

U.S. EPA. December 1991. "Medical and Institutional Waste Incineration: Regulations, Management, Technology, Emissions, and Operations." EPA/625/4-91/030.

U.S. EPA. January 1991. "Treatment Technology Background Document." Office of Solid Waste.

Zanowiak, P. 1982. "Pharmaceuticals." *Kirk-Othmer Encyclopedia of Chemical Technology, Vol. 17.* 3rd Edition. New York: John-Wiley.

9

Chemical Waste Streams

Dana Acton-Irvin

Upon completion of this chapter, you will be able to meet the following objectives:

- Provide an overview of the regulations that apply to the generation of waste streams in the chemical industry.
- Describe the general waste characteristics of several common waste streams found in the chemical industry.
- Discuss methods and procedures that can be applied to monitoring a majority of chemical effluents.
- Address pollution prevention methods.
- Discuss health and safety concerns associated with waste streams.
- Discuss cost benefits of reducing the amount of hazardous chemicals within waste streams. This procedure is referred to as "source point reduction."

INTRODUCTION

In 1992 the chemical industry reported 1.54 billion pounds[1] of chemicals released into the environment. Most chemicals reported are used to manufacture commercial products and some chemicals are used as intermediates in the production of products. The majority of chemicals are being disposed of responsibly. However, releases of chemicals into the environment still occur. According to the 1992 report, less than 7 percent of chemicals are released to the environment. In 1992, chemical companies had reduced releases of toxic chemicals to the environment by 4 percent from 1991 and by 35 percent since 1988. Off-site transfers of waste for treatment and disposal fell by 19 percent since 1991 and by 39 percent since 1988. The chemicals released into the environment originate from chemical waste streams.

This chapter will introduce the reader to an overview of regulations applying to the chemical industry's generation of waste streams, describe the general characteristics of waste streams as a result of chemical operations, and discuss methods and procedures that can be used to monitor the majority of industrial effluents.

REGULATORY OVERVIEW

In 1976 Congress enacted the Resource Conservation and Recovery Act (RCRA). For the first time the federal government was authorized to establish a comprehensive program for the cradle-to-grave management of hazardous waste. Regulations applicable to facilities that treat, store, or dispose (TSD) of hazardous waste are set out in 40 Code of Federal Regulations (CFR) Parts 260–265. The procedural system covering the RCRA permit program is found in 40 CFR Parts 124, 270, and 271. All facilities that treat as hazardous waste (e.g., incinerate), store a hazardous waste (e.g., drum storage yards), or dispose of a hazardous waste (e.g., off-site disposal to a hazardous waste landfill) are required to comply with RCRA regulations.

In addition to federal RCRA regulations, many individual states have passed their own hazardous waste laws. In many cases, the state regulations "mirror" the federal RCRA regulations with minor exceptions. For example, the Tennessee hazardous waste regulations mostly mirror the federal regulations, but take a few exceptions such as substituting the terminology "division director" for "EPA regional administrator." However some states, such as California, have much more stringent regulations than RCRA. In any case, it is necessary to not only be knowledgeable of the federal RCRA regulations, but of individual state hazardous waste regulations, too.

The federal government has also enacted several other regulations that govern the chemical industry. The Clean Air Act (CAA) regulates chemical by-products emitted into the air. The Clean Water Act (CWA) and the Safe Drinking Water Act (SDWA) regulate chemicals that enter water. Once a hazardous material is transported offsite, the Hazardous materials Transportation Act becomes applicable.

WASTE STREAMS

According to statistics, the chemical industry produced over 729 billion pounds of organic and inorganic chemicals in 1994.[2] The top ten chemicals produced are listed in Table 9–1.

The chemical industry produces such a varied amount of products that it would be difficult, if not impossible, to describe each specific waste stream. Therefore, the more common types of waste streams will be discussed.

There are four major categories of waste streams: gaseous (gases, vapors, airborne particulates), wastewater (aqueous waste), liquid waste streams (nonaqueous

TABLE 9–1
Top ten chemicals produced by industry in 1994.

Chemical	Billions of Pounds Produced
Sulfuric acid	89.20
Nitrogen	67.54
Oxygen	49.67
Ethylene	48.52
Lime	38.35
Ammonia	37.93
Propylene	28.84
Sodium hydroxide	25.83
Phosphoric acid	25.26
Chlorine	24.20

waste), and solid waste streams (including sludges and slurries). **Gaseous** waste streams can be a result of such operations as fiber spinning, spraying operations, and processing. **Wastewater streams** are the result of raw water usage. **Liquid waste streams** can result from operations such as solvent recovery systems, vessel washings, and condensate. **Solid waste streams** can be the result of filter cakes, distillation fractions, spent catalysts, vessel and tank residues, bag house dusts, and sludge from an on-site wastewater treatment facility.

Each waste stream within the facility should be identified and quantified. At a minimum, the following information should be documented for each waste stream: waste ID and name, source and origin, component or property of concern, annual generation rate for the component of concern and the overall waste, and cost of disposal for each pound generated and a yearly cost. This information is obtained from product information such as formulations and mass balances.

The following is an example of a common waste stream formed from the production of explosives: Within the manufacturing area there are catch basins. A catch basin is a depression that allows a solid to settle to the bottom and liquid to continue downstream. In this example the solid that has settled to the bottom is removed for disposal. Because this solid is from the manufacturer of explosives, the solid is considered an explosive contaminated material. The type of explosive depends upon the type of product being manufactured. In this case, the solid is contaminated with trinitrotoluene and RDX, which is a common explosive used to manufacture weapons. Based on this information the waste stream is identified as follows:

Waste ID/Name: *Explosive contaminated wastewater*
Source/Origin: *Building F-9 catch basin*
Component/or Property of Concern: *Trinitrotoluene, RDX (common explosive)*
Annual Generation Rate
  Component(s) of Concern: *13 pounds*
  Overall: *159 pounds*
Cost of Disposal
  Unit Cost ($ per pound): *22 for open burning: 250 for off-site disposal*
  Overall cost (per year): *3,500 for open burning: 40,000 for off-site disposal*

WASTE-MONITORING PROGRAM

The majority of chemicals released into the environment are the by-products of chemical operations carried along through waste streams. Most chemical manufacturing processes produce some quantities of waste. This waste can be liquid, gas, or solid in form. Unfortunately, the disposal of these wastes can cause detrimental effects on the environment as is evidenced by the continued hazardous waste cleanup effort throughout the United States.

Due to the continued growth of industry, it has become clear that current methods of chemical waste disposal have be reevaluated. In addition, methods to eliminate waste disposal need to be identified. Our air, water, and land's ability to retain the waste is approaching its maximum holding potential. Further industrial expansion, or even continued operations at existing levels of pollution, could result in severe health and social degradation.

The end result of waste streams will be either as on-site releases and waste management or off-site waste management. **On-site releases** are made to the air, water, land, or to underground injection wells. On-site waste management includes inplant recycling, energy recovery, and treatment. Off-site waste management includes publicly owned treatment works (POTWs), treatment facilities, disposal such as in landfills, recycling, and energy recovery.

The latest statistics indicate that approximately 1.5 billion pounds of waste were reported as being released on-site. An additional 1.2 billion pounds of waste were reported as being transferred off-site. Table 9–2 shows a breakdown of wastes released both on-site and off-site.[3]

TABLE 9–2
On-site and off-site waste releases.

Disposal Method	Pounds (Millions)
On-Site	
Fugitive or nonpoint air emissions	157.53
Stack or point air emissions	399.93
Surface water discharges	224.34
Underground injection	684.64
Releases to land	69.48
Total on-site releases	1536.93
Off-Site	
Transfers to recycling	455.37
Transfers to energy recovery	349.82
Transfers to treatment	166.69
Transfers to POTWs	203.60
Transfers to disposal	46.55
Other off-site transfers	1.64
Total off-site releases	1223.68

An initial step to begin reducing waste-stream generation is to implement a program to identify and monitor the waste streams and their constituents. A waste monitoring program is necessary for the following reasons:

1. To assure responsible regulatory agencies of the manufacturer's compliance with effluent requirements and the implementation schedule set forth in the discharge permit. A discharge permit is a permit issued from the EPA under its National Pollution Discharge Elimination System (NPDES) program. This permit gives the holder the ability to discharge a limited amount of wastewater at given intervals into public waterways. Under the permit system, the party creating the discharges is responsible for monitoring the waste stream's compliance with the permit.
2. To maintain sufficient control of inplant operations to prevent violations of permit specifications. Material losses or reduced performance of process equipment will result in increased waste loads. Analysis of the waste streams can often pinpoint malfunctions and result in prompt corrections. If the inplant control system is carried out effectively, there will be a minimum of regulatory involvement in plant production operations.
3. To develop necessary data for the design and operation of wastewater treatment facilities. The monitoring program will also provide basic data that will be valuable in the design of a wastewater treatment system to meet regulatory requirements.
4. To insure product and material losses to sewer lines. Monitoring can eliminate inaccurate accusations of illegal or harmful waste discharge practices. Adequate monitoring records can document that a facility was operating in conformance with permit requirements.

A waste-monitoring system should become an integral portion of the manufacturing process and be used as a measure of efficient operation. Once incorporated into the production system, it will be an invaluable check on the overall efficiency of plant operations, as well as aid in meeting legal requirements.

DEVELOPING THE MONITORING PLAN

Five stages are necessary to develop a monitoring plan. They include

- Management awareness
- Task team selection
- Process analysis
- Waste survey
- Initiation

Each stage is described here. Figure 9–1 is an example of the steps involved in establishing an effluent monitoring program.

Management Awareness

The first stage in developing a monitoring plan is to have management awareness. Management must be understanding and cooperative to begin initiation of the program. As the program is being designed and implemented, it is essential that management

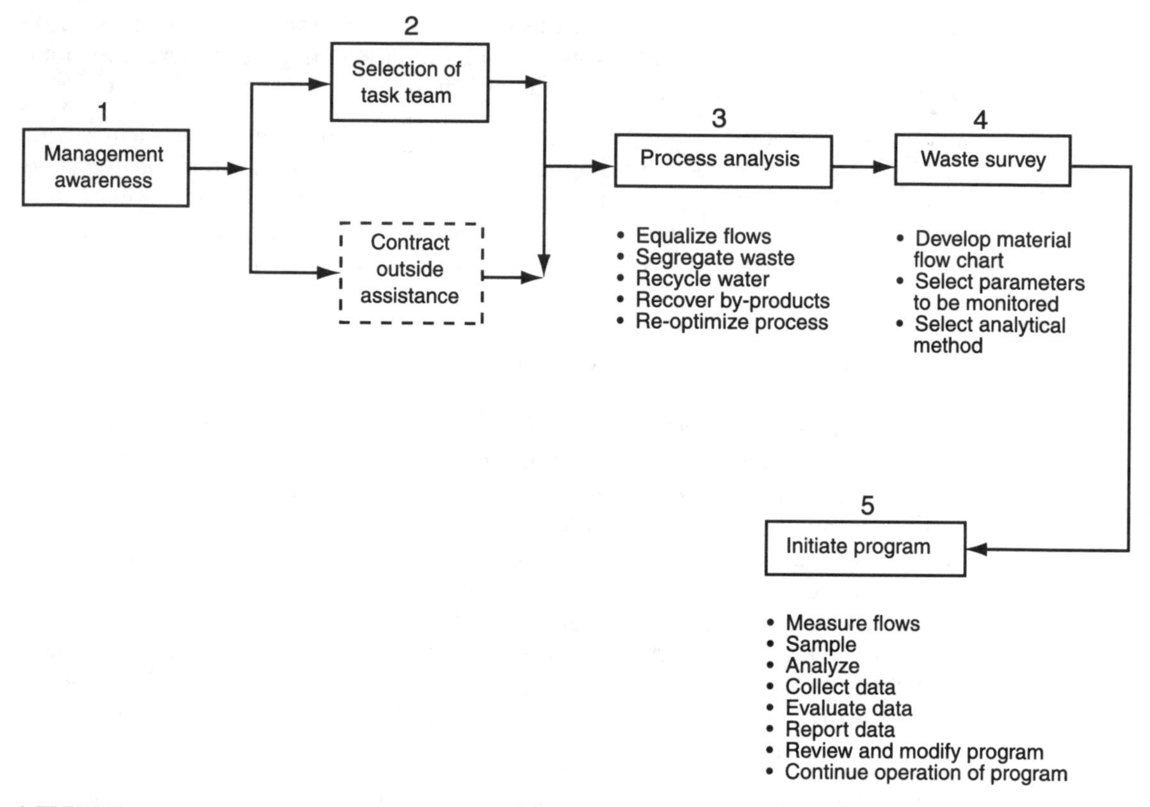

FIGURE 9–1
Steps involved in establishing a waste-monitoring program.

guarantees that various groups such as production, analytical laboratory, and engineering will cooperate fully. Management must also be aware of the legalities and final overall cost reduction to the operation.

Task Team Selection

The second stage is the selection of a task team to develop the monitoring program. In many cases, an outside consultant is hired to develop the monitoring program. This decision is one the manufacturer must make based upon a judgment of inhouse capabilities. When an inhouse task team is chosen, it is imperative that the project team leader be at a high enough management level to guarantee full cooperation of the supporting departments. In either case, however, a thorough knowledge of the manufacturing facility, its operations, and the analytical techniques required for characterization of the waste streams is essential.

Process Analysis

The third stage is to analyze the process. During this stage of program development, water usage and waste streams generated will be characterized. In many cases, simple processes such as equalizing flows, segregating wastes, recycling water, recovering by-products, and re-optimizing processes can reduce waste-stream generation prior to implementing a monitoring program.

Waste Survey

The fourth stage is to survey the waste produced. The waste survey provides a material balance of the flow of pollutants through a facility. Since the savings to be realized from the waste survey almost always exceeds the cost, the majority of industries will find that survey expenditures yield an excellent return.

The first consideration in the development of an industrial waste-stream survey is a review of the entire production process. A complete picture may be acquired by a **material flow chart** of the entire plant for each operation, all raw materials, additives, end products, by-products, and liquid and solid wastes. The waste-stream survey plan should take into consideration the following factors: seasonal variations, peak pollution loads, intermittent discharges, and continuous discharges.

Following the construction or the understanding of an existing flow chart, the next step is to define the amounts of raw materials, additives, products, and wastes for each operation. After the materials are known it is possible to establish a **mass balance** for each operation.

Mass balances are important for many projects since they allow for quantifying losses or emissions that were previously unaccounted for. Also, mass balances assist in developing the following information:

1. Determining a baseline for tracking the progress of your efforts.
2. Gathering data to estimate the size and cost of additional equipment and other modifications.
3. Gathering data to evaluate economic performance.

In its simplest form, the mass balance is represented by the mass conservation principle:

$$\text{Mass In} = \text{Mass Out} + \text{Mass Accumulated}$$

The mass balance should be made individually for all components that enter and leave the process. An example of estimating releases of particulates to the air from a grinding/blending operation using mass balance follows.

A facility grinds and blends chromium (III) oxide (Cr_2O_3) for use as a pigment in a water-based paint. The air around the grinding operation is exhausted to a filter with a particulate collection efficiency of 99 percent. According to the facility measurements, the annual amount of chromium (III) oxide solids collected from the filter is 2,000 pounds. What is the estimated amount of particulate chemical released into the air? To find the solution, take the amount of solid chemical processed per year minus the amount of solid dissolved or suspended in final product minus the amount of solids in filter cake minus the amount of solids in wastewater sludge.

$$\text{Amount of } Cr_2O_3 \text{ released to air} = 2{,}000 \text{ pounds} \times [(1-0.99) \div 0.99] = 20 \text{ pounds.}$$

To report as a release of parent metals (Cr), adjust as follows: The amount of metal released equals the amount of metal compound released times the molecular weight of the parent metal portion divided by the molecular weight of the metal compound.

$$\text{Molecular weight of Cr} = 52$$
$$\text{Molecular weight of } Cr_2O_3 = 152$$
$$\text{Amount of Cr released to air} = 20 \text{ pounds}$$
$$Cr2O3 \times (52 \times 2) \div 152 = 14 \text{ pounds}$$

Mass balances can assist in determining the concentrations of waste constituents where analytical test data is limited. They are particularly useful where there are points in the production process that make it difficult or uneconomical to collect analytical data, such as an inaccessible pipe. A mass balance can help determine if fugitive losses are occurring. For example, the evaporation of solvent from a parts cleaning tank can be estimated as the difference between solvent put into the tank and solvent removed from the tank.

By definition, the mass balance includes both materials entering and leaving a process. Mass balances are easier, more meaningful, and more accurate when they are done for individual units, operations, or processes. For this reason, it is important to define the mass balance area properly. The area should be drawn around the specific operation of concern, rather than a larger group of areas or the entire facility. An overall mass balance for the facility can then be constructed from individual unit balances. An example of estimating releases of solvent to the air from a solvent cleaning tank using mass balance follows.

A solvent solution is used to clean metal parts containing 80 percent by volume of methylene chloride. Purchasing and inventory records indicate that 1,300 gallons of this solvent solution were used throughout the year. Facility operational records show that 17 drums (55 gallons each) of spent cleaning solvent was shipped off-site for recovery during the same period. Assume that the methylene chloride concentration in the spent solvent is 80 percent. Calculate the mass balance of methylene chloride released into the air. The volume of methylene chloride released to air equals:

amount used (1,300 gal × 0.80)– amount recovered (17 drums x 55 gal/drum × 0.80) = 292 gallons

There are 11.05 pounds of methylene chloride per each gallon of methylene chloride. The amount of methylene chloride released to air equals:

292 gallons × 11.05 pounds methylene chloride per/gal. of methylene chloride = **3,227 pounds of methylene chloride released to the air**

There are several drawbacks to mass balances. One is that the analytical data and flow measurements may not allow an accurate measure of the waste stream. In particular, in processes with very large inlet and outlet waste streams the absolute error in measurement of these quantities may be greater in magnitude than the actual waste stream itself. The time span is also an important factor. Mass balances constructed over a shorter time span require more accurate and more frequent waste-stream monitoring to close the balance. Mass balances performed over the duration of a complete production run are typically the easiest to construct and are reasonably accurate. The EPA's Office of Toxic Substances has prepared a guidance manual,

Estimated Releases of Waste Treatment Efficiencies for the Toxic Chemical Inventory (Form EPA 560/4-88-02). This manual contains additional information for the developing mass balances for the listed toxic chemicals.

If an up-to-date one is not already available, at this stage it will be necessary to develop a **sewer map** showing water, wastewater, sanitary, storm, and drain lines. The difficulty of locating sewer lines for wastes fed to each outfall becomes a time-consuming and complex problem in older facilities. Piping diagrams are seldom updated as changes are made over the years and these drawings must be accepted with this understanding. The details of the map should be specific for pipe size, location, type of supply and drain connections to each processing unit, and direction of flow, with location of roof and floor drains, manholes, catch basins, and control points defined. To determine the sources of wastewater in sewers, it is frequently convenient to add a tracer to the waste stream in the outlet of a production unit. By plotting the flow of the tracer, it is possible to establish a sewer map. Commonly used tracers are dyes, floats, and smokes.

The next phase of the monitoring plan is to establish **sampling stations.** Sampling stations are used to gather information regarding the waste stream, such as peak time releases and concentration of chemicals being released. Important factors to be considered in selecting the sampling stations area are:

1. The flow of the waste stream is known or can be estimated or measured.
2. The sampling station is easily accessible with adequate safeguards.
3. The location should be where the waste stream is well mixed.

When sampling, both the quantity and quality of what enters and exists the facility are of major importance to assess the operation's overall performance. However, sampling should be performed on specific unit operations. This will assure consistent operations of the facility.

In many cases, regulatory agencies determine what analysis must be conducted on the waste streams. Additionally, analysis is required to assure compliance with permit requirements. Specific analysis will be dependent upon the type of operation. The following are examples of wastewater parameters for selected chemical manufacturing processes.

Inorganic Chemical Manufacturing

Acidity/alkalinity	BOD_5	Fluoride
Total solids	COD	Silicates
Total suspended solids	TOD	Phosphorous
Total dissolved solids	Chlorinated benzenoids	Cyanide
Chlorides	Polynuclear aromatics	Metals
Sulfates	Phenol	Temperature

Organic Chemical Manufacturing

BOD_5	TOC
COD	Organic chloride
pH	Total phosphorus
Total suspended solids	Metals
Total dissolved solids	Phenol
Oil and grease	Cyanides
	Total nitrogen

Plastic and Synthetics Manufacturing

BOD	Total dissolved solids
COD	Sulfates
pH	Phosphorus
Total suspended solids	Nitrate/organic nitrogen/ammonia
Oil and grease	Cyanides
Phenols	Polynuclear aromatics
Toxic additives	Zinc
Chlorinated benzenoids	Mercaptan

After the parameters of analysis have been determined, the next step is to collect a sample. Proper collection is of the utmost importance, as the results of the sampling will lead to all subsequent decisions. Obtaining good results will depend upon certain details. Among these are the following:

1. Insuring that the sample taken is truly representative of the waste stream. Obtaining a sample that is truly representative of the waste stream may be the source of significant errors. Obtaining a representative sample should be of major concern in a monitoring program.
2. Using proper sampling techniques. The two most common types of samples are known as grab samples and composite samples and either may be obtained manually or automatically.
3. Protecting the samples until they are analyzed. Personnel must assure that the samples are preserved properly and analyzed in the time period specified for the method. For example, COD analysis must be conducted within a 24-hour period from sample collection time.

Initiate the Program

The final task is to initiate the monitoring program. This phase will involve all of the previous steps, from identification of the process, to the waste survey and characterization, sampling, and analysis. Following analysis it is then necessary to evaluate the data and report the findings. The data will allow the operation to be evaluated to tell if there is waste being generated and where the waste originates and into what waste stream the waste is going, such as gaseous, liquid, or solid.

POLLUTION PREVENTION

One of the greatest challenges facing business today is maintaining industrial competitiveness while ensuring a healthy environment. To remain competitive in an environmentally driven world market, American industry must develop new and improved process technologies designed to prevent pollution at the source. We can no longer rely upon end-of-the-pipe pollution control solutions.

In 1992, the EPA enacted the federal Pollution Prevention Act (PPA). This policy states that pollution should be prevented or reduced at the source whenever feasible. Pollution that cannot be prevented should be recycled in an environmentally safe manner, whenever possible. Pollution that cannot be prevented or recycled should be treated in an environmentally safe manner, whenever feasible. Disposal or other release to the environment should be employed only as a last resort and should be conducted in an environmentally safe manner. The two major categories of pollution prevention are source reduction and recycling. Source reduction techniques avoid the generation of hazardous wastes, thereby eliminating the problems associated with handling these wastes, whereas recycling techniques allow hazardous materials to be put to beneficial use.

Source Reduction

Source reduction is simply using techniques to reduce or eliminate the amount of toxic chemicals that are carried through the waste stream and released into the environment. Source reduction techniques are characterized as good operating practices, technology changes, material changes, or product changes. Each of these categories of source reduction is described here.

Good Operating Practices

Good operating practices are procedural, administrative, or institutional measures that a company uses to minimize waste generation. These practices can often be implemented with little cost and, therefore, have a high return on investment. These practices can be implemented in all areas of a chemical facility, including production, maintenance operations, and raw material and product storage. Examples of good operating practices include waste minimization, management and personnel practices, material handling and inventory practice, loss prevention, waste segregation, cost accounting practices, and production scheduling.

Examples of good management and personnel practices include employee training, incentives and bonuses, and other programs that encourage employees to conscientiously strive to reduce waste.

One example of good material handling is the institution of programs that reduce loss of input materials due to mishandling, expired shelf life of time-sensitive materials, and proper storage conditions.

Loss prevention is primarily avoiding leaks from equipment and spills. Some simple techniques include covering solvent tanks when not in use, placing equipment so as to minimize spills and losses during transport of parts or materials, and using drip pans, secondary containerization, and splash guards.

Waste-stream segregation is the technique of reducing the volume of hazardous wastes by preventing the mixing of hazardous and nonhazardous wastes by segregating them to avoid cross-contamination of waste streams.

Allocating waste treatment and disposal cost directly to the departments or groups that generate waste, rather than charging them to general company overhead accounts is an example of a cost accounting practice.

By using proper production scheduling of batch runs, the frequency of equipment cleaning and the resulting waste-stream generation can be reduced.

Technology Changes

Technology changes normally involve process and equipment modifications that reduce waste, primarily in a production setting. They range from minor changes that can be implemented in a short period of time to the replacement of processes involving large capital costs and duration, such as changes in the production process, equipment changes, implementing automation, and changes in process operating conditions (e.g., flow rates and pressure). One technique is treating waste streams at the source instead of the outlet. Often a particular compound can be treated at the source so that it will not react in the sewer to form less-desirable regulated compounds. One example is chlorine, which reacts to form chlorinated phenols. Volatile constituents and aquatic toxicity are best treated at the source whenever possible. Another example of a technology change is the reuse of water within the production area. Raw, pretreated, or final treated wastewater streams can frequently be reused within a given production area. Often piping modifications are necessary to achieve this "gray-water" reuse. These pretreated wastewater streams can then be used for cooling tower and scrubber makeup.

Material Changes

Input material changes accomplish waste-stream minimization by reducing or eliminating the hazardous materials that initially enter the production. Also, changes in input materials can be made to avoid the generation of hazardous wastes within the production processes. Examples include avoiding the use of heavy-metal pigments, using a less-hazardous or toxic solvent for cleaning, and purchasing raw materials that are free of trace quantities of hazardous or toxic impurities. Material Safety Data Sheets or product formulations obtained from the manufacturer can be used to assess the toxic characteristics of raw materials prior to their purchase.

Product Changes

Product changes are performed by the manufacturer of a product with the intent of reducing waste streams. They can include product substitution, product conservation, and changes in product composition. For example, in the paint industry water-based coatings are finding increasing applications where solvent-based paints were used before. These products do not contain toxic or flammable solvents that make solvent-based paints hazardous when they are disposed of. Also, cleaning the applicators with solvent is no longer necessary. The use of water-based paints instead of solvent-based paints also greatly reduces volatile organic compound emissions to the atmosphere.

Recycling Techniques

Recycling techniques are characterized as use/reuse techniques and resource recovery techniques. Recycling via use and/or reuse involves the return of waste material either to the originating process as a substitute for an input material, or to another process as an input material. The recycling process is not considered a waste stream and is, therefore, not discussed in detail here.

HEALTH AND SAFETY CONCERNS

Waste streams can be a source of health and safety concerns. Even in an enclosed piping system, there is a potential of exposure to the waste stream through sampling, leaking, or even pressure release. In many cases the materials in the waste stream are hazardous and can be toxic. For instance, exposure to long periods of particulates in air emissions can cause health problems, depending on the type and size of the particulate. Information regarding safe levels of exposure can be obtained from a variety of sources, including the American Conference of Governmental Industrial Hygienists. This organization evaluates and determines safe exposure levels to a number of common chemicals.

Another health and safety issue involves the exposure to bacterial sources possibly present in sewage waste streams and wastewater treatment facilities. In many cases, personal protective clothing may be necessary to handle materials and to perform sampling of these waste streams.

COST BENEFITS OF WASTE STREAM REDUCTION

The economic evaluation is carried out using standard measures of profitability, such as payback period, return on investment, and net present value. This information is normally obtained from the business unit within the corporation. In performing the economic evaluation, various costs and savings must be broken down into capital costs and operation costs. For smaller facilities, this procedure will be much less formal. Several obvious options, such as the installation of controls and good operating practices, may be implemented with little or no economic evaluation. In these instances, no complicated analyses are necessary to demonstrate the advantages of adopting P_2 techniques. For larger facilities, direct capital costs can include costs for designing, purchasing, and installing equipment and indirect costs of permitting, training, start-up, and finance charges.

In the past, companies have tended to ignore costs associated with waste treatment, storage, and disposal. However, regulatory requirements imposed on facilities have caused the costs of waste management to increase to the point where it is becoming a significant factor in a company's overall cost structure.

SUMMARY

- Waste stream generation in the chemical industry has become a significant concern.
- The cost of disposing of wastes is forever increasing as our natural resources are depleting.
- In addition, the regulatory requirements regarding waste stream generation are becoming more complicated and harder to comply with.
- It has therefore become necessary for the chemical industry to become more responsible for chemical waste stream generation and disposal.

QUESTIONS FOR REVIEW

1. What type of facilities does the Resource Conservation and Recovery Act regulate?
2. What are the four types of waste streams? Give an example of the source of each.
3. List two reasons with explanations of why a waste monitoring program is necessary.
4. What are the five steps involved in developing a monitoring plan?
5. List three factors to be considered in selecting a sampling station.
6. The two major categories of pollution prevention are source reduction and recycling. Give an example of each of these types of pollution prevention techniques.

ACTIVITIES

1. Interview the manager of a local facility (e.g., waste water treatment plant, chemical manufacturer) to find out the discharge points of the facility.
2. Contact your state environmental department regarding discharge permits for your local area. Find out the amount, type, and location of discharge allowable under the permit for a particular facility.
3. Contact the EPA for the most recent edition of the Toxics Release Inventory. Which state has the most discharges? Which state receives the most hazardous waste? Is the trend of hazardous waste discharges increasing or decreasing?

NOTES

1. 1992 Toxics Release Inventory. April 1994. EPA 745-R-94-001.
2. *Chemical & Engineering News*, June 26, 1995.
3. 1992 Toxics Release Inventory. EPA 745-R-94-001.

10

Pulp and Paper Industry

Tisha Wilkinson

Upon completion of this chapter, you will be able to meet the following objectives:

- List the major sources of effluent pollution in the pulp and paper industry.
- List the major sources of air pollutants in the pulp and paper industry.
- Identify the waste component associated with selected components of wood.
- List a number of small-volume hazardous waste streams produced by the pulp and paper industry.
- Identify the types of waste generated in the chemical pulping process.
- Identify the most dominating compunds found in pulp mill effluents.
- List the minimum analysis requirements for major effluent waste streams.
- Identify air emissions sampling practices.

INTRODUCTION

Concerns over the impacts of pulp and paper mill effluent on the receiving environment have been inherent since the industry first started. Pulp and paper mill effluent discharged into water causes oxygen depletion, distributes non-filterable residues, and adds known and unknown compounds that often have adverse effects on aquatic organisms. If present in sufficient quantities these can even be lethal. Other environmental impacts from pulp mill effluent are associated with odor, fish flesh tainting, acidity, slime growth, thermal effects, foam, scum and color.

Air pollution abatement efforts in the pulp and paper industry are concerned with controlling gaseous and particulate emissions. Although all types of emissions may be controlled at least in part by the production process, some type of external treatment of discharging gas volumes is also required. Although not dangerous in airborne concentrations, odorous materials (particularly reduced sulfur gases) are the most difficult to eliminate (Smook, 1992).

GENERAL PROCESS

The basic process in the production of pulp and paper products is the conversion of logs into bleached and dried pulp or wood fibers. Logs are debarked and reduced to wood chips. The chips are introduced to a digester where they cook in chemical liquors to free the wood fibers by dissolving the lignin and nonfiber materials binding the fibers together. After the chips have finished cooking in the liquor, the impure pulp is screened and washed to remove the liquor chemicals, dissolved lignins, and other impurities. The pulp is then bleached to increase brightness (increase the degree of whiteness), and then washed again to remove the remaining chemicals. Once washed, the pulp is dried and baled. The pulp is then ready to be converted into a variety of paper products in the paper mill.

Pulping Methods

The purpose of pulping is to free cellulose fibers from the other wood components. There are numerous pulping processes, with the specific process used determining the type of waste produced. Pulping methods in the industry include chemical, semi-chemical, mechanical, thermo-mechanical, chemi-therom mechanical, and bleached chemi-thermo mechanical.

Chemical Pulping

Chemical pulping uses chemicals to separate the lignin, hemicellulose, and extractives from wood fibers. Kraft and sulfite processes are the two major types of chemical pulping.

In the kraft pulping process, logs are debarked and then put through a chipper. The wood chips are cooked in a white liquor solution of sodium sulfide and sodium hydroxide. The treatment dissolves 90–95 percent of the lignin, nearly all hemicellulose and wood extractives, and a small amount of cellulose-derived polysaccharides. Approximately 55 percent of the original wood is dissolved in what is now termed the "black liquor." Byproducts are recovered, and the liquor evaporated to high concentration and then burned for recovery of energy and inorganic chemicals (Murray, 1992). Inorganic sulfur is trapped in a condenser during the evaporated phase. Highly volatile emissions from the condenser, and volatile compounds and gases from the burnt concentrated black liquor, are released to the atmosphere. Air contaminants released from pulping include particulate, sulfur dioxide, and total reduced sulfur (TRS) compounds. (These compounds contribute to the foul odors associated with pulp mills.)

The less common of the two chemical pulping processes is the sulfite process. The wood chips are reduced to their component parts by cooking in a pressurized vessel using a liquor composed of alkaline oxides and sulfurous acid (H_2SO_3) as the primary or secondary delignification chemicals.

Chemical recovery is an integral part of the kraft process. BOD_5 is much higher in the sulfite process because there is no chemical recovery. In either process, suspended solids and color are problems in the effluent. Of the two chemical pulping processes, toxicity and foam problems are greater in the kraft effluent.

Semi-Chemical Pulping

The semi-chemical pulping process is a two-step pulping process that uses a mild liquor, such as neutral sodium sulfite/sodium carbonate solution, for partial delignification of the wood. The final separation of fibers is achieved by mechanical action.

There is no chemical recovery in the semi-chemical pulping process. The BOD_5/tonne of pulp is between that of the kraft and sulfite chemical pulping methods. Effluent problems are suspended solids and color, to a lesser degree than chemical pulping.

Mechanical Pulping

The pulp fibers in mechanical pulping are released by mechanically grinding the wood. Wastewater streams have a low BOD_5 because most of the lignin and hemicellulose remains with the pulp fibers.

Mechanical pulping requires high energy costs. Suspended solids are a major problem.

Thermo-Mechanical Pulping

In thermo-mechanical pulping (TMP), the wood chips are steamed prior to and during the initial pulping process. Secondary atmospheric refining follows the initial mechanical pulping stage. TMP effluent has lower suspended solids and higher BOD_5 than mechanical pulping.

Chemi-Thermo Mechanical Pulping

Pulp made by the chemi-thermo mechanical (CTMP) process uses wood chips that are pretreated with a chemical (usually sodium sulfite) either prior to or during presteaming to aid in subsequent mechanical processing in refiners. CTMP requires a lower volume of water than chemical pulping and, therefore, yields higher BOD_5 and toxicity in the effluent. Suspended solids are a problem.

Bleached-Chemi-Thermo-Mechanical Pulping

A bleaching stage is added to the CTMP process to brighten the pulp in bleached-chemi-thermo-mechanical pulping. Color of the effluent is lower than in chemical pulping, but the BOD_5 and toxicity are higher. Suspended solids are a problem in the effluent.

Secondary fiber plants recycle wastepaper that is either mixed paper, old newspapers, old corrugated containers, pulp substitutes, or high-grade deinked wastepaper. The wastepaper is repulped, and the contaminants and ink are removed. The stock is then thickened before being formed into a sheet, pressed, dried, and baled.

MAJOR SOURCES OF POLLUTION FROM PULP AND PAPER OPERATIONS

The major categories of water pollution that are of concern to the pulp and paper industry are effluent solids, oxygen demand, toxicity, and color. The major categories of air pollutants from pulp and paper facilities include particulates, gases/vapors (odors), and liquid droplets (mists).

The major sources of effluent pollution in a pulp and paper mill include:

- Wastewater from wood handling/barking and chip handling
- Condensate from the digester and evaporators
- White waters (water containing fiber fines and/or additives) from pulp screening, cleaning, and thickening
- Filtrate from the bleach plant washer
- White water from the paper machine
- Fiber and liquor spills from all sections

The major sources of air pollutants in pulp and paper industry include:

- Fine particulates from the kraft recovery furnace
- Course particulates from hog fuel and coal-fired boilers
- Sulfur oxides, particularly from sulfite mill operations
- Nitrogen oxides from all combustion processes
- Reduced sulfur gases from kraft pulping and recovery operations
- Volatile organic compounds from digester relief and spent liquor evaporation

INGREDIENTS AND MATERIALS BALANCES

Woody Materials

Wood is composed of 45 percent cellulose (fiber), 25 percent lignin, 25 percent hemicellulose, and 5 percent extractives. Each component produces different types of waste as shown in Table 10–1 (Kringstad and Lindstrom, 1984).

Cellulose is glucose in a tightly bound molecular structure. It is susceptible to degradation by microorganisms, but not easily attacked by bacteria and molds because of the protective hemicellulose and lignin covering its fibers.

Hemicellulose, which surrounds the cellulose fibers, is composed of a random arrangement of five-carbon sugars. It forms a chemical bond with the lignin and acts as a chemical bonding agent between the cellulose and lignin.

TABLE 10–1
Wood composition and waste components.

Wood Component	% Composition	Waste Component
Cellulose	45	Fiber (suspended solids)
Lignin	25	Chromophores (color)
Hemicellulose	25	BOD_5
Extractives	5	Toxicity

Lignin surrounds and gives strength to the cellulose-hemicellulose framework. Lignin is very difficult to degrade. When slow degradation takes place, phenolic compounds, which are generally toxic, are released. The phenolic compounds released from lignin during the chlorine bleaching of pulp are responsible for much of the toxic compounds released in pulp mill effluents (Murray, 1992).

Extractives are the wood components that may be extracted by organic solvents. Although extractives are of much lower concentrations than phenolic materials in mill wastes, some are relatively toxic and can significantly reduce water quality and marine habitat in lakes and rivers receiving mill effluents (Murray, 1992).

HAZARDOUS WASTE STREAM IDENTIFICATION AND CHARACTERIZATION

The large-volume wastes produced by the paper industry are generally not classified as hazardous; however, there are small-volume wastes that if generated are classified as hazardous and must be transported and managed according to the Resource Conservation and Recovery Act (RCRA). These include:

- **Spent halogenated solvents** used in degreasing
- **Corrosive waste** generated from the use of strong acids and bases
- **Paint waste** containing solvents and paint waste with heavy metals
- **Ink waste,** which can include solvents, metals, or ignitable materials
- **Petroleum distillates** from cleanup operations

In the chemical pulping process, acids and alkalis, lime, sulfurous acid, sodium hydroxide, and sodium sulfide are typically used. The process generates acid and/or alkaline waste.

Chlorine bleaches, sulfate bleaches, and chloroform solvents are used in the bleaching process. This produces toxic wastewater and wastewater treatment sludge, as well as acid/alkaline waste.

Acidic compounds are among the most dominating in pulp mill effluents. Resin acids are naturally occurring compounds in wood resins. Chlorinated resin acids may be found in spent bleach liquors, particularly from softwood kraft pulping. The acidic compounds may contribute to acute toxicity in untreated effluent.

Chloroorganic compounds may be formed by the reaction of elemental chlorine or hypochlorite with lignin residues from the cooking process (COWIconsult, 1989).

Chlorophenolics may be found in different spent bleach liquors depending on the raw material, the pulping process, and the amount of chlorine used for bleaching.

Chloroform is formed in the bleach plant.

Dioxins (polychlorinated dibenso-p-dioxins) and furans (polychlorinated dibenzoi-p-furans) have been identified in emissions from the pulp industry; however, their concentration in stack gases from burning condensed black liquor is very low. Dioxins and furans are also produced during pulp bleaching; however, research and development has led to a substantial reduction to near elimination of these compounds in effluents (Murray, 1992).

Pigments used in the paper-making process contribute to general hazardous wastewater treatment sludge.

The waxes, glues, synthetic resins, and hydrocarbons used in sizing and starching contribute to toxic wastewater and sludges.

Ink, solvent, paint, ignitable, and toxic waste are generated from the inks, paints, solvents, rubbers, and dyes used in the coating, coloring, and dying operation.

Cleaning and degreasing practices using tetrachloroethyline, trichloroethyline, methylene chloride, trichloroethane, and carbon tetrachloride generate toxic rinse water and solvent waste that is considered hazardous.

Used antifreeze is potentially toxic because of its ethylene glycol content. Due to the nature of its use, antifreeze can become contaminated with dissolved metals. Primary lead, mercury, cadmium, chromium, copper, and zinc are all possible contaminants of used antifreeze. Antifreeze may also become mixed with other hazardous fluids such as gasoline or solvents.

Sandblasting grit may become contaminated during the sandblasting operation. Sandblasting lead-based paints may contaminate the used grit with enough soluble lead to require the material to be regulated as a hazardous waste.

REGULATORY IMPACT

Under the federal Clean Water Act (CWA), pollution control limits may be either technology-based or water-quality based. At a minimum, all dischargers must meet national technology-based limits, and individual states may use a water quality approach to set more stringent effluent limits as required. Dischargers are issued permits with discharge limits and compliance schedules that meet the federal technology-based standards as well as state ambient water-quality standards. The specific requirements of a permit are determined on a site-specific basis. The permits are usually valid for five years unless the mill is undergoing major process changes, in which case short-term permits may be issued.

There are distinct advantages to enforcing the regulations aimed at reducing the amount of pulp and paper mill pollutants. Certainly improving the quality of the environment and of human life are among those advantages. As new, more strict environmental regulations are introduced, the industry is forced to modernize and to adopt technologies that are more environmentally sound. However, the installation of pollution-control systems or changes to production methods may reduce the competitiveness of some mills. As a result, these mills will likely not be able to invest the capital required to conform to the stricter environmental standards, forcing them to close, either partially or completely, or be sold.

Water-quality guidelines in the United States are typically administered by the permit system, which lays down the specific criteria and schedule of compliance for each industrial facility. Guidelines are normally applied uniformly, with an awareness of the site's specifics. In more environmentally sensitive locations, more stringent limitations may be imposed. Likewise, more relaxed limitations may be tolerated on mill discharges in less-environmentally sensitive areas.

If a mill is found to be out of compliance of its permit requirements, the state environmental officials will issue a waste discharge requirement, which is essentially a request that the mill fix the problem (Odendahl and Holloran, 1993). If the mill does not fix the problem, the authorities will follow up with a waste discharge order demanding that the mill fix the problem. Following this, the authorities will issue a cease and desist order by the state, which demands that the mill remedy the situation or else be fined a specified amount. At this point, the state may involve the federal authorities. After consultation with the mill, the EPA will issue a consent decree detailing the terms and deadlines for the mill to be in compliance. The decree will include a specific list of fines for not meeting the deadlines (Odendahl and Holloran, 1993).

Limitations on industrial air emissions fall under government regulation. The basic statutory framework is set forth in the Clean Air Act (CAA). The provisions of the act are administered by the EPA through a permit system.

Since solid waste disposal of sludge is a large-volume waste in pulp and paper operations, more stringent regulations can potentially impact the viability of an operation. Traditionally, pulp and paper mill sludges were readily disposed of in on-site and off-site landfills. The primary environmental concern, however, is the potential for landfilled waste sludge leachate to contaminate groundwater and surface water. Leachate control and groundwater monitoring is now a regulated requirement at solid waste disposal facilities.

The federal environmental efforts are being supplemented by state and local enforcement activities. Every state has a functioning environmental bureaucracy that implements federal policies and supplements those with its own.

TREATMENT AND CONTROL TECHNOLOGY

Effluent Treatment and Control Technology

Internal Systems to Treat Mill Effluent

Changes to the production process reduce adverse environmental effects. The industry has reduced the quantity of organochlorinated compounds by substituting chlorine dioxide for chlorine in the bleaching process. The second method, using hydrogen peroxide, totally eliminates lethal compounds. The third method, oxygen delignification, supplements the chemical process to reduce the quantity of chemicals needed at the bleaching stage (Madore, 1992). Using chlorine dioxide and oxygen delignification can reduce the quantity of organochlorines by about 50 percent (Sinclair, 1990).

External Systems to Treat Mill Effluent

Primary treatment of pulp and paper effluent generally refers to the process for removing suspended solids from the effluent. This stage is followed by a secondary treatment stage that biologically treats the effluent to reduce BOD and toxicity. Tertiary treatment to further clarify and remove effluent color may follow the secondary treatment stage, but is not common in the industry.

The two main methods of wastewater primary treatment in the pulp and paper industry are sedimentation and dissolved air flotation. Utilizing a gravity clarifier, consisting of a circular tank equipped with a sludge removal mechanism, sedimentation can remove up to 95 percent of settleable solids, 70–90 percent suspended solids, and approximately 10 percent of the BOD_5 in the mill effluent. Dissolved air flotation methods of primary treatment introduce pressurized air through an upflow unit as the effluent enters the bottom of the tank. Air bubbles carry suspended solids to the surface, where a scraper separates the particles from the effluent overflowing from the unit.

Preliminary screening to either method is often used to remove large particles from the effluent steam.

Of the two primary treatment methods, sedimentation is the more popular method because it is less sensitive to variations in flow and solids concentration, and it requires little attention and maintenance (Smook, 1992).

After primary treatment, the wastewater still contains dissolved and suspended material that must be removed. Secondary treatment is a biological process that uses microorganisms to consume organic wastes in the effluent. Essentially, the objective of the secondary treatment is to reduce the BOD_5 in the effluent by 70–95 percent and to reduce the effluent toxicity. Living microorganisms commonly referred to as "bugs" are used to remove the contaminating materials present in the mill wastewater. The first step involves transferring dissolved and suspended materials from the liquid carrying them into the living mass of bugs. The living bugs use the waste organic material in the effluent as their food source. The organic materials transferred to the bugs are either dissolved or suspended. The second step involves the removal of the bugs from the liquid by means of separation, leaving the liquid substantially cleansed of contaminants.

One method of secondary treatment is the aerated stabilization basin (ASB). The ASB depends upon mechanical aeration in lagoons with a six- to ten-day retention time to achieve acceptable effluent quality. Aerators provide the necessary oxygen for relatively high biological activity. Nutrients added to the ASB in the form of urea and phosphoric acid increase the biodegradation efficiency.

Another secondary treatment method is the activated sludge system (AS). The AS is considered a high-rate method of secondary treatment. Its essential element is the microbial floc suspended in an aeration tank or mixing chamber. The tank is normally sized for three to eight hours of retention. Clarified wastewater is continuously fed to the tank. The activated sludge solids multiply as dissolved organic waste is metabolized. At the same time, effluent is drawn off to a clarification unit. Some of the sludge is recirculated to the aeration tank to maintain a high-density floc. The remainder of the sludge is disposed of by landfill or incineration (Smook, 1992).

Another high-rate method of secondary treatment is the biological filter. Biological filters promote contact between a free-flowing wastewater and a stationary growth of microorganisms in the presence of atmospheric air (Smook, 1992). The trickling filter is designed so that wastewater flows down through a bed of plastic modules. Microorganisms are cultivated on the plastic modules. The entire system is maintained in an aerobic condition by the free passage of air through the unit. The microorganisms consume the soluble organic components of the wastewater, primarily to produce new biomass. When the biomass reaches a certain size, it sloughs off and is carried out of the trickling filter and into the secondary clarifiers (Smook, 1992).

Environmental Monitoring of Pulp and Paper Mill Effluent

Environmental monitoring of pulp and paper mill effluent should, at a minimum, include an analysis of each major waste stream for suspended solids (TSS), dissolved solids, BOD, pH, and toxic effects. Standard tests for each of these components are used. Biological monitoring is necessary to determine long-term impacts. The regulations require that all pulp and paper mills be in compliance with the prescribed discharge limits as outlined in the mill's operating permit. As such, mills are required to conduct regular testing according to specified sampling and analytical protocols and procedures to ensure that they meet or exceed the operating permit.

Emission Treatment and Control Technology

Emission control efforts in the pulp and paper industry are concerned with the control of gaseous and particulate emissions. All types of airborne emissions may be partly controlled by operating strategies or by modifications to the production process. More difficult to manage are the discharging gas volumes, particularly reduced sulfur gases, that may not be dangerous in airborne concentrations but are odorous.

Much of the gaseous air pollutants from pulp and paper mill operations, including organic materials and total reduced sulfur (TRS), can be removed through combustion. The lime kiln and power boiler are typical thermal-incineration sources that, once adapted, are effective in destroying pollutants.

Noncondensible gas (NCG) systems collect gases from various mill process areas. The concentrated NCGs are transferred to a wet scrubber to reabsorb some of the sulfur. The remaining concentrated NCGs and the dilute stream are fired in the kiln with the power boiler as a backup.

Wet scrubbers remove dust or gaseous contaminants from a gas stream through contact with an absorbing liquid. The gas contacts an absorbing or wet liquid (usually water). The dust laden water droplets are drawn into a separator where the water is separated from the gas. The gas can then be drawn to the combustion area.

Electrostatic precipitators (ESP) can be used to collect particulate. Typically 99.9 percent of dust particles down to 0.1 microns can be removed with an electrostatic precipitator. The ESP uses high voltage to create a negatively charged field through which dust-laden gas passes. Positively charged collection plates attract the negatively charged particles, thereby drawing the particles out of the gas stream. A periodic mechanical action "raps" the plates to discharge the accumulated particles into a dry bottom hopper or into a wet bottom pool of liquor (Smook, 1992).

Fabric filters are also used to collect solid particulates. The gas passes through cloth. The particles cannot penetrate the fabric filter. As the filter builds up with the collected material, the pressure differential increases and the filter must be cleaned. Cleaning the filter is achieved by mechanical shaker, reverse flow of air, or flushing jet of air. The dust that is discharged falls into a hopper and is removed by some form of transportation device (Smook, 1992).

Environmental Monitoring and Testing of Pulp and Paper Mill Emissions

Environmental monitoring of pulp and paper mill emissions generally requires both source sampling and ambient sampling. Source sampling of stack emission streams measures specific emission rates and determines if regulatory limitations are being met. Continuous emission monitors are used to measure the gaseous contaminants in emissions including TRS, sulfur dioxide, nitrogen oxides, hydrocarbons, carbon dioxide, and carbon monoxide (Smook, 1992).

Ambient sampling and testing confirms the local air quality and determines whether ground-level concentrations of specific contaminants are within allowable limits. Measuring ambient pollution in the local area is normally concerned with settleable particulates (dust), suspended particulates, reduced and oxidized sulfur gases, and corrosion effects (Smook, 1992).

Chemical Control and Disposal

Hazardous chemical containers that contain acutely hazardous waste require "triple rinsing." Depending on the hazardous chemical being rinsed out of the container, an appropriate solvent must be used to complete the rinsing procedure. The solvent must then be collected, managed, and disposed of as a hazardous waste. The triple-rinsed container can then be disposed of as a nonhazardous solid waste.

STATE-OF-THE-ART POLLUTION PREVENTION

Over the past two decades, pulp and paper mill design and technological changes have been driven by environmental considerations. Pulp and paper facilities of the future will become even more efficient and have substantially less of an impact on the environment. Improved efficiency will continue to be achieved through advanced process control of existing technology. Specifically, digester control will be the key focus to improve pulp quality, reduce demand for pulp bleaching agents, and to increase energy recovery. It is expected that presses will replace traditional filters as standard bleach plant-washing equipment. This will further reduce water and steam consumption as well as chemical consumption (Meadows, 1995).

Emerging technology in the pulp and paper industry is leading toward zero effluent operations. **Zero effluent** is the elimination of liquid process effluents from a pulp or paper mill without transferring an unreasonable load to the airshed or soil. Cooling water that has not been in contact with the process, sanitary effluents, and storm water run-off would still exist in the zero-effluent mill (Ontario Ministry of the Environment, 1992). The zero-effluent design modifies the production processes to reduce the effluent volume and then evaporate the effluent, incinerate the concentrate, and reuse the evaporator condensates as mill water supply (Ontario Ministry of the Environment, 1992).

Ozone bleaching has been adopted rather slowly in the pulp and paper industry; however, the elimination or reduction of chlorine-containing chemicals offers significant potential for environmental gains. Furthermore, the elimination of chlorides in the bleach plant facilitates the recycle of all bleachery effluents back to the chemical recovery system (Smook, 1992).

Advances in the disposal of pulp and paper mill sludge have developed substantially in the last ten years. The increasing cost of landfilling sludge and more strict environmental regulations have led to more diverse methods of disposal. Fiber transfers, whereby, one mill's sludge can be another mill's raw material feedstock, are attracting favorable attention in the industry. Similarly, developments in the area of manufacturing soil from sludge have begun (Coburn and Dolan, 1995).

ENVIRONMENTAL HEALTH & SAFETY ISSUES AND CONCERNS

Effluent/Water Quality

In high concentrations, conventional water pollutants can cause health and environmental problems. Without enough oxygen in the water, fish and aquatic life will suffocate.

Suspended solids problems include interference with the transmission of light in the water, adverse affects on fish growth rate and resistance to disease, and filling in of gravel spawning beds.

Fecal coliform bacteria can cause gastrointestinal illnesses in people who swim in waters that have too much of this bacteria. Water that is too acidic will kill fish and their eggs, while elevated levels of oil and grease can cause death, deformities, and impair cellular and physiological processes in aquatic life. Since petroleum products and organochlorines bioaccumulate, that is, accumulate over time in tissue and concentrate going up the food chain (biomagnification), potential health problems exist for humans also.

Air Quality

Small particles can build up in lungs over time and impair breathing capacity. High levels of PM_{10} (pollution particulate matter at 10 microns or less—1 micron is 1 millionth of a meter) can produce an array of adverse health effects, including aggravating pre-existing respiratory problems, and causing a temporary reduction in lung capacity. Children are especially vulnerable.

Since carbon monoxide reduces oxygen levels in the blood stream, elevated levels of pollution can be especially harmful to individuals with heart diseases. Healthy individuals exposed to high levels of carbon monoxide can be impacted by reduced mental alertness and impaired visual perception and manual dexterity. High concentrations of lead and carbon monoxide emissions have been attributed to learning disabilities, especially in young children. The corrosiveness of ozone has a very detrimental impact on building structures and landmarks, causing rusting, cracking, and fading.

Most negative perceptions of industrial activity are the result of particulate and odors. Particulate can be visible from a long distance, create a persistent haze, and leave deposits on cars, windows, and painted surfaces.

Solid Waste Disposal

Landfilling waste sludge has been the most common method of solid waste disposal for pulp and paper mills. Incinerating solid wastes is also practiced; however, this is normally considered a first stage of disposal to reduce the volume prior to landfilling.

The potential for leachate contamination of ground and surface waters is the primary concern with landfilling pulp and paper mill sludges. For this reason federal and state regulatory agencies have become involved in developing criteria for solid waste disposal facilities, and developing and enforcing regulations pertaining to them.

Hazardous Chemicals

Many solvents are so hazardous that they must be used with personal protective equipment like respirators, gloves, caps, and glasses. Exposure to solvents can cause a number of detrimental health effects. Hazardous solvents can get into the food chain and bioaccumulate in tissue.

SUMMARY

- The high organic content of pulp and paper mill waste, together with the presence of bleaching chemicals results in the production of toxic process waste.
- Internal process changes that replace or partially replace chlorine pulp bleaching have resulted in a reduction of toxic emissions.
- External effluent treatment processes are more effectively treating mill effluents through both primary and secondary treatment stages.
- A heightened awareness of hazardous chemical handling practices has resulted in better control and handling of hazardous materials, thus reducing the risk to human health.
- Tougher environmental standards have necessitated, for many companies, installing and operating complex pollution-abatement equipment, process changes, and heightened analysis and monitoring of emissions. In this respect, the industry reaction to more stringent environmental standards has meant, at least initially, retrofitting process technology in an attempt to improve the technical performance of the operation *vis a vis* their environmental performance.

QUESTIONS FOR REVIEW

1. What are the major sources of effluent pollution in the pulp and paper industry?
2. What are the major sources of air pollutants in the pulp and paper industry?
3. Match the waste component associated with each component of wood:

Wood Component	**Waste Component**
Lignin	BOD_5
Extractives	Fiber (suspended solids)
Cellulose	Chromophores (color)
Hemicellulose	Toxicity

4. List three small-volume hazardous waste streams produced by the pulp and paper industry.
5. What type of wastes are generated in the chemical pulping process?
6. What are the most dominating compounds found in pulp mill effluents?
7. Environmental monitoring of pulp and paper mill effluent should, at a minimum, include an analysis of each major waste stream for _________.
8. Environmental monitoring of pulp and paper mill emissions generally requires both _________sampling and _________ sampling.

REFERENCES

Coburn, R., and G. Dolan. 1995. "Industry/Government Partnership Works to Solve New York Environmental Issues." *Pulp & Paper*. October 1995.

COWIconsult. 1989. "The Technical and Economic Aspects of Measures to Reduce Water Pollution Caused by the Discharges from the Pulp and Paper Industry." Denmark.

Kringstad, K.P., and K. Lindstrom. 1984. "Spent Liquors from Pulp Bleaching." *Environmental Science and Technology,* Vol. 18.

Meadows, D.G. 1985. "The Pulp Mill of the Future: 2005 and Beyond." *Tappi Journal.* October 1995.

Madore, O. 1992. "The Pulp and Paper Industry: The Impact of the New Federal Environmental Regulations." Supply and Services Canada. June 1992.

Murray, W. 1992. "Pulp and Paper: The Reduction of Toxic Effluents." Supply and Services Canada. April 1992.

Odendahl, S., and M. Holloran. 1993. "International Approaches to Pulp and Paper Regulation: How Do We Compare?" Thunder Bay, Ontario: Paper presented at the CPPA Environment Conference. October 1993.

Ontario Ministry of the Environment. 1992. "Best Available Technology for the Ontario Pulp and Paper Industry." Toronto: Queen's Printer for Ontario.

Sinclair, W.F. 1990. "Controlling Pollution for Canadian Pulp and Paper Manufacturers: A Federal Perspective." Ottawa: Environment Canada.

Smook, G.A. 1992. *Handbook for Pulp & Paper Technologists,* 2nd Edition. Vancouver: Angus Wilde Publications.

11

Wood Preserving

Les Lonning and Edward Smith

Upon completion of this chapter, you will be able to meet the following objectives:

- Have a basic understanding of the history of the wood-preserving industry.
- Understand the primary wood-treatment processes currently used in the United States.
- Understanding of benefits of wood treatment.
- Identify the basic characteristics and chemical constituents of the most widely used wood preservatives in use today.
- Identify the potential sources of hazardous waste generation.
- Recognize the primary regulatory programs applicable to the wood-preserving industry.
- Understand the scope and objectives of the various pollution prevention and hazardous waste management efforts employed by the wood-preserving industry. The reader should also gain an understanding of the relationship of these efforts.
- Develop an understanding of the toxicity and potential health effects of the chemicals employed in the wood-preserving industry.
- Develop an understanding of the potential environmental impacts associated with chemicals employed in the wood-preserving industry.
- Recognize the available treatment technologies used in the wood-treating industry.

INTRODUCTION

Forests cover about 30 percent of the land area of the world. They range from equatorial swamps to mountain rain forests, from small plantations containing a few tree species to luxurious, vibrant jungles with hundreds of trees and species. Approximately one-third of this area is being harvested, with another one-fifth available for potential harvesting. However, most of the remaining forest areas cannot be commercially utilized because of physical or economic constraints. Growth and decay are essential processes that help maintain the balance of nature. Trees are attacked by insects and fungi, and in the process complex substances such as cellulose manufactured in the growing plant are broken down to more simple materials. Without decay, forests would become choked with old trees and nothing else would be able to grow. Trees that have been cut into wood for various construction purposes can be protected from attack by insects, fungi, and marine borers by the application of wood preservatives.

For a wood preservative to be considered effective, it should exhibit certain characteristics. The American Wood Preservers' Association (AWPA) Technical Committee has prepared the following set of qualifications for wood preservatives:

1. It must be toxic to wood-destroying organisms.
2. Its value as a preservative must be supported by field and/or service data.
3. It must possess satisfactory physical and chemical properties that govern its permanence under the conditions for which it is recommended.
4. It must be relatively free from objectionable qualities and handling.
5. It must be subject to satisfactory laboratory and plant quality-control methods.
6. It must be available under provisions of current patents.
7. It must be in actual commercial use.

GENERAL PROCESSES

Since the early- to mid-1800s there have been a number of wood preservatives that have been used, both in Europe and the United States. One of the early ones, which was patented as a wood preservative by John Bethel in 1838, was creosote. In addition, there have been a number of inorganic toxic wood preservatives that were comprised of copper, zinc, arsenic, and/or mercury.

The wood-preserving industry has settled on a few basic materials and formulations for wood preservatives. These basically fall into four classes:

1. Creosote
2. Oil-borne preservatives, such as pentachlorophenol
3. Water-borne preservatives
4. Petroleum products

The petroleum products are not necessarily wood preservatives, but are used as a carrier system for the oil-borne preservatives and play an integral role in the protection of wood.

Coal-Tar Creosote

Coal-tar creosote is a residual product from the distillation and processing of coal tar. It is a blend of naphthalene, drain oil, wash oil, anthracene drain oil, and heavy distillate oil. The oil is blended to meet the physical requirements of the specifications noted by the AWPA.

Coal-tar creosote is a mixture of literally hundreds of various compounds. The bulk of them are aromatic hydrocarbons. The principal constituents in creosote are phenanthrene, naphthalene, flouranthene, fluorene, acenaphthene, pyrene, and dibenzofuran.

Pentachlorophenol

Pentachlorophenol as commercially specified for wood preservation must contain a minimum of 85 percent pentachlorophenol and less than 1 percent matter insoluble in a NaOH solution. Pentachlorophenol is most commonly shipped to the treatment facility as a solid, in block form or pelletized, although it can also be shipped as a concentrated solution. At the treatment facility the solid penta, or concentrated liquid, is usually diluted in a carrier oil comprised of either a No. 2 diesel or what is known as a heavy-aromatic oil. The working solution strength normally runs between 5.0 and 7.5 percent by weight (w/w).

Water-Borne Preservatives

Water-borne preservatives consist of compounds such as arsenic, chromium, copper, zinc, and fluorides. The idea of using water as a carrier for this type of preservative has long been one that has been economically appealing. However, the problem in early formulations was a means whereby the preservative metals could become "fixed" in the wood while still retaining their preservative effectiveness. (Fixed or fixation refers to the reaction between the preservative solutions and the wood.) During this process, the water-soluble metal oxides are chemically reduced to form less toxic, water-insoluble metal complexes. Fixation of the preservative solutions increases the leach resistance of the metals and leads to increased customer safety.

The most common water-borne preservative used today is CCA, or chromated copper arsenate, because of its economic incentives and excellent service records. In CCA solutions, arsenic works as the insecticide, copper is a fungicide, and hexavalent chrome is the fixation agent.

Petroleum Products

There is a relatively small consumption of petroleum products by the wood-preserving industry. The oils that fall into this category are primarily those used for blending with creosote or oils that are used as carriers (or solvents) for pentachlorophenol solutions. Although they are not considered wood preservatives they are part of the wood-preservative system.

The primary use of petroleum products is in the pentachlorophenol preservative system. In the southern United States, these petroleum oils are primarily No. 2 diesel, while in the West and Midwest the carrier systems for pentachlorophenol are generally a heavier oil, which has high aromaticity.

For treatment of wood products with preservatives to be successful, the preservative should penetrate evenly into the wood in sufficient depth and concentration to protect it from wood destroyers in the environment for which it is to be used. The timber has to be in a suitable condition for treatment for this to be possible, and the preservative applied according to the relevant process specification. The most important and successful method for applying these wood preservatives is the **pressure-impregnation process.** In this normal practice, timber is enclosed in a steel pressure vessel and preservative is forced into the wood cells under hydraulic pressure of somewhere between 125 and 150 pounds per square-inch, the pressure period lasting anywhere from one to three hours. The steel pressure vessel is commonly referred to as a "retort cylinder."

The typical oil-borne treating cycle is as follows:

1. An initial vacuum period, usually lasting somewhere between thirty minutes to one hour. This step combined with step No.2 "conditions" the wood by removing moisture from the wood cells.
2. A seasoning period in which the wood is heated and/or moisture is driven off during a cycle that lasts generally between twelve and thirty-six hours. This step assures that the wood is of the proper moisture content for treatment. Moisture removal is required to allow room for the impregnation of preservative solutions into the wood cell structure.
3. A preliminary air-pressure period, usually lasting approximately thirty minutes. This step in the process expands the wood cell structure, which aids the impregnation of preservative solutions.
4. A pressure period lasting typically between one and four hours impregnates the wood with the preservative solution.
5. Final vacuum, which usually lasts somewhere between one and two hours. This vacuum step draws excess preservative from the wood, minimizing drippage and making the final product safer to handle.

For water-borne treatments, the pressure period requires much less time and there is usually no seasoning period so the overall treating cycles are considerably shorter. Moisture from the wood is typically achieved by air drying or kiln drying the wood products prior to treatment in the retort.

Following the treatment process, the treated wood is removed from the treatment vessel and placed on a containment pan or drip pad. The containment pans and drip pads are designed to collect preservative solutions that are either present on the wood surface or are released from the wood cells as they return to atmospheric temperature and pressure. The recovered treating solutions are commonly returned to the treatment process and re-used. Generalized schematics of the CCA and pentachlorophenol pressure wood-treatment processes are provided in Figures 11–1 and 11–2. The regulatory issues related to these containment structures will be discussed in greater detail later in this chapter.

The primary benefit of wood preserving is the increase in service life. Chemical treatment of wood products can increase service life by five or more times over untreated wood. For example, it is not uncommon for creosote and pentachlorophenol treated utility poles that were installed in the 1940s to still be in service. Untreated wood exposed to moisture and in contact with the ground can show signs of decay within three years. The treatment of wood products also conserves forest resources.

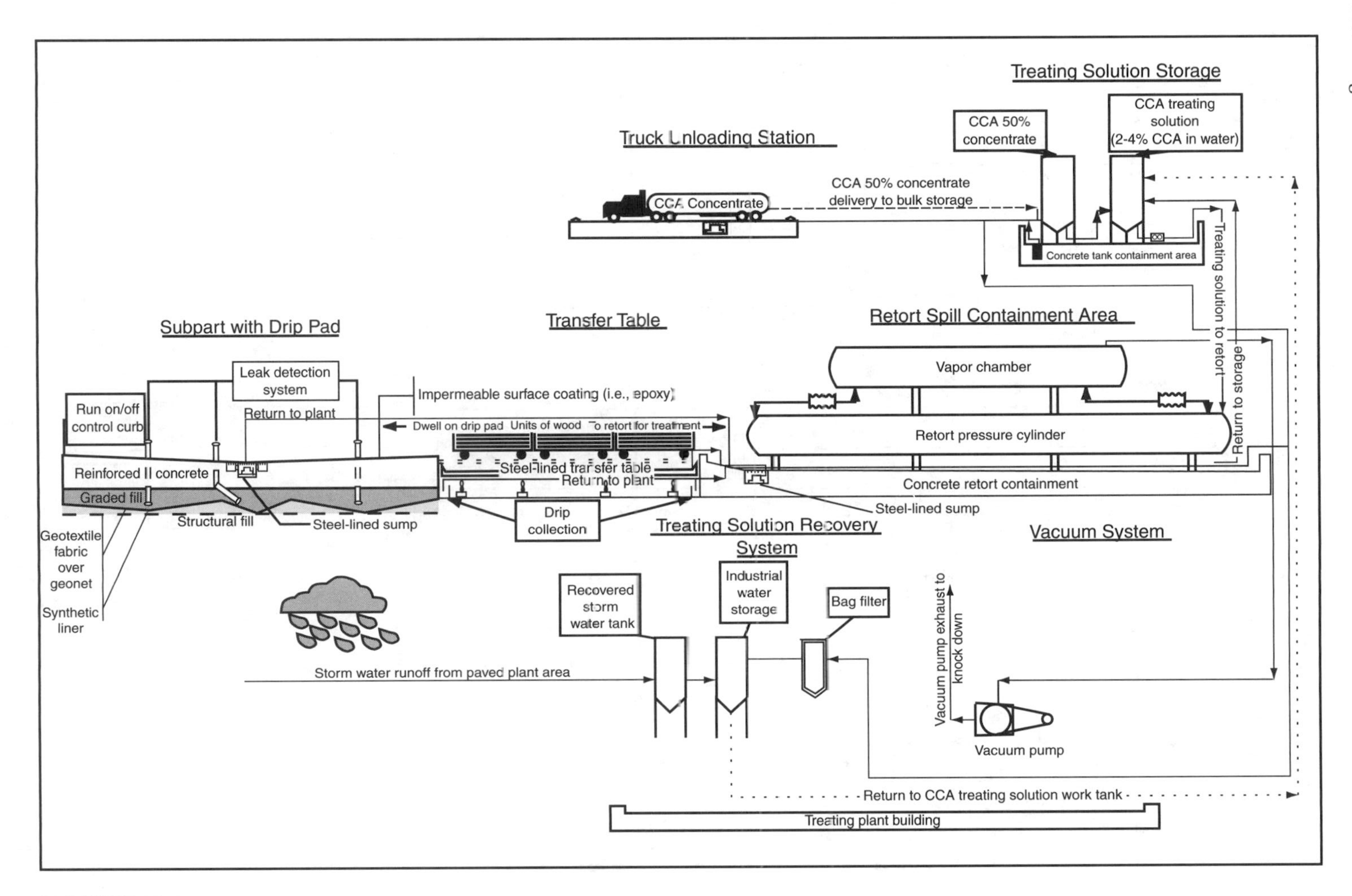

FIGURE 11–1
A generalized schematic of the CCA wood-treatment process.

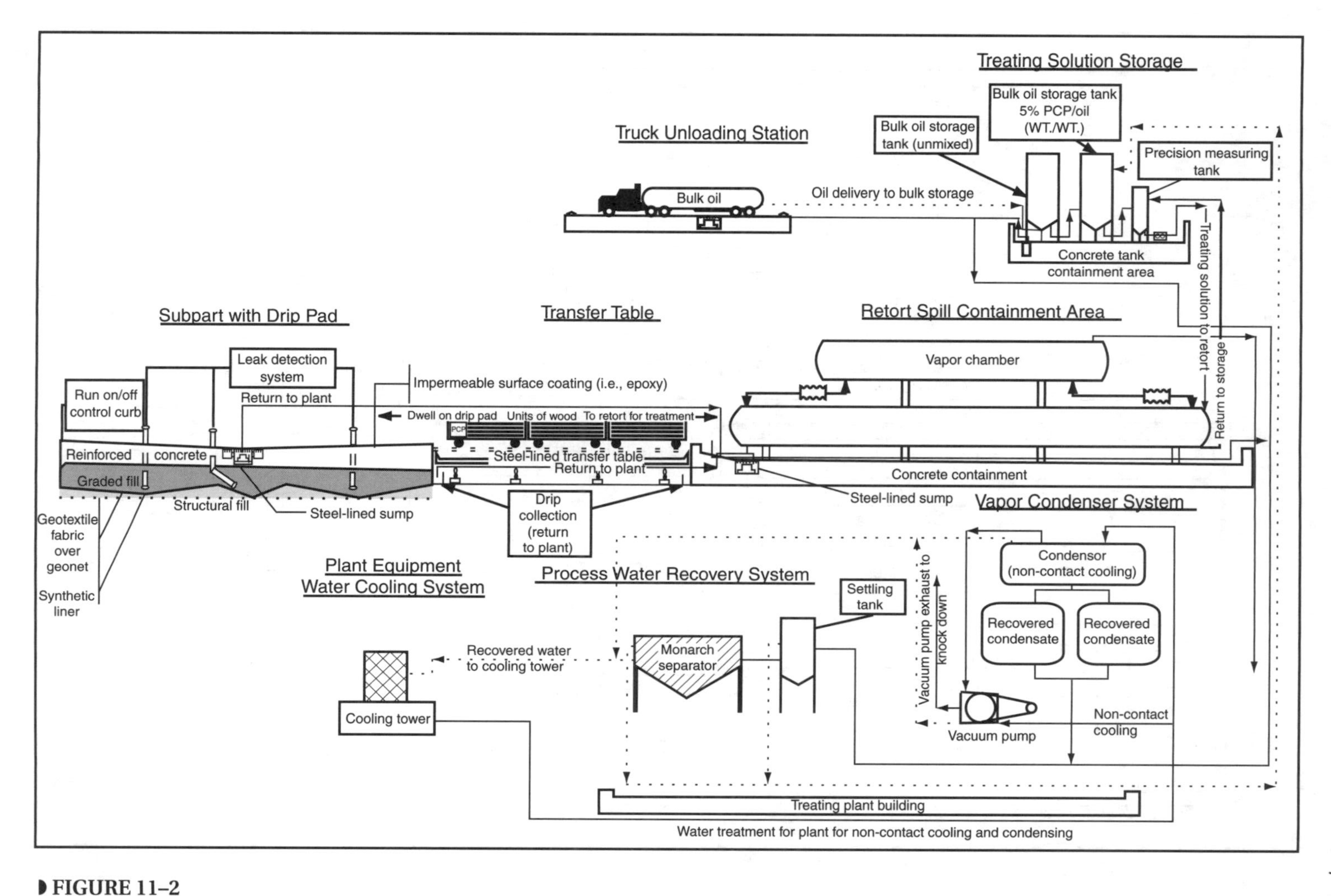

FIGURE 11–2
A generalized schematic of the pentachlorophenol wood-treatment process.

INGREDIENTS AND MATERIAL BALANCE

Coal-Tar Creosote

Creosote is a complex mixture of over 250 individual organic chemicals that are products of a fractional distillation of coal tar. Approximately 35 percent of creosote by weight is aliphatic hydrocarbon (oil) and the remaining 65 percent consists of polynuclear aromatic hydrocarbons (PAHs).

Coal tar is a byproduct of coking of bituminous coal. Practically all coal tar produced in the United States results from this high-temperature process. Commercial creosote is a blend of several coal-tar distillation fractions, selected to provide specific physical characteristics to the resulting product. Almost all creosote is blended with tar acids and is processed with fatty acids to form a water-emulsifiable product that is most commonly used as a wood preservative, insecticide, disinfectant, and/or an animal repellent.

A listing of the primary compounds found in creosote is provided on Table 11–1. This table also provides the approximate percentage of each of the primary constituents by weight.

Pentachlorophenol

Pentachlorophenol (penta) is a widely used wood preservative that is normally carried in a petroleum solvent. The commercial synthesis of penta is readily accomplished by the direct chlorination of phenol. This process also results in the formation of other chlorinated phenols, chlorinated ether compounds, and trace levels of dioxin/furan compounds. The primary constituents found in technical grade penta, typically used in the wood-treatment industry, are provided on Table 11–2. The approximate ranges for each constituent or constituent group are also provided.

TABLE 11–1
Primary creosote components.

Creosote Components	Composition (% by Weight)
Phenanthrene	19.4
Naphthalene	17.0
Fluoranthene	11.8
Pyrene	8.4
Acenaphthene	7.8
2-Methyl naphthalene	6.5
Fluorene	6.0
Dibenzofuran	5.2
Carbazole	5.1
1, 2-Benzanthracene/chrysene	4.2
1-Methyl naphthalene	3.5
Anthracene	2.5
Biphenyl	1.9
Acenaphthylene	0.5

Source: U.S. EPA, EPA/625/7-90/001, November 1990.

TABLE 11–2
Primary components of technical grade penta.

Technical Grade Pentachlorophenol	Composition Range (% by Weight)
Pentachlorophenol	90–94
2, 3, 4, 6-Tetrachlorophenol	0–1.5
Hydroxypolychlorodibenzo ethers	4–7
Dioxin/furan compounds (Octa, traces of hepta- and hexachlorodioxin)	< 0.1

Source: Vulcan Chemicals, Material Safety Data Sheet, Form 3239-710, September 1995.

Penta is considered to be a broad-spectrum biocide and is effective against bacteria, fungi, and insects. This characteristic has resulted in the use of this compound in a number of applications, ranging from utility poles to slimacidal usage in paper mills and cooling towers.

Water-Borne Preservatives (Inorganic Arsenicals)

Arsenic is produced as a byproduct of the nonferrous smelting industry, mainly from copper smelters but also from gold, zinc, and lead smelters. During the smelting process, white arsenic trioxide (As_2O_3) fumes are formed and then condensed in brick buildings as the gasses are cooled. Arsenic trioxide is used as a rodenticide and as a starting material for all other arsenical pesticides. The crude As_2O_3 is collected and further refined for use in the wood preserving industry. The arsenic compound typically used for wood treating is arsenic pentoxide (As_2O_5).

Copper compounds used in wood treating are manufactured from copper ore and recycled copper from the metals manufacturing and chemical industry. These copper products are typically produced as bivalent copper compounds, copper carbonate, cupric hydroxide, cupric oxide, and copper sulfate.

The source of chromium compounds used for wood treating is from the importation and processing of chromite ore or from the reclamation of recycled chromium products. The chromium products used in the wood treating industry are manufactured as potassium chromate, potassium dichromate, chromium trioxide, and chromic acid.

REGULATORY CONSIDERATIONS

The wood-treatment industry in the United States has been subject to intensive evaluation by environmental and occupational health agencies, and the environmental standards to which the industry must adhere are continually becoming more stringent. This level of review and evaluation is primarily due to the toxic nature of the chemicals used by the industry, and the increased understanding of the potential impact that these chemicals may have if released into the environment in an uncontrolled manner. The increased degree to which these and other industrial chemicals have been regulated is also a result of technological advances that allow for the detection of chemicals at increasingly lower levels, and the completion of additional toxicological studies.

The following section outlines the primary federal regulations that currently affect the wood-preserving industry. Federal regulations that are addressed here affect preservative use and application, and the storage, handling, and disposal of hazardous waste. State and local environmental regulations are not discussed.

Federal Insecticide Fungicide and Rodenticide Act

Under the Federal Insecticide Fungicide and Rodenticide Act (FIFRA), the EPA regulates the use and application of creosote, pentachlorophenol, and the inorganic arsenicals (water-borne preservatives). Only certified applicators or persons under their direct supervision may purchase or use these preservatives.

The majority of wood preservatives are subject to special review, formally known as Rebuttable Presumption Against Registration (ARPAR, under FIFRA). Each preservative that is registered under FIFRA has certain requirements that it must adhere to with respect to the label and its language. This information applies only to the pesticide and not to the treated-wood product.

Most commercially used wood preservatives, including creosote, pentachlorophenol, and CCA, were classified as restricted-use pesticides in 1987; currently there is a request for information to fill in data gaps on preservatives. In response to potential data gaps identified by regulatory agencies, some manufacturers have formed task forces to provide the necessary funding and to retrieve and submit supplemental information.

Resource Conservation and Recovery Act

The Resource Conservation and Recovery Act (RCRA) regulates the storage, handling, and disposal of hazardous waste. The waste may be considered hazardous if it is a listed waste or exhibits hazardous characteristics. RCRA regulations list more than 400 hazardous wastes, including waste derived from nonspecific sources, which are labeled F-wastes, and wastes from manufacturing processes and discarded chemical products, which are labeled K-wastes. If a waste is not listed, it may be considered a hazardous waste if it exhibits hazardous characteristics or fails the **Toxics Characterization Leachate Procedure (TCLP)** test. The TCLP test measures the levels of hazardous chemicals that are contained in an acidified water sample that is passed through the waste material. The hazardous characteristics, as defined by RCRA regulations, are ignitability, corrosivity, reactivity, and toxicity.

The following listed wastes are related to the wood-preserving industry.

- K001: the bottom sediment sludge from the treatment of waste waters from wood preserving processes that use creosote and/or pentachlorophenol
- F032: pentachlorophenol process residuals and preservative drippage, including wastewaters
- F034: creosote process residuals and preservative drippage, including wastewaters
- F035: inorganic arsenical process residuals and preservative drippage, including wastewaters

RCRA regulations as they apply to the management of hazardous waste also directly impact the manufacturing processes within wood treatment facilities. Those with the most notable impact are the RCRA regulations, finalized in 1992, that require wood-treatment facilities to maintain and operate drip pads and treated-wood product storage yards consistent with Subpart W. These regulations are detailed within Sections 264 and 265 of the Code and Federal Regulations, and specify conditions under which wood-treatment facilities must operate in order to avoid the requirement for specific hazardous waste permits. Such hazardous waste permits are generally reserved for facilities that are primarily designed to treat, store, and/or dispose of hazardous wastes, including hazardous waste landfills and incinerators.

A physical description of a Subpart W drip pad and the operational requirements of drip pads and treated wood storage yards will be addressed in the Pollution Prevention (P_2)section of this chapter.

Note: The EPA, with input from the wood preserving industry, published the "Wood Preserving Resources Conservation and Recovery Act (RCRA) Compliance Guide, A Guide to Federal Regulation" (Document No. EPA 305-B-96-001) in June 1996. This comprehensive document is an excellent reference for those wishing additional details on RCRA regulations as they apply to the wood preserving industry. This document is available electronically on the World Wide Web at http: //es.inel.gov./oeca/metd/guide.

Clean Air Act

The primary regulations controlling air pollution in the United States were implemented pursuant to the federal Clean Air Act (CAA). These regulations are administered by the EPA, but the regulatory lead can also be taken by individual states following federal approval of its State Implementation Plan (SIP).

The CAA does not currently have board applicability in the wood-treating industry. This is primarily due to the fact that standards for hazardous air pollutants (HAPs) have not been established for common wood-preserving chemicals (arsenic, copper chromium, and pentachlorophenol). In addition, most wood-treating facilities are relatively small and are therefore not drawn into the CAA's major source operating permit requirements, based on excessive emissions of criteria pollutants such as volatile organic compounds, nitrogen oxides, sulfur oxides, carbon monoxide, lead, and particulates. Wood-treating facilities may provide calculated emission estimates to the regulating authorities in order to confirm their nonmajor (or minor) air pollution source status.

POLLUTION PREVENTION

Pollution prevention at wood-treatment facilities is addressed through the development and implementation of contingency and management plans. Each plan is typically directed at the protection of a specific route for pollution migration or release into the environment, and most of the plans are required by federal, state, and/or local regulations. For instance, spill contingency plans are designed to prevent the accidental release of chemicals to surface water. Specific P_2 plans are also developed to outline best management practices in the production of treated wood products that are specifically intended to reduce the toxicity and quantity of the hazardous waste that is generated.

Summary of Plans

Management plans detail procedures for the management of wood-treatment chemicals and associated equipment and also outline contingency and response procedures in the event of a release to the environment. The training steps taken to protect and inform on-site workers and contractors, and the reporting requirements to educate the surrounding community and emergency responders, are also included in these site plan documents. A summary of each plan document typically in use at wood treatment facilities is detailed here and the statutory basis for each is also provided where applicable.

Spill Prevention Containment and Counter Measure Plan (40 CFR 112)

Spill Prevention Containment and Counter Measure (SPCC) plans were first required by the EPA for facilities handling bulk quantities of oil in 1973, and are directly applicable to wood-treatment facilities using oil-borne preservative solutions. SPCC or similar spill response plans are also widely used by wood-treatment plants utilizing water-borne treatment solutions.

A SPCC plan describes the operations at the facility and identifies conditions under which hazardous materials and wastes are stored and managed. This plan also details inspection procedures for plant equipment, including tanks, piping, valves, retorts, and hazardous waste storage containers. The containment structures are described and detailed response plans for potential leaks or spills are included. These plans must also be reviewed and certified by an independent professional engineer and be updated every three years, or when significant plant operation changes occur. The professional engineer must also verify that the containment structures are of adequate structural integrity and size to contain 110 percent of the volume of the single largest tank within the given containment area for a period of seventy-two hours.

Employee education on SPCC plan implementation is an integral part of the regulation. The plan must designate a response team leader and specify the duties and responsibilities of each team member. SPCC plans, and all plan updates, are also sent to the emergency response teams within the host communities for the applicable wood treatment facilities. These SPCC regulations are intended to work in concert with local Emergency Planning and Community Right To Know programs required under Title III of the SuperFund Amendments and Reauthorization Act of 1986 (commonly referred to as SARA 313).

Hazardous Waste Operations and Emergency Response Plan (40 CFR 1910.120)

This plan instructs plant employees on the proper methods for characterizing, handling, and storing hazardous waste at the facility. The plan also identifies the required labeling, manifesting, and shipping requirements for the transportation of hazardous wastes offsite. All hazardous waste generated at wood-treating facilities must be disposed of and/or treated at permitted treatment, storage, and disposal facilities (TSDFs). This plan also identifies procedures for hazardous waste removal from facility equipment (i.e., tanks) and includes inspection requirements and recordkeeping for hazardous waste storage areas. Employee training on this plan is generally scheduled to run concurrent with other safety or hazard communication training, such as the OSHA Hazard Communication Program.

Storm Water Pollution Prevention Plan (SWP_3)

Storm Water Pollution Prevention (SWP_3) plans are generally a requirement of National Pollutant Discharge Elimination System (NPDES) permits held for storm water discharges from wood-treatment facilities. The plans are generally prepared or reviewed by an independent certified professional engineer and are also subject to review and approval by the regulating authority that issues the NPDES permit.

This plan is similar to the SPCC plan; however it focuses on the Best Management Practices (BMPs) for storm-water pollution prevention instituted at the wood treating facility. The BMPs identify facility improvements and changes in plant operating procedures that will improve storm water quality. The BMPs at wood-treatment facilities can include the paving of the treated products storage yards and the installation of the storm-water treatment systems. The treatment of storm water leaving a wood-treating facility is often required to meet state and federal surface water quality standards. Technologies used in the treatment of storm water are similar to those discussed in the Treatment and Control Technologies section; however the design of the treatment system must also consider the intermittent nature of storm water flows. Generally this is accomplished by balancing the capacity of the treatment system with the ability and cost to store storm water during periods of high flow.

Drip Pad Management Program (40 CFR 264/265)

According to Subpart W of the RCRA regulations, a drip pad must be "constructed of nonearthen materials, excluding wood and nonstructurally supported asphalt." It is designed to be sloped in order to free the drain drippage of treatment solutions from freshly treated wood. Drip pads are required at all active wood-treatment facilities using the pressure treatment process, pursuant to the finalization of the EPA's December 24, 1992, modified regulations for hazardous waste management. A drip pad is typically a steel-reinforced concrete containment structure that is either constructed with an impermeable surface or is equipped with a secondary containment-and-leak-detection monitoring system. The integrity of the structure is insured by the annual certification of surface impermeability by an independent professional engineer, or by the regular monitoring of the secondary containment system. Impermeable surfaces, as defined by the regulation, must have a demonstrated hydraulic conductivity of less than 1×10^{-7} centimeters per second.

Subpart W drip pads are designed to prevent the release of preservative drippage from newly treated wood products. The design and construction oversight must be directed by an independent professional engineer, and the completed drip pads must be certified by a professional engineer to meet federal regulations.

The drip pad operational requirements of these regulations must be detailed in site specific drip pad management program plan documents in place at all wood treating facilities. These plans provide specific instruction to plant personnel on the required maintenance, operation, inspection, and reporting applicable to drip pads. Inspection records must be maintained (for a minimum of five years) as part of the operating log for the drip pad, and are subject to review by regulatory inspectors.

Contingency Plan for Incidental and Infrequent Drippage in the Treated Products Storage Yard (40 CFR 264/265)

Pursuant to the EPA's December 24, 1992, modification of the regulations for hazardous waste management applicable to wood treatment facilities, contingency plans must be developed and maintained for the treated products' storage yard at wood-treatment facilities. The plans are required by the EPA due to concerns that freshly treated wood removed from the drip pad may release small quantities of preservative in response to changes in climatic conditions.

These plans detail the procedures and frequency of the storage yard inspections made in response to infrequent and incidental preservative drippage. The objective of the plans is to insure the removal of potential drippage from the storage yard so that preservative residue will not accumulate and cause contamination to the underlying soil or environmental media. Institution of these plans by the wood-treating industry also convinced the EPA that potential drippage of treatment solutions in the treated wood storage yard does not constitute disposal of a hazardous waste and thereby allows the industry to operate exempt from specific permits required for the treatment, storage, and disposal of hazardous wastes. Paving of the storage yards and the installation of storm water monitoring and controls, to a large extent, addresses the potential impacts from the storage of treated products.

OSHA Written Hazard Communication Program

This program details the employee and contractor training conducted at wood treatment facilities in response to the OSHA (federal and state) Hazard Communication (Right-To-Know) regulations. This training educates employees and contractors on the potential occupational hazards related to working with hazardous materials at the plant and provides information on how to avoid potential health-related problems. The training also addresses the warning signs of possible accidental exposure and identifies specific steps to take if an accident occurs. The potential chronic and acute effects of exposure to wood preserving chemicals are discussed in the following section of this chapter. It should also be noted that wood-treatment facilities are subject to regular state or federal OSHA compliance inspections. These inspections include exposure monitoring of plant employees.

ENVIRONMENTAL HEALTH AND SAFETY CONSIDERATIONS

The toxic nature of wood-preserving chemicals presents significant health and safety considerations for the industry. Toxicological information for the typical wood preserving chemical is presented here for both the short-term (acute) and the potential long-term (chronic) effects. Note that the majority of the potential health effects attributed to wood-preserving chemicals are based on studies completed on animals involving chemical concentrations typically higher than those encountered in an industrial setting. Appropriately, the regulatory exposure and/or contaminant levels derived from these types of studies are designed to be conservative and protect the most sensitive individuals.

Information regarding the permissible exposure limits to wood-preserving chemicals in the work place and the federal water quality standards is also summarized on Table 11–3. This summary provides information on the relative toxicity

TABLE 11–3
Summary of permissible exposure limits and water-quality standards.

	Permissible Exposure Limit (PEL) OSHA 8-Hour TWA[1]	EPA Drinking Water Maximum Contaminant Level	EPA Chronic Surface Water (Fresh Water) Quality Standard	EPA Acute Surface Water (Fresh Water) Quality Standard
Pentachlorophenol	0.5 mg/m^3	1 ug/l [5]	13 ug/l [3]	20 ug/l [3]
Creosote	0.2 mg/m^3 [2]	Not applicable	Not applicable	Not applicable
Copper	1.0 mg/m^3	1,300 ug/l [6]	11 ug/l [4]	17 ug/l [4]
Chromium VI	0.1 mg/m^3	50 ug/l [7]	10 ug/l [4]	15 ug/l [4]
Arsenic	0.010 mg/m^3	50 ug/l [7]	190 ug/l [4]	360 ug/l [4]

[1] National Institute for Occupational Safety and Health (NIOSH), Pocket Guide to Chemical Hazards, June 1994.
[2] Creosote–OSHA PEL for coal tar pitch volatiles.
[3] Code of Federal Regulations (40 CFR Section 131.36, 1993).
[4] Federal Register–V l. 60, No. 86, May 4, 1995.
[5] Code of Federal Regulations (40 CFR Section 141.61, 1993).
[6] Code of Federal Regulations (40 CFR Section 141.51, 1993).
[7] Code of Federal Regulations (40 CFR Section 141.11, 1993).

among the wood-preserving chemicals and can also be used to compare relative toxicity with chemicals used outside of the industry. A review of water-quality standards will also be helpful during the discussion of the treatment and control technologies applicable to the wood-treating industry.

Pentachlorophenol

Acute Toxicity

Inhalation of concentrations of 0.3 milligrams per cubic meter (mg/m^3) of pentachlorophenol can cause nose irritation. Concentrations of 1 mg/m^3 can cause irritation of the upper respiratory tract, resulting in sneezing and coughing. Persons acclimated to penta can tolerate levels above 2 mg/m^3. Symptoms of overexposure include rapid heart beat and respiration, elevated body temperature and blood pressure, muscular weakness, excessive sweating, dizziness, and nausea. Air purifying respirators fixed with organic vapor cartridges are typically worn by workers potentially exposed to elevated levels of penta fumes, dusts, or mists.

Penta is also readily absorbed through skin contact, particularly when in solution. Absorption through the skin can result in systemic poisoning and the acute exposure symptoms described previously. Penta is considered a poison and the single-dose toxicity is high.

Workers at wood-treatment facilities are required to wear protective, chemically resistant clothing and gloves when working with penta to prevent exposure to the skin. Eating and smoking are also generally prohibited in the treating area to minimize unintentional ingestion. Respirators are routinely worn, particularly when working in confined spaces and when opening the retort door at the end of a treatment cycle.

Chronic Toxicity

Technical-grade penta has been evaluated for cancer-causing effects in laboratory animals, and researchers have attempted to understand the potential for human toxicity and carcinogenicity by evaluating exposure effects on employees from the wood preserving industry. Toxicological interest in penta has been, in part, due to the presence of trace levels (parts per million) of hexa, hepta, and octachlorodibenzo-p-dioxin. The most toxic of the dioxin compounds, 2,3,7,8 TCDD, is not normally associated with penta.

According to the U.S. EPA Integrated Risk Information System (IRIS) toxicologic data base, numerous studies have been conducted on laboratory animals to determine the acute (short-term) and chronic (long-term) health effects of exposure to penta by the oral, inhalation, and dermal absorption routes. These studies concluded that penta causes a statistically significant increase in the incidence in tumors in male mice (oral) and concluded that penta is a “probable human carcinogen” (B2 Classification).

A National Toxicity Program (NTP 1989) fed up to 400 parts per million (ppm) technical-grade penta to mice and up to 600 ppm purified penta to mice, five days a week for 106 weeks. A statistically significant increase in the incidence of liver tumors and endocrine tumors was observed in male mice for both tests involving technical-grade and purified penta. A significant increase in liver tumors was only observed in female mice that were fed the highest concentrations of purified penta.

Studies on human carcinogenicity and effects on mortality have, so far, been inadequate and inconclusive. The authors of one study (Gilbert, et al., 1990) concluded that their test results do not suggest any clinically significant adverse health effects nor any increased cancer morbidity or mortality from exposure to penta from wood preserving chemicals. However, the narrative section of the IRIS database seriously questioned this conclusion and cites potential problems on the selection criteria of the test population and a possible observational bias of the study’s examiner.

The IRIS carcinogenicity assessment concluded that penta is a probable human carcinogen (Classification B2), based on inadequate human data and the sufficient evidence of carcinogenicity in animals. This classification is based on data from a single species.

Creosote

Acute Toxicity

Inhalation of creosote can result in moderate irritation of the respiratory tract. Repeated and prolonged exposure can result in breathing difficulties, central nervous system effects, and cardiovascular collapse. Exposure to the skin can result in severe irritation and, when accentuated with sunlight, photo toxic burns. Ingestion of creosote can result in gastrointestinal irritation, nausea, vomiting, abdominal pain, and in extreme cases, death.

Chemically resistant clothing and gloves should be worn by workers in the wood treatment process. Air purifying respirators are also used by workers to reduce potential exposure from inhalation of dust, fumes, and vapors, particularly when working in confined spaces.

Chronic Toxicity

Creosote is a coal tar distillate and is a complex mixture of over 250 hydrocarbons, some of which are known to cause cancer in laboratory animals. A listing of the major components is provided in Table 11–1.

CCA

Acute Toxicity

CCA (copper chromated arsenate) is generally purchased from a chemical supplier and delivered to the wood-treatment facility as a 50 percent concentrate (weight by volume – metals/water). In this form, the solution presents a serious health concern as it is very acidic (pH = 2) and can cause serious burns and tissue damage to skin. At the wood-treatment facility, the concentrate is generally diluted to 1–2 percent working solution, at which point the acute risks of exposure are lowered significantly.

CCA treating solution (1–2 percent) can cause irritation when exposure to eyes and skin occurs. Repeated and/or prolonged exposure to skin can also result in systemic effects as the solution can be absorbed through the skin. Inhalation of dusts and mists can cause irritation to the throat and lungs when levels exceed the recommended exposure limits. The exposure limits are summarized on Table 11–3.

Workers at wood-treatment facilities generally wear chemically resistant clothing and gloves when working with the CCA treating solution. Workers handling the 50 percent concentrate wear protective face shields, splash aprons, and acid-resistant clothes and gloves.

Chronic Toxicity

Two of the three primary constituents of the CCA treating solution, chromic acid (containing hexavalent chrome) and arsenic pentoxide, are classified as heavy metals and are considered human carcinogens based on laboratory animal studies.

ENVIRONMENTAL CONSIDERATIONS

Wood-preserving chemicals released in an uncontrolled manner can present a serious risk to human health and the environment, due to the toxic nature of the chemicals and their potential to migrate. Some of the physical characteristics that can impact the ability of chemicals to migrate into the environment are summarized on Table 11–4.

Wood-treating facilities represent approximately 5 percent of the sites listed on the EPA's SuperFund National Priorities List (NPL), and additional sites are being investigated and/or conducting remedial action within the United States under state-run environmental cleanup programs. The high percentage of wood-treating sites is primarily due to the toxic nature of the wood-treating chemicals and the lack of scientific knowledge regarding the potential effects of wood-preserving chemicals in the early history of the industry.

TABLE 11–4
Physical characteristics of wood-preserving chemicals.

	Solubility	Vapor Pressure	Molecular Weight
Pentachlorophenol	14 mg/l@20C	0.001 mmHg @ 20 C	266
Creosote (Phenanthrene)[1]	1.0 mg/l @ 25 C	0.00068 mmHg @ 20 C	178
Copper as CuO	Soluble[2]	Not applicable	80
Chromium as CrO_3	Soluble[2]	Not applicable	100
Arsenic as As_2O_5	Soluble[2]	Not applicable	230

{1} Phenanthrene constitutes approximately 20% of whole creosote.
{2} Copper, chromium, and arsenic solubility is based on metals in typical oxide formulations used in the wood-treating industry.

Currently, researchers, environmental officials, and the wood-treating industry are attempting to identify innovative and cost-effective means of addressing the cleanup or remediation of wood-treating chemicals that have been found contaminating environmental media, including soil, groundwater, and surface water. Although a complete discussion of the remediation of soils, groundwater, and surface water is well beyond the scope of this section, a brief discussion of this topic will assist the reader in understanding the range of issues faced by the wood-treating industry today. A summary discussion of environmental remediation also provides an excellent incentive for the development of innovative operation controls and complete execution of environmental controls that are presently available. Consideration of the remedial treatment technologies also complements the discussion of treatment and control technologies in use at active facilities.

A review of the EPA's presumptive remedy selection process will provide students wishing to understand the complexity of issues relating to environmental cleanup processes with a concise outline of the available technologies. The EPA's presumptive remedy/technology selection guide provides specific performance information that has been digested from the evaluation of cleanup attempts at similar sites. By reviewing past experiences, successful and not so successful, the EPA intends to stream-line the selection process and improve the likelihood of success of cleanup actions at wood-treating sites. It should be noted that the EPA has utilized the "presumptive remedy" approach at sites contaminated with volatile organic compounds and polychlorinated biphenols. Presumptive remedies have also been developed for municipal landfills, grain storage facilities, coal gasification plants, and hazardous waste sites with contaminated groundwater.

It was the EPA's objective for the presumptive remedy process to move the cleanup actions more quickly through the administrative process and lead to earlier implementation of remediation. This approach also focuses on the identification of early actions (interim actions) that can be taken prior to the implementation of a site-wide remedy. Interim actions are generally designed to control source areas of contamination at the site and can ultimately result in a shorter cleanup time frame. The presumptive remedy process is also thought to benefit the responsible party by consolidating technologies that have become common, thereby limiting the number of remedial alternatives that must be evaluated in the feasibility study. This approach can direct more of the responsible party's financial resources to the cleanup actions instead of toward fulfilling administrative requirements. Another potential benefit of the presumptive remedy process is the greater level of consistency in the selected remedies among similar types of sites. Consistency in the selection of remedial actions can prevent the development of a financial advantage among business owners within competing markets.

TREATMENT AND CONTROL TECHNOLOGIES

EPA documents on the presumptive remedy for wood-treating sites outline the process that EPA site managers should take in developing a list of options for remediating process residuals and environmental media (soil, sediment, groundwater, and surface water). These documents also identify the treatment technologies that are appropriate for treating sludge, soil, sediment, groundwater, and surface water impacted by wood-preserving chemicals. This listing is also an appropriate summary of the control technologies that are currently in use or are being considered for use within active wood-treating facilities.

The summary tables (Tables 11–5 and 11–6) from an EPA fact sheet on treatment technology selection are presented here. These tables address the treatability of typically encountered wood preservative chemicals and combinations of chemicals. The tables also include the effective ranges of contaminant reduction that can be achieved by the various available control technologies. The far right-hand columns also present typically utilized treatment trains, or groups of technologies that have been used sequentially to achieve reductions in chemical concentrations. The tables have been edited to assist students not familiar with terms used in the field of environmental remediation and treatment technologies.

TABLE 11–5
Treatment technologies for sludge, soil, and sediment.

Chemical/ Containment	Treatment Technologies	Treatability/ Effectiveness RREL Database (% Reduction)	Treatment Trains (Combination of Treatment Steps)
CCA	Immobilization[1,2]	80–90% TCLP (B, P, F)	Soil washing immobilization[3]
PCP	Incineration[2]	90–99% (B, P, F)	—
	Other thermal treatment[3]	—	—
	Biotreatment[3]	—	Soil washing/ biotreatment [3]
	Dechlorinization[3]	—	—

TABLE 11–5
(continued)

Creosote	Incineration[2] Other thermal treatment[3] Biotreatment[3]	90–99% (B, P, F) — —	— — Soil washing/ biotreatment [3]
PCP + creosote	Incineration[2] Other thermal treatment[3] Biotreatment[3]	95–99% (B, P, F) — —	— — Soil washing/ biotreatment [3]
Creosote + CCA	Not applicable	See notation 4	Incineration/immobilize ash[2] Soil washing/biotreatment immobilization[3]
PCP + CCA	Not applicable	See notation 4	Incineration/immobilize ash[2] Soil washing /biotreatment/ immobilization[3] Dechlorination/immobilization[3]

Source: U.S. EPA A540-F-93-020, May 1993. Selected text edited.

Notes:

[1] Arsenical compounds are often problematic for stabilization; however, the levels of arsenical compounds typically found at wood-treating sites are low enough to not present concerns.

[2] These technology recommendations assume that no site-specific constraints exist. Technology selection should ensure that the specified treatment efficiencies will meet established cleanup goals.

[3] These other technologies may warrant site-specific evaluations because they lack full-scale performance data. Site-specific conditions also may favor a subset of the major technology. Bench scale and/or pilot studies may be necessary to refine the selection of the most appropriate specific treatment method.

[4] Performance data are from the Risk Reduction Engineering Laboratory (RREL). The database is derived from bench scale (B), pilot-scale (P), or full-scale (F) demonstration projects. Dashes indicate insufficient data. The RREL is updated on a regular basis and is available through the Alternative Treatment Technology Information Center (ATTIC). The numbers represent total treatment efficiency, not destruction/removal efficiency (DRE). Additional technologies and their treatment efficiencies are added as more information becomes available.

[5] Performance efficiency for treatment trains is a function of contaminant concentration, matrix, and volume. It can generally be presumed that the performance of treatment trains will equal or exceed that of individual treatment technologies.

TABLE 11–6
Treatment technologies for surface water and groundwater.

Chemical/ Contaminant	Treatment Technologies	Treatability/ Effectiveness RREL Database (% Reduction)	Treatment Trains (Combination of Treatment Steps)
CCA	Precipitation Reverse osmosis (RO) Ion exchange	97–99% (B, P, F) 99% (P)	Precipitation/immobilization Precipitation/RO/immobilization Precipitation/ion Exchange/immobilization
PCP	Carbon treatment Biotreatment Oxidation	95–99% (P, F) 99% (B, P, F) 99% (B, P, F)	Phase separation/carbon Phase separation/biotreatment Phase separation/oxidation

TABLE 11–6
(continued)

Creosote	Carbon treatment Biotreatment Oxidation	82–99% (P, F) 99% (P, F) 99% (B, P)	Phase separation/carbon Phase separation/biotreatment Phase separation/oxidation
PCP + creosote	Carbon treatment Biotreatment Oxidation	82–99% (P, F) 99% (B, P, F) 99% (B, P)	Phase separation/carbon Phase separation/biotreatment Phase separation/oxidation
Creosote +CCA	Carbon treatment Oxidation Precipitation	—	Phase separation/treat organic/ treat metals
PCP + CCA	Carbon treatment Oxidation Precipitation	—	Phase separation/treat organic/ treat metals

Source: U.S. EPA A540-F-93-020, May 1993. Selected text edited.

KEY: Treat organic = carbon treatment or chemical (O_3, ClO_2, H_2O_2) or ultraviolet oxidation

Treat metals = reverse osmosis or ion exchange or chemical precipitation and immobilization of residues

*Performance data from RREL (Risk Reduction Engineering Laboratory). Database is derived from bench scale (B), pilot-scale (P), or full-scale (F) demonstration projects. Dashes in the table indicate insufficient data.

Primary Treatment Technologies

The following treatment technologies are the main ones used in wood-preserving facilities. They are described briefly.

- **Immobilization.** A process by which contaminants in a solid phase are mixed with binding agents to limit leaching. Binding agents typically used include fly-ash and cement.
- **Incineration and other thermal treatment.** Incineration and other thermal processes are primarily aimed at the destruction of organic chemicals and wastes by elevating the temperature to a level at or above the combustion temperature of the given waste material. This process can also involve the recovery of thermal energy from materials with acceptably high BTU values. The residual ash from the incineration process may require additional treatment or evaluation prior to disposal.
- **Biotreatment.** A treatment technology that uses microorganisms to degrade contaminants. Primarily applicable to organic contaminants; however, innovation processes have been developed recently that may prove effective in reducing inorganic (metals) contamination.
- **Soil washing.** This process uses water and mechanical action to remove hazardous constituents from soil particles. The wash water is often augmented with chemical surfactants, leaching agents, pH adjustments, and chelating agents to improve the contaminant removal efficiency. The wash water is generally reused throughout the washing process.
- **Dechlorination.** A process applicable to pentachlorophenol waste designed to reduce the toxicity of chlorinated compounds by breaking the chemical bonding of the chlorine molecules and removing the chlorine atoms.

- **Phase separation.** This simple process relies on differences in the specific gravity of waste materials, and is particularly useful in wood-treating operations for managing pentachlorophenol/oil solutions and creosote. Pentachlorophenol solutions are lighter than water with a specific gravity that ranges from 0.85 to 0.95. Creosote is more dense than water and has a specific gravity of approximately 1.15. In phase separation systems, mixtures of wood-preserving chemical and water are separated through a series of baffles placed at the top or bottom of the water stream. Material lighter or heavier than water is skimmed off the top of the water surface or removed from the bottom of the water column.
- **Precipitation.** This is a process primarily applicable to wood-preserving waste streams containing arsenic and hexavalent chromium. The process typically involves the addition of chemicals to the waste stream to raise the pH and cause the metals to precipitate due to reduced solubility. Additional chemicals, such as polymers and fluculants, may also be added to accelerate the separation of the precipitated metals.
- **Reverse osmosis (RO).** This process involves the removal of contaminants solubilized in a wastewater stream through the use of selective membrane filters. RO may be applicable for the removal of the metals from wastewater streams from wood-preserving facilities; however, flow rates through these systems are generally very low and operational costs are generally high.
- **Ion exchange.** In this process chemical contaminants in a wastewater stream are removed through substitution with calcium or sodium typically contained in a media of resinous beads. This process is most applicable to the wood-treating industry for the removal of arsenic and copper.
- **Carbon treatment.** This treatment process relies on the natural process (London Dispersion Forces) by which chemicals in liquid or gaseous streams are attracted to, and then held, on the surface of the solid carbon media. At wood-treating facilities carbon adsorbtion systems are commonly used to remove organic chemicals (pentachlorophenol and creosote compounds) from wastewater streams.
  The ability of carbon to adsorb chemicals is greatly enhanced through the activation process, which develops a large internal pore volume (surface area) per unit volume of the media. According to product literature from Calgon Carbon Corporation, 1 pound of activated carbon contains an effective total area of 100 acres.
- **Oxidation.** This process involves the destruction or alteration of contaminants in the waste stream through the addition of chemicals (oxidizers) or the introduction of an ultraviolet light source. Through this process the organic chemical contaminants, such as pentachlorophenol and creosote, are oxidized and the resultant chemical products are primarily carbon dioxide and water. Total mineralization is said to occur if the destruction of the contaminants is complete and the only resultant products are carbon dioxide and water.

SUMMARY

- Since the early to mid-1800s there have been various wood preservatives used in Europe and the United States.
- Wood preserving can extend the useful life of wood 8 to 10 times that of untreated wood. However, for this extended longevity they must be toxic to decay organisms and insect attack.

- The treatment of wood today is accomplished by two primary processes: pressure impregnation and thermal treatment. The pressure process employs pressures of between 125 and 150 pounds per square inch.
- Three main wood preservatives are currently in use in the United States: creosote, pentachlorophenol, and CCA. CCA is the most common of the three.
- Due to the inherently toxic nature of the chemicals selected by the wood-treating industry coupled, with an increased understanding of the potential environmental impact of these chemicals when released to the environment in an uncontrolled manner, the industry has been subjected to intensive evaluation by the environmental and industrial health agencies.
- These federal regulations control the use and application of the preservatives as well as the storage, handling, and disposal of any hazardous waste generated from their use.
- Wood-preserving chemicals that have been released in an uncontrolled manner can present a serious risk to human health and the environment. When a uncontrolled releases occur, it becomes necessary to remediate the chemicals within the soil, groundwater, and surface water. There is a continuing effort by EPA and industry to identify and implement cost-effective remedial actions at active and inactive wood-preserving sites within the United States. A number of these technologies are currently presented in EPA's presumptive remedy for wood-preserving sites.

QUESTIONS FOR REVIEW

1. Name the three main types of wood preservatives employed in the United States today?
2. What are the three most common chemicals found in water-born wood preservations?
3. List the top five major chemical components of creosote?
4. Name the two most common process types used in the wood-preserving industry today?
5. Name three primary regulations that currently effect the wood-preserving industry?
6. Identify four management plans used by the wood-preserving industry to control pollution.
7. Identify a symptom of creosote exposure to the skin.
8. What are the two major objectives of the EPAs presumptive remedy process?
9. Of the three major wood preservatives, which are considered to be biodegradable?
10. Incineration and thermal treatment of organic wood-preserving wastes can derive what beneficial side-effect?

ACTIVITIES

1. Compare and contrast the concentrations of chemical constituents in the primary wood-preserving solutions with the EPA drinking water and surface water standards. Consideration of the solubility of these chemicals should also be considered during this exercise.
2. Research and review the registration requirements for new wood preservatives under FIFRA.
3. Research and review new technologies employed for the reduction or destruction of wood-preserving chemicals.

REFERENCES

American Wood Preservers Institute. Summer 1995. "Answers to Often-Asked Questions About Treated Wood." Vienna, VA, pp. 23–24 and 33.

Beazer East, Inc. December 1990. "A Technology Overview of Existing and Emerging Environmental Solutions of Wood Treating Chemicals." National Environmental Technology Applications Corporation, Pittsburgh, PA: University of Pittsburgh Allied Research Center, pp. 7–15.

Beazer East, Inc. December 1990. "A Technology Overview of Existing and Emerging Environmental Solutions of Wood Treating Chemicals." National Environmental Technology Applications Corporation. Pittsburgh, PA: University of Pittsburgh Allied Research Center, pp. 16–89.

Calgon Carbon Corporation. December 1993. *Operating Manual—Model 8 Granular Carbon Adsorption System.* Pittsburgh, PA, pp. 6.7–6.8.

Environmental Protection Agency. July 1, 1993. Part 264. *Code of Federal Regulations, Protection of the Environment, 40, Parts 260 to 299.* Rockville, MD: Government Institutes, Inc., pp. 290–294.

Environmental Protection Agency. July 1,1993. Part 265. *Code of Federal Regulations, Protection of the Environment, 40, Parts 260 to 299.* Rockville, MD: Government Institutes, Inc., pp. 425–430.

Environmental Protection Agency. July 1, 1993. Part 131. *Code of Federal Regulations, Protection of the Environment, 40, Parts 100 to 149.* Rockville, MD: Government Institutes, Inc., pp. 288–293.

Environmental Protection Agency. July 1, 1993. Part 141. *Code of Federal Regulations, Protection of the Environment, 40, Parts 100 to 149.* Rockville, MD: Government Institutes, Inc., pp. 599–692.

Environmental Protection Agency. May 1993. "Technology Selection Guide For Wood Treater Sites." *Publication 9360.0-46FS, EPA540-F-93-020.* Washington DC: Office of Emergency and Remedial Response Emergency Response Division 5202G, pp. 1–6.

Environmental Protection Agency. November 1990. *Approaches for Remediation of Uncontrolled Wood Preserving Sites.* Cincinnati, OH: Center for Environmental Research Information, pp. 3–4.

Environmental Protection Agency. June 1996. "Wood Preserving Resource Conservation and Recovery Act (RCRA) Compliance Guide, A Guide to Federal Regulation. "*Publication EPA 305-B-96-001.* Section 8. Washington DC: Office of Compliance, Office of Enforcement and Compliance Assurance, pp. 5–6.

Fuller, B., R. Holberger, D. Carstea, J. Cross, R. Berman, and P. Walker. June 1977. "The Analysis of Existing Wood Preserving Techniques and Possible Alternatives." METREK Division/The MITRE Corporation, pp. 9–41.

Grimes, M. (writer), and P.J. Marrer (pesticide training coordinator). 1992. *Wood Preservation, Pesticide Application Compendium 3.* Oakland, CA: University of California Statewide Integrated Pest Management Project Division of Agriculture and Natural Resources, pp. 39–54.

Hickson. June 1993. "Diluted Wolmanac® Solution." *Material Safety Data Sheet.* Conley, GA: Hickson Corporation, pp. 1–4.

Koppers Company. September 1985. "Creosote." *Material Safety Data Sheet.* Pittsburgh, PA: Koppers Company, Inc., pp. 1–4.

National Library of Medicine. November 28, 1995. "Copper (II) Oxide." Specialized Information Services, TOXNET®Files, Hazardous Substance Data Bank. Washington DC: Government Printing Office.

National Library of Medicine. November 28, 1995. "Pentachlorophenol." Specialized Information Services, TOXNET®Files, Integrated Risk Information System (IRIS). Washington DC: Government Printing Office.

Nyer, E.K. 1992. *Groundwater Treatment Technology, Second Edition.* New York: Van Nostrand Reinhold, pp. 91–94 and pp. 218–247.

Puszcz, S.G., and M.V. Tumulty, PE., CGWP, and H2M Group. 1994. "Groundwater Pollution Control Technologies." *Environmental Science and Technology Handbook.* Rockville, MD: Government Institutes, Inc., pp. 334–345.

Romano, A.M. Summer 1994. "Recent Developments in Cleanup Technologies." *Remediation.* pp. 363–371.

Vulcan Chemicals. September 1, 1995. "Pentachlorophenol." *Material Safety Data Sheet.* Birmingham, AL: Vulcan Chemicals Company, pp. 1–5.

12

Electroplating Waste Streams

Mark Aronson

Upon completion of this chapter, you will be able to meet the following objectives:

- Understand the electroplating process.
- Describe the waste streams that are produced by the electroplating process.
- Appreciate the impact of government regulation on the management of electroplating waste streams.
- Describe available technologies for waste minimization and treatment.
- Understand the concept of industrial ecology.

INTRODUCTION

Electroplating is one of the most important techniques used in the process of metal finishing. It is estimated that there are 2,300 job-plating shops in the United States and an additional 5,000 to 7,000 captive shops and facilities performing plating and finishing operations.[1] **Job-plating** shops own 50 percent or less of the area of the work undergoing finishing; **captive shops** own more than 50 percent of their work. Industries that may use electroplating during the manufacturing process include: aerospace, appliance, automotive, electronics, hardware, heavy equipment, jewelry, musical instrument, and telecommunications.[2]

The techniques used to prepare metals for electroplating and the electroplating process itself may generate wastewater, solid wastes, and air emissions. Pollutants contained in these waste streams may degrade water and air quality, thus posing a danger to the environment and human health. In addition, many of these same pollutants corrode equipment, generate hazardous gases, cause treatment plant malfunctions, and make sludge disposal more difficult.[3] Reducing the generation of wastes or recycling wastes benefits the industry. Effective management of waste streams can reduce use of raw materials, lower disposal costs, and reduce the liabilities associated with waste production and disposal.[4]

The potential environmental impact of electroplating waste streams on the environment has caused the industry to be one of the most regulated in the United States. The two federal agencies charged with the creation of regulations and their enforcement for the metals finishing industry are the EPA and the Department of Labor (Occupational Safety and Health Administration, or OSHA).[5] The specific regulations applying to the electroplating industry are reviewed later in this chapter. Federal regulation has encouraged development of technologies that avoid and minimize waste production in the electroplating process.

ELECTROPLATING: THE INDUSTRIAL PROCESS

Pretreatment Technologies

The process of electroplating requires preparation of the materials being plated. This procedure is often called "pretreatment" to indicate the steps that precede the actual plating process. The purpose of pretreatment is to remove soils (dirt, oil, and grease), oxides, and nonmetallic and metallic inclusions that would prevent adhesion of the metal to be deposited on the work piece during electroplating.[6,7]

Pretreatment chemicals used to clean metals for plating can be divided into two categories: solvents and cleaners. A **solvent** could be defined as a liquid that can dissolve another substance. In most of the industrial literature, solvent refers to nonaqueous substances; often these are hydrocarbons. A **cleaner** is a substance that is water-based or is water soluble. Solvents are used to dissolve soils (dirt, oil, and grease). Cleaners work by displacing, dissolving, or chemically altering a contaminant.[8]

Solvent Cleaning

Oils, fats, and waxes have been removed from metals using chlorinated hydrocarbon solvents for many years. The most commonly used ones are 1,1,2-trichloro-1,2,2-trifluoroethane (CFC-113), 1,1,1-trichloroethane (TCA, also called methyl chloroform or MCF), trichloroethylene (TCE), tetrachloroethylene (also called perchloroethylene or PERC), and dichloromethane (methylene chloride or METH).[9] These solvents have been very popular because they are good cleaners and are nonflammable.

CFC-113, TCA, TCE, and PERC are often used in vapor degreasing. The vapor degreasing process uses a spray of heated solvent vapor. The vapor condenses on the cooler metal parts dissolving the contaminants, which run off the part. The parts are warmed during the cleaning process and vapor will no longer condense on them. The degreased parts then can be removed and quickly dry in the air due to the solvent evaporating.

TCA and CFC-113 are also used in cold cleaning and spot cleaning. In cold cleaning, parts are placed in a tank containing the solvent. The cleaning of the parts is facilitated by agitation of the solvent or parts. Ultrasonic waves may also be used in this process. Spot cleaning refers to the localized application of a solvent to clean an area on the work piece. The solvent is applied to a cloth and wiped across the portion of the workplace to be cleaned.

Cleaners (Aqueous Cleaning)

Aqueous cleaners work by displacing dirt and debris rather than dissolving them like the organic solvents. Strongly alkaline cleaners often contain builders (sodium phosphates, carbonates, hydroxides, and silicates), surfactants (detergents and soaps), and sometimes cyanide compounds. Alkaline cleaners are used to remove paint coatings and soil from the metal parts.

Acid cleaners often contain mineral acids (hydrochloric, nitric, or sulfuric acid), organic acids, and detergents. These cleaners remove **oxidation** and metal contaminants (smut) on the workplace. Later in this chapter, we will see that the traditional solvents and cleaners are being replaced by new technologies that are environmentally friendly and safer in the workplace.

The Chemistry of Electroplating

Electroplating is done by passing an electrical current through a solution containing dissolved metal ions. The part to be plated is placed in the solution and acts as the **cathode** (negative electrode), thus attracting the positively charged metal ions to its surface. An electrode attached to the **anode** (positive electrode) is placed into the solution as a source of ions for the plating process (see Figure 12–1). The voltage supply, or rectifier, provides the direct current (DC voltage) for the electroplating process. The actual chemistry behind the process of electroplating is simple. The chemical reaction for silverplating a part is shown below (Ag is the chemical symbol for the element silver):

$$Ag^{+1} + e^{-} \longrightarrow Ag \text{ (silver metal)}$$

Silver ion + electron —> silver metal

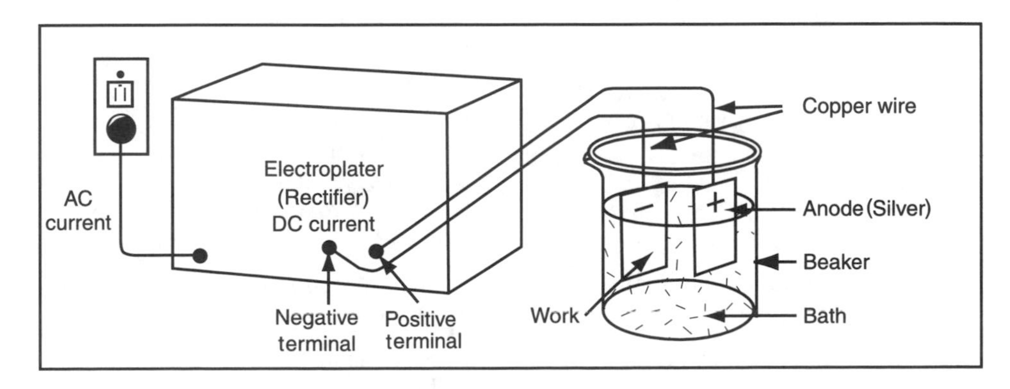

FIGURE 12–1

Source: Grobet File Company of America, Inc. Sal-Hyde Plating Manual No.62.552-1. Carlstadt, NJ.

The electrons necessary for this reaction are available on the surface of the part to be plated; it is attached to the negative electrode (cathode) of the rectifier. Silver ions dissolved in the plating solution possess a positive charge and are attracted to the surface of the part. When the silver ion touches the part, the ion reacts with the electron forming a deposit of silver metal. The electroplating process is continued until the desired coating thickness is produced on the part.

Ferrous and **nonferrous metals** can be electroplated with a variety of metals and their alloys including: aluminum, brass (alloys of copper and zinc), bronze (alloys of copper and tin), cadmium, chromium, copper, iron, lead, nickel, tin, and zinc. Precious metals are also used for plating such as gold, platinum, and silver.[10] The electroplating solutions usually contain a water-soluble salt of the metal to be deposited in the plating process (see Table 12–1). The chemistry occurring in the electroplating tank is only one step of the electroplating process.

TABLE 12–1
Common electroplating bath compositions.

Electroplating Bath Name	Composition
Brass and bronze	Copper cyanide Zinc cyanide Sodium cyanide Sodium carbonate Ammonia Rochelle salt
Cadmium cyanide	Cadmium cyanide Cadmium oxide Sodium cyanide Sodium hydroxide
Cadmium fluoroborate	Cadmium fluoroborate Fluoroboric acid Boric acid Ammonium fluoroborate Licorice
Copper cyanide	Copper cyanide Sodium cyanide Sodium carbonate Sodium hydroxide Rochelle salt
Copper fluoroborate	Copper fluoroborate Fluoroboric acid
Acid copper sulfate	Copper sulfate Sulfuric acid
Copper pyrophosphate	Copper pyrophosphate Potassium hydroxide Ammonia

TABLE 12–1
(continued)

Electroplating Bath Name	Composition
Fluoride-modified copper cyanide	Copper cyanide Potassium cyanide Postassium fluoride
Chromium	Chromic acid Sulfuric acid
Chromium with fluoride catalyst	Chromic acid Sulfate Fluoride

Source: *EPA Guides to Pollution Prevention: The Fabricated Metal Products Industry.* EPA/625/7-90/006. July 1990. P. 10.

The Electroplating Process

The electroplating process refers to the sequence of operations from cleaning of the parts to the finished plated product. This process is responsible for the creation of the waste streams that are of interest to us. Therefore, to understand the origin of these waste streams and the technologies used to manage them we need a working knowledge of the complete electroplating process. The overall process is summarized in Figure 12–2.

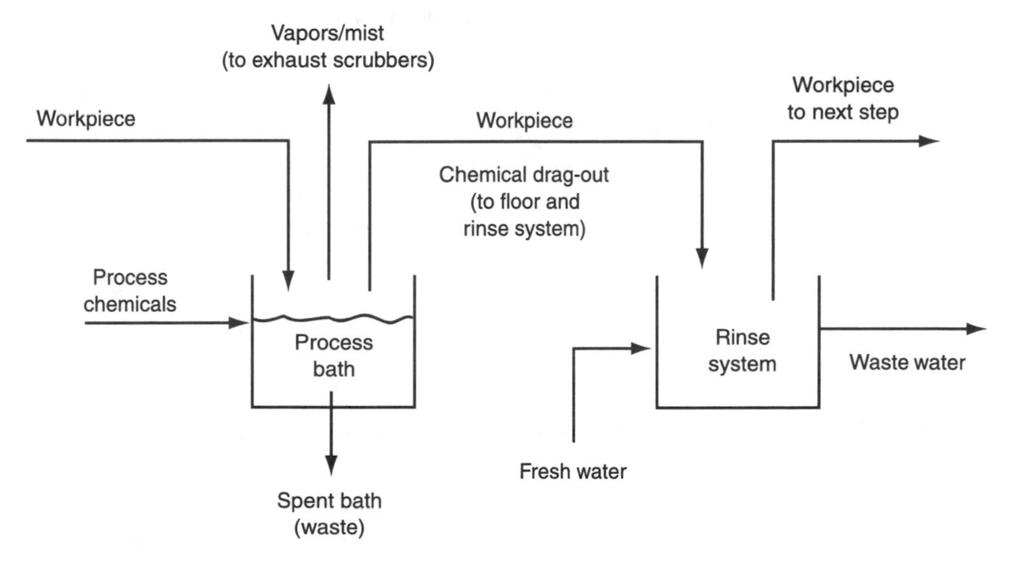

FIGURE 12–2

Source: *EPA Guides to Pollution Prevention: The Metal Finishing Industry.* EPA/625/R-92/011. October 1992. P. 5.

The pretreatment chemical processes of cleaning, **degreasing, pickling,** etching, and electroplating are typically done in tanks (baths) and then are followed with a rinsing cycle. Some potential pollutants can be released to the air via vapors and mists. The chemicals from the cleaning or electroplating steps form the **dragout.** Dragout refers to the chemicals adhering to the work piece that are transferred to the next tank during rinsing or fall to the floor between the tanks. It is easy to see that potential pollutants can enter the air, plant floor, or wastewater. The most important source of waste generation in the electroplating process occurs in the rinsing operation.[11]

Electroless Plating

It is possible to deposit metal coatings on other metals without the use of electrical current. **Electroless plating** produces coatings that are harder and more wear and corrosion resistant than electroplated coatings. Electroless plating chemicals are available for copper, gold, nickel, silver, and tin metal plating. The chemical basis of the system relies on the use of specific chemical reducing agents that can provide the necessary electrons for the reduction of the metal ions to be plated.

To illustrate the chemistry involved in electroless plating, we can examine the chemical reactions taking place in electroless nickel plating below:

1. $H_2PO_2^- + H_2O \longrightarrow H^+ + HPO_3^{2-} + 2H$ (reactive)
2. $Ni^{2+} + 2H$ (reactive) $\longrightarrow Ni + 2H^+$

In the first reaction, the hypophosphite ion is used to form reactive hydrogen. The reactive hydrogen is the reducing agent (provides electrons) for the formation of the elemental nickel metal in the second reaction.[12] There are two side reactions that also occur:

3. $H_2PO_2^- + H$ (reactive) $\longrightarrow H_2O + OH^- + P$
4. $2H$ (reactive) $\longrightarrow H_2$

Reaction 3 forms phosphorus that is incorporated into the nickel deposit and gives electroless plating its unique qualities. Reaction 4 leads to the formation of hydrogen gas while the plating occurs.

The chemical nature of this system allows the production of composite coatings. Diamond or silicon carbide can be added to the electroless plating baths and will be incorporated into the coatings. This increases the wear resistance of the coating. Chemicals that have lubricating properties can be added to the electroless plating solutions to produce coatings with considerable lubricity.

IDENTIFICATION AND CHARACTERIZATION OF WASTE STREAMS

Identification and characterization of industrial waste streams are the first steps in waste minimization. Identification of the waste stream refers to identifying the process, the waste stream, and chemical compounds present in the waste stream. Characterization refers to identifying the physical and chemical nature of the compounds in the waste stream. The EPA has produced waste minimization worksheets to assist managers with waste stream identification and characterization. The worksheets pertaining to waste stream identification and characterization are included in Appendix G.

HAZARDOUS WASTE STREAMS IN THE ELECTROPLATING PROCESS

Our examination of the electroplating process demonstrated that there are two processes that can generate waste streams:

1. Pretreatment technologies (cleaning, degreasing, stripping, etc.).
2. Electroplating activities (dragout from electroplating solutions).

Table 12–2 summarizes the waste streams, wastes, and hazards produced by the electroplating process. It is important to note that water is the primary waste stream carrier in the electroplating process, either in the form of wastewater or rinse water. Some air emissions are possible during the electroplating process but are primarily limited to the pretreatment stage when solvent degreasing technologies are utilized.

TABLE 12–2
Summary table of metal finishing industry waste.

Waste	Potential Hazards	Waste Stream	Process
Alkali (hydroxide)	Corrosivity	Wastewater	Cleaning, etching
Acid (nitric, sulfuric, hydrochloric, hydrofluoric)	Corrosivity	Wastewater	Cleaning, pickling, etching, bright dipping
Surfactants	Aquatic toxicity	Wastewater	Cleaning
Oil and grease	Aquatic toxicity	Wastewater, spent solvent	Cleaning
Cadmium, zinc, nickel, copper, other metals	Toxicity	Plating bath, dragout, rinse water, spent filters, sludge	Plating
Perchloroethylene, trichloroethylene, other solvents	Inhalation, dermal	Spent solvent (liquid or sludge), air emissions	Cleaning
Cyanide	Toxicity	Plating bath, dragout, rinse water, other wastewater	Plating, tumbling, stripping, heat treating, desmutting
Chromates	Toxicity	Plating bath, dragout, rinse water, sludge, other wastewater, mist	Plating, chromating, etching
Water	—	Rinse water, dragout, process bath, air emission (evaporation), cooling water, boiler blowdown	Various

Source: *EPA Guides to Pollution Prevention: The Metal Finishing Industry.* EPA/625/R-92/011. October 1992. P. 7.

FEDERAL REGULATION OF THE ELECTROPLATING INDUSTRY

The purpose of this section is to summarize the legislation and regulations affecting the electroplating industry. The primary legislation affecting the electroplating industry has been passed since 1972 (see Table 12–3). Most of the regulatory activity concerns water pollution. This should not be surprising since wastewater is the most important waste stream in the electroplating process.

TABLE 12–3
Summary of environmental legislative activities affecting the electroplating industry.

Year	Legislation	Requirements
1972	Federal Water Pollution Control Act (FWPCA) Amendments (Public Law 92-500)	Required all industries discharging into waterways to meet technology-based standards of pollution control: Best Practicable Technology (BPT)—By July 1, 1977 Best Available Technology (BAT)—By 1983 (later revised to 1984) New Source Performance Standards (NSPS)—If source begins construction after publication of the applicable proposed regulations. Required all industries discharging into municipal systems to attain industry-specific effluent limitations (pretreatment standards). Required periodic review and updating of technology-based requirements. Established National Pollutant Discharge Elimination System (NPDES) permit program. Required self-monitoring by plants discharging to navigable waters. Established federal control over municipal systems.
1976	National Resource Defense Council (NRDC) Consent Decree (NRDC et al. vs. Train)	Committed EPA to a schedule for developing BAT effluent limitations for 21 major industries covering 65 recognized toxic substance classes (129 specific compounds). This schedule was later incorporated into the Clean Water Act.
	Resource Conservation and Recovery Act (RCRA) (Public Law 94-580)	Established controls for disposal of all solid wastes. Defined hazardous solid wastes.

TABLE 12–3
(continued)

Year	Legislation	Requirements
		Established tests to determine which wastes are covered. Established standards for solid waste generators, storage facilities, and disposal sites. Established manifest system for transportation of hazardous wastes.
1977	Clean Water Act (Public Law 92-217)	As an amendment to FWPCA, revised FWPCA deadlines. Defined classes of pollutants as toxic, conventional, and nonconventional with major emphasis on the toxic compounds associated with the NRDC Consent Decree. Linked pretreatment standards to BAT guidelines for toxics. Established BCT (Best Conventional Technology) level of compliance for industrial discharges of conventional pollutants (e.g., oil and grease, pH, suspended solids, biochemical oxygen demand) based upon the cost to municipalities to treat conventional pollutants and industries' incremental treatment costs. Authorized municipal systems to relax pretreatment standards under certain conditions for individual dischargers.
1984	Hazardous and Solid Waste Amendments of 1984 (Public Law 98-616)	As an amendment to RCRA, brought small-quantity generators (100 to 1,000 kg/month) under RCRA requirements. Small-quantity shipments must be accompanied by manifests; small quantities can be stored on site for 180 days. Required certification by generators after 9/1/85 that the generation of hazardous wastes has been minimized. Required new underground storage tanks of petroleum and hazardous wastes to be constructed to prevent leaks, and existing tanks to be monitored for leaks.

TABLE 12–3
(continued)

Year	Legislation	Requirements
		Prohibited landfilling of bulk or non-containerized liquids after 5/8/85. Banned certain wastes from land disposal unless they can be demonstrated not harmful to human health and human environment.
		Required certain surface impoundments receiving hazardous wastes to be retrofitted with double liners, groundwater monitoring capabilities, and leachate collection systems.
		Required EPA to report to Congress on hazardous wastes not addressed by RCRA because they are sent through municipal sewers.

Source: *EPA Environmental Regulations and Technology: The Electroplating Industry.* EPA/625/10-85/001. P. 3.

The EPA is charged with the development of regulations mandated by legislation. The other federal agency involved in promulgating regulations for industry is OSHA. The following discussion examines the impact of federal regulation on the electroplating industry.

Federal Water Pollution Control Act as Amended by the Clean Water Act of 1987

The Clean Water Act (CWA) has a stated objective to "restore and maintain the chemical, physical, and biological integrity of the nation's waters."[13] The EPA is the federal agency given the job of administering the CWA's programs. The EPA has developed standards for the electroplating and metal-finishing industry. These standards are found under Title 40, Code of Federal Regulations (CFR). The applicable parts of the CFR are listed below; each will be discussed separately:[14]

- 40 CFR 413—EPA Electroplating Standards
- 40 CFR 433—EPA Metal-Finishing Standards
- 40 CFR 403—EPA General Pretreatment Regulations
- 40 CFR 122—EPA National Pollution Discharge Elimination System (NPDES)

40 CFR 413—EPA Electroplating Standards

Electroplating shops that discharge wastewater to publicly owned treatment works (POTW) are subject to monitoring requirements listed under 40CFR 413.14, pretreatment standards for existing sources (PSES). These standards must be met before discharge to a POTW (see Table 12–4, discharge concentration limits). Note that shops having a discharge of less than 10,000 gallons per day (gpd) have fewer restrictions than shops that discharge more than 10,000 gpd. Shops plating precious metals (silver, gold, platinum) are covered by limits under 40 CDR 413.24. Shops that are involved in procedures similar to electroplating are also subject to PSES (40 CFR 413 subparts D–H).[15] These procedures include: chromating, phosphating, chemical etching and milling, electroless plating, and printed circuit-board manufacturing.

40 CFR 433—EPA Metal-Finishing Standards

The metal-finishing standards are more restrictive than the PSES and apply to all electroplaters that are direct dischargers (facilities discharging directly into a waterway). Direct dischargers must apply for an NPDES permit specifying what pollutants may be discharged and a schedule for compliance, monitoring, and reporting.

The regulations concerning the applicability of 40 CFR 413 and 40 CFR 433 to specific metal-finishing operations are complicated and are beyond the scope of this review. In general, job shops own 50 percent or less of the area of the materials undergoing metal finishing and will be subject to 40 CFR 413. Job shops are also subject to 40 CFR 433 if they are direct dischargers. Captive shops, which own more than 50 percent of the area of the materials undergoing plating, will be subject to both 40 CFR 413 and 40 CFR 433 if they are indirect dischargers and 40 CFR 433 as direct dischargers.

TABLE 12–4
Discharge concentration limits for PSES to a POTW from common metal EP facilities.

	Concentration (mg/liter)	
Pollutant	**Daily Maximum**	**4 Days Average**
<10,000 gpd flow		
Cadmium	1.2	0.7
Cyanide, (A)	5.0	2.7
Lead	0.6	0.4
TTO[1]	4.57	—
>10,000 gpd flow		
Cadmium	1.2	0.7
Chromium, (T)	7.0	4.0
Copper	4.5	2.7
Cyanide, (T)	1.9	1.0
Lead	0.6	0.4
Nickel	4.1	2.6
Zinc	4.2	2.6
Metals (T)*	10.5	6.8
TSS[2]	20.0	13.4
TTO[1]	2.13	—
pH	7.5-10.0 su	7.5-10.0 su

A = Amenable to chlorination
T = Total concentration
1 = Total toxic organics
2 = Total suspended solids
* = Total metals is defined as the sum of Cu, Ni, Zn, and Cr (both trivalent and hexavalent)

Source: Mabbett, A. 1994. EPA/OSHA Government Regulations: Summary for the Finishing Industry: Products Finishing: 59 (2A), P. 274. Cincinnati, OH: Gardner Publications, Inc.

TABLE 12–5
EPT effluent discharge concentration limits for PSES to a POTW from MF.

Pollutant	Concentration (mg/liter) Daily Maximum	Monthly Average
Cadmium	0.69	0.26
Chromium, (T)	2.77	1.71
Copper	3.38	2.07
Cyanide, (A)	0.86	0.32
Cyanide, (T)	1.20	0.65
Lead	0.69	0.43
Nickel	3.98	2.38
Silver	0.43	0.24
Zinc	2.61	1.48
TTO1	2.13	—

A = Amenable to chlorination
T = Total concentration
1 = Total toxic organics

Source: ditto #4 , Mabbett p. 276.

The effluent standards listed in Table 12–5 are the result of applying the best practicable control technology (BPT, part 433.433.13) for PSES. Metal-finishing shops are also subject to effluent limitations for existing point source discharges using best available technology (BAT, part 433.14) and new source performance standards (NSPS, part 433.16). There are also standards for pretreatment from new sources (PANS, part 433.17).

40 CFR 403—EPA General Pretreatment Standards

40 CFR 403 establishes the responsibilities of federal, state and local government, industry, and the public to implement national pretreatment standards to control pollutants that pass to publicly owned treatment works (POTWs) or that may contaminate sewer sludge. Electroplaters must comply with the General Pretreatment Standards summarized below:[16]

- Effluents cannot create a fire or explosion hazard.
- PH of the effluents must be >5.0.
- Effluents cannot obstruct flow to a POTW.
- Effluents cannot interfere with sewage plant operations.
- Effluents cannot be excessively hot.
- Effluents cannot contain petroleum, mineral, or nonbiodegradable oils.

Part 403 also allows industries to be categorized under the National Pollutant Discharge Elimination System (NPDES), which are subject to the National Categorical Pretreatment Standards (CPS). Electroplating shops that discharge effluents that do not flow to a POTW are covered under 40 CFR 122.

40 CFR 122—EPA National Pollution Discharge Elimination System (NPDES)

Part 122 of the NPDES requires facilities with point source discharge (storm sewers and direct discharge into waterways) to obtain a permit for their discharges. The EPA or authorized states give the permittee the right to discharge specified pollutants from specified point sources for a period of five years.[17] Local POTWs can apply for control authority status and establish their own effluent limitations as long as they are at least as stringent as the federal standards.

Resource Conservation and Recovery Act (40 CFR 260-281, 42 USC 6901)

Industries that generate, store, transport, or dispose of hazardous wastes must comply with the standards authorized in the act. RCRA requires "cradle to grave" management of hazardous wastes. This act regulates the handling, storage, treatment, disposal, and transportation of hazardous wastes.

Industries that produce hazardous wastes during their operations are required by law to notify the EPA. They become a waste generator and have responsibility forever for the wastes produced. Permits must be obtained if the generator is going to store or treats the waste on-site (40 CFR 270). Generators that produce less than 1,000 kilograms per month are designated as small-quantity generators. Small-quantity generators may store up to 6,000 kilograms of hazardous waste on-site for up to 180 days without a permit (see Table 12–6). Large-quantity generators are allowed to store wastes ninety days without a permit (see Table 12–7). Generators producing less than 100 kilograms of hazardous waste are exempt from RCRA but are responsible for some minimum standards (see Table 12–8).

If hazardous wastes are transported, the transporter must apply for an EPA transporter number. This number allows tracking of the hazardous waste.

Electroplating shops are involved in the generation of hazardous wastes. To be classified as a hazardous waste the waste stream chemicals would be listed as hazardous

TABLE 12–6
U.S.EPA requirements for small-quantity generators.

Produce between 100 Kg (220 lbs) and 1,000 Kg (2,200 lbs) in any calendar month of hazardous waste.
Do not accumulate more than 1,000 Kg (2,200 lbs) waste on-site at any one time.
Never store more than 6,000 Kg (13,200 lbs).
Notify EPA and obtain U.S.EPA I.D. number.
Store waste no more than 180 days (possible 270 days).
Properly store wastes in containers or tanks.
Post emergency numbers near telephones.
Informal employee training.
Disposal only at a RCRA permitted site.
Use only transporters with U.S.EPA I.D. numbers.
Use proper DOT packaging and labeling.
Use Hazardous Waste Manifest.
Keep records for three years.
Report missing shipments.

Source: Kushner, J., and A. Kushner. 1994. *Water and Waste Control for the Plating Shop*. Cincinnati, OH: Gardner Publications Inc. P. 270.

wastes by the EPA (40 CFR 261-Subpart D) or meet one of the characteristics of a hazardous waste (see Table 12–9).

TABLE 12–7
U.S.EPA requirements for large-quantity generators.

Generates more than 1,000 Kg (2,200 lbs) of hazardous waste in any calendar month.
Notify EPA and obtain U.S.EPA I.D. number
Store waste no more than 90 days.
Properly store wastes in containers or tanks.
Prepare and retain a written Contingency Plan.
Prepare and retain a written training plan with annual training.
Prepare and implement a written Waste Minimization Plan.
Disposal only at a RCRA permitted site.
Use only transporters with EPA I.D. numbers.
Use proper DOT packaging and labeling.
Use Hazardous Waste Manifest.
Keep records for three years.
Report missing shipments in writing.
Submit biennial report of your hazardous waste activity.
Submit a biennial Waste Minimization Report.

Source: Kushner, J., and A. Kushner. 1994. *Water and Waste Control for the Plating Shop*. Cincinnati, OH: Gardner Publications Inc. P. 270.

TABLE 12–8
U.S.EPA requirements for conditionally exempt small quantity generators.

Generate no more than 100 Kg (220 lbs) in any calendar month of hazardous waste.
Do not accumulate more than 1,000 Kg (2,200 lbs) on-site at any one time.
Never store more than 6,000 Kg (13,200 lbs).
At least, send waste to a state approved sanitary landfill.

Source: Kushner, J., and A. Kushner. 1994. *Water and Waste Control for the Plating Shop*. Cincinnati, OH: Gardner Publications Inc. P. 270.

TABLE 12–9
U.S.EPA hazardous waste characteristics summary (1) common to electroplating.

Characteristic	Waste Code	Contaminant	Description
Ignitability	D001	—	Liquid with flashpoint <140°F
Corrosivity	D002	—	pH ≤ 2.0, pH ≥ 12.5, or pH < 2.0, pH 12.5, or corrodes steel 6.35 mm (0.250 inches per year)
Reactivity	D003	Cyanide Sulfide	Sufficient to generate toxic gases at 2<pH <12.5

TABLE 12–9
(continued)

Characteristic	Waste Code	Contaminant	Description
Reactivity	D003	Cyanide Sulfide	Sufficient to generate toxic gases at 2<pH <12.5
Toxicity characteristic			
	Waste Code	**Contaminant(2)**	**Max. Allowable Level (mg/l)***
	D004	Arsenic	5.0
	D005	Barium	100.0
	D018	Benzene	0.5
	D006	Cadmium	1.0
	D019	Carbon tetrachloride	0.5
	D021	Chlorobenzene	100.0
	D022	Chloroform	6.0
	D007	Chromium	5.0
	D026	Cresol	200.0
	D027	1, 4-Dichloro-benzene	7.5
	D028	1, 2-Dichloro-ethane	0.5
	D029	1, 1-Dichloro-ethylene	0.7
	D008	Lead	5.0
	D009	Mercury	0.2
	D035	Methyl ethyl ketone	200.0
	D036	Nitrobenzene	2.0
	D037	Pentachloro-phenol	100.0
	D010	Selenium	1.0
	D011	Silver	5.0
	D039	Tetrachloro-ethylene	0.7
	D040	Trichloro-ethylene	0.5
	D041	2, 4, 5-Tri-chlorophenol	400.0
	D042	2, 4, 6-Tri-chloroplenol	2.0

*Extract form Toxicity Characteristic Leaching Procedure (TCLP) per 40CR261, Appendix II.
1) See 40CFR261, Subpart C, for full language and explanation.
2) See 40CFR261.24 for entire listing of contaminants.

Source: Kushner, J., and A. Kushner. 1994. *Water and Waste Control for the Plating Shop*. Cincinnati, OH: Gardner Publications Inc. P. 270.

Solids collected from wastewater treatment are a listed hazardous waste except from the following processes:[18]

- Sulfuric acid anodizing aluminum
- Tin plating on carbon steel
- Zinc plating (segregated basis) on carbon steel
- Aluminum or zinc-aluminum plating on carbon steel
- Cleaning/stripping associated with aluminum, tin, and zinc plating on carbon steel
- Chemical etching and milling of aluminum
- Sulfuric acid

Used cleaning solutions, stripping solutions, plating solutions, filters, and residues from tank bottoms are not listed wastes unless cyanides are present. Before these materials can be eliminated from consideration as being hazardous, they must not exhibit any of the characteristics found in Table 12–9. Hazardous materials should be identified during the identification and characterization of the waste stream.

Comprehensive Environmental Response, Compensation and Liability Act (40CFR 300, 42 USC 9601-9675)

CERCLA, or the "SuperFund" program, was created to correct deficiencies in RCRA. RCRA deals with regulating the generation of hazardous wastes but did not provide a legal remedy for dealing with hazardous waste sites that were uncontrolled or abandoned.[19] CERCLA empowered the EPA to legally force the cleanup of abandoned waste sites when there is an "imminent and substantial danger to public health, welfare, or the environment." CERCLA also provides for recovery of cleanup costs by the EPA when the responsible party can be determined. One of the most important provisions of CERCLA is the creation of the hazardous substance SuperFund. This fund is used by the EPA for cleanup of hazardous waste sites. The fund is supported by tax revenues from the petroleum and chemical industries and an environmental tax on corporations.[20]

Clean Air Act

Air pollutants from electroplating operations could result from cleaning operations where volatile organic compounds (VOCs) are used or in plating baths where acid or alkaline vapors are produced. Permits may be required under an EPA-approved State Implementation Plan (SIP). Emissions are usually managed through a Best Available Control Technology (BACT).

Under Title III of the Clean Air Act (CAA) amendments of 1990, 189 hazardous air pollutants were regulated (only 188 are currently regulated). The EPA published a list of industry groups or "source categories" that emit one or more of these air pollutants. One of the listed categories targets hard and decorative chromium electroplating and chromium anodizing procedures. Major sources are ones emitting 10 tons/year or more of a listed pollutant or 25 tons/year or more of a combination of listed pollutants.

Maximum Achievable Control Technology (MACT) standards appeared in the January 25, 1995, edition of the *Federal Register* (Vol. 60, p. 1,948). These regulations pertain to all electroplating facilities engaged in hard and decorative chromium electroplating and/or chromium anodizing. According to the regulations, decorative plating operations were to be in compliance by January 25, 1996, and hard chromium and chromium anodizing shops by January 25, 1997. Also under Title III, industries storing any of the 162 listed chemicals above specified thresholds must prepare a risk management plan for accidental release prevention.[21]

The MACT regulations for chromium electroplating are too detailed to be included here. The EPA has published a book explaining the regulations entitled *A Guidebook on How to Comply with the Chromium Electroplating and Anodizing National Emission Standards for Hazardous Air Pollutants* (EPA-4531B-95-001). The MACT regulations can be downloaded from the EPA's Technology Transfer Network (TTN); call (919)541-5742 for more information. Also of interest for monitoring of plating emissions is the EPA-produced video, "Construction and Operation of the EPA Method 306A Sampling Train and Practical Suggestions for Monitoring of Electroplating and Anodizing Facilities." This video is available from North Carolina State University at (919)515-5875. The Resources section at the end of this chapter lists some Internet sites that are useful in obtaining further information on the above regulations.

Title VI of the 1990 amendments is also applicable to plating operations, providing for the control and phase-out of chlorofluorocarbons (CFCs). The EPA lists two classes of chemicals known to destroy stratospheric ozone:

1. Class I—chlorofluorocarbons, halons, carbon tetrachloride, and methyl chloroform
2. Class II—hydrochlorofluorocarbons (HCFCs)

Electroplating and other metal-finishing operations have used these compounds for pretreatment degreasing and cleaning. Fortunately, newer technologies are replacing these harmful chemicals in the metal-finishing industry.

Occupational Safety and Health Act (29 CFR 1900–1910)

This act was created to provide national standards for health and safety in the workplace. Since the metal-finishing industry commonly uses chemicals and technologies that could be detrimental to employees, knowledge of some of the specific standards is useful. A summary of some relevant standards is given here:

Occupational Injuries and Illnesses Reporting (29 CFR 1904) In general, employers are required to keep records on job-related injuries and illnesses. Specific recordkeeping requirements include the posting of an annual summary in the workplace where other employee information is located.

Hazard-Communication Standard (29 CFR 1910.1200) This standard is known as the "Worker Right-To-Know." The Hazard-Communication Standard (HCS) requires specific labeling of products by chemical companies and distributors. Employers are required to inform the employees of hazard information about hazardous chemicals by use of labels, Material Safety Data Sheets, and training. Employers must develop a hazard communication program that must include a chemical inventory, assessment of hazard potential, Material Safety Data Sheets for all chemicals, container labeling, as well as a employee training program.[22]

Cadmium (29 CFR 1910.1027) The standard requires employers to monitor the amount of cadmium exposure experienced by employees. This information must be given to the employees in writing and they must be posted in an area accessible to them. The standard also lowers the eight-hour time-weighted average (TWA) Permissible Exposure Limit (PEL) to 0.75 ppm eight-hour TWA in air.

WASTE TREATMENT AND CONTROL TECHNOLOGIES

The discussion of waste control and treatment technologies will be divided into two parts: Pretreatment waste control and treatment, and electroplating waste control and treatment.

Pretreatment Waste Control and Treatment

Pretreatment wastes arise in the plating shop primarily from cleaning and stripping operations that prepare the workplace for plating. By far, the most urgent problem in pretreatment waste streams is air pollution from the use of chlorinated cleaning solvents. The following reasons are given by the EPA for application of new cleaning technologies:[23]

1. RCRA regulations make using processes that generate significant solvent waste streams unattractive to the industry. Disposal of hazardous solvent wastes is costly and involves continued liability concerns. Waste generators must maintain a waste minimization program. Elimination and reduction of solvent use help industries demonstrate they are taking steps to minimize hazardous waste streams.
2. Hazardous solvent emissions have to be reported under Title III of the Super-Fund Amendment and Reauthorization Act (SARA). Industry is required to monitor and report the release of toxic chemicals in the Toxics Release Inventory (TRI). Cleaning solvents 1,1,1-trichloroethane (TCA), trichloroethylene (TCE), methylene chloride (METH), and perchloroethylene (PERC) are toxics that must be reported. Industries that can reduce or eliminate toxic solvent use below the threshold of the regulations do not need to complete a TRI report for that solvent.
3. Many solvents used in cleaning are part of a list of seventeen priority toxic chemicals in the 33/50 program targeted for early reduction.
4. OSHA regulations for solvent air emissions in the workplace are becoming increasingly stringent.
5. Title III of the CAA amendments requires adoption of Maximum Achievable Control Technologies (MACT) for 189 hazardous air pollutants (HAPS). Solvents

used for cleaning and vapor degreasing processes are major sources of HAPS and are subject to MACT standards.

6. The Pollution Prevention Act establishes source reduction as the preferred method for pollutant management. The EPA can grant a six-year extension on the MACT compliance date to any existing source of airborne toxic chemicals that reduces emissions voluntarily by 90 percent (95 percent for particulates) below 1987 levels before January 1, 1994.
7. Under the CAA amendments, MACT standards are expected for halogenated solvent cleaners.
8. Class I ozone layer-depleting substances (OLDS) must be phased out by 1996. The OLDS Class I list includes CFCs and halons. CFC-113 and TCA, which are important cleaners, are included in this list.

Cleaner Technology for Pretreatment Processes

According to the EPA, cleaner technology is using a source reduction method or recycling to eliminate or significantly reduce hazardous waste generation. The EPA also suggests that source reduction methods should precede recycling in the hierarchy of pollution prevention options. This means recycling should be used to minimize or reduce the need to treat wastes that remain after source reduction methods are implemented.

There are two main strategies in cleaner technologies for cleaning and degreasing operations:

1. Replace toxic solvents with alternative cleaning solutions (aqueous-based cleaners).
2. Process changes that use different technologies for cleaning or eliminate the need for cleaning.

Table 12–10 summarizes the current available technologies for alternative cleaning solutions, while Table 12–11 summarizes the available technologies for process changes.

Electroplating Waste Control and Treatment

Source reduction of pollutants is the first step in the waste minimization process, followed by effective recycling and recovery technologies. Also, alternative treatment systems can be used to further reduce generation of wastes. A brief discussion of selected source reduction, recycling, and recovery techniques follows; these are summarized in Table 12–12.

Source Reduction Processes

1. Eliminate processes using hexavalent chromium and cyanide whenever possible because special equipment is needed to detoxify these chemicals.
2. Use nonchelated process chemicals. Chelating agents are used in cleaning baths and in electroless plating. Chelating agents lower the concentration of specific metal ions in the bath. The problem is that chelating agents interfere with treatment procedures that depend on precipitation. Additional precipitants may need to be added, increasing the volume of hazardous sludge after treatment.

TABLE 12–10
Available technologies for alternatives to chlorinated solvents for cleaning and degreasing: descriptive aspects.

Technology Type	Pollution Prevention Benefits	Reported Application	Operational Benefits	Limitations
Aqueous cleaners	• No ozone depletion potential • May not contain VOCs • Many cleaners reported to be biodegradable	• Excellent for removing inorganic and polar organic contaminants • Used to remove light oils and residues left by other cleaning processes • Used to remove heavy oils, greases, and waxes at elevated temperatures (>160°F)	• Remove particulates and films • Cleaner performance changes with concentration and temperature, so process can be tailored to individual needs • Cavitate using ultrasonics	• Nonflammable and nonexplosive; relatively low health risks compared to solvents; consult Material Safety Data Sheet (MSDS) for individual cleaner • Contaminant and/or spent cleaner may be difficult to remove from blind holes and crevices • May require more floor space, especially if multistage cleaning is performed in line • Often used at high temperatures (120–200°F) • Metal may corrode if part not dried quickly; rust inhibitor may be used with cleaner and rinsewater • Stress corrosion cracking can occur in some polymers
Semi-aqueous cleaners	• Some have low vapor pressure and so have low VOC emissions • Terpenes work well at low temperatures, so less heat energy is required • Some types of cleaners allow used solvent to be separated from the aqueous rinse for separate recycling or disposal	• High solvency gives cleaners good ability for removing heavy grease, waxes, and tar • Most semiaqueous cleaners can be used favorably with metals and most polymers • NMP used as a solvent in paint removers and in cleaners and degreasers	• Rust inhibitors can be included in semiaqueous formulations • Nonalkaline pH; prevents etching of metals • Low surface tension allows semiaqueous cleaners to penetrate small spaces • Glycol ethers are very polar solvents that can remove polar and nonpolar contaminants • NMP used when a water-miscible solvent is desired	• NMP is a reproductive toxin that is transmitted dermally; handling requires protective gloves • Glycol ethers have been found to increase the rate of miscarriage • Mists of concentrated cleaners (especially terpenes) are highly flammable; hazard is overcome by process design or by using as water emulsions

TABLE 12–10
(continued)

Technology Type	Pollution Prevention Benefits	Reported Application	Operational Benefits	Limitations
			• Esters have good solvent properties for many contaminants and are soluble in most organic compounds	• Limonene-based terpenes emit a strong citrus odor that may be objectionable • Some semiaqueous cleaners can cause swelling and cracking of polymers and elastomers • Some esters evaporate too slowly to be used without including a rinse and/or dry process • May be aquatic toxins
Petroleum hydrocarbons	• Produce no wastewater • Recyclable by distillation • High grades have low odor and aromatic hydrocarbon content (low toxicity) • High grades have reduced evaporative loss	• Used in applications where water contact with parts is undesirable • Used on hard-to-clean organic contaminants, including heavy oil and grease, tar, and waxes • Low grades used in automobile repair and related service shops	• No water used, so there is less potential for corrosion of metal parts • Compatible with plastics, most metals, and some elastomers • Low liquid surface tension permits cleaning in small spaces	• Flammable or combustible, some have very low flash points, so process equipment must be designed to mitigate explosion dangers • Slower drying times than chlorinated solvents • The cost of vapor recovery, if implemented, is relatively high
Hydrochlorofluorocarbons (HCFCs)	• Lower emissions of ozone-depleting substances than CFCs • Produce no wastewater	• Used as near drop-in replacements for CFC-113 vapor degreasing • Compatible with most metals and ceramics, and with many polymers • Azeotropes with alcohol used in electronics cleaning	• Short-term solution to choosing an alternative solvent that permits use of existing equipment • No flash point	• Have some ozone depletion potential and global warming potential • Incompatible with acrylic, styrene, and ABS plastic • Users must petition EPA for purchase, per Section 612 of CAAA

Source: *EPA Guide to Cleaner Technologies: Alternatives to Chlorinated Solvents for Cleaning and Degreasing.* EPA/625/R-93/016. February 1994. Pp. 7–8.

TABLE 12–11
Available technologies for cleaning and degreasing: descriptive aspects.

Cleaning/ Degreasing Technology	Pollution Prevention Benefits	Reported Application	Operational Benefits	Limitations
Add-on controls to existing vapor degreasers	• Reduce solvent air emissions	• Retrofitted on existing vapor degreasers	• Allow gradual phase-in of emission controls • Major process modifications not required • Cleaning principle remains the same • Relatively inexpensive	• Reduce but cannot eliminate air emissions • Performance depends on other features of existing degreaser • Dragout on parts cannot be eliminated
Completely enclosed vapor cleaner	• Virtually eliminates solvent air emissions	• Same as conventional open-top vapor degreasers	• Virtually eliminates air emissions and workplace hazards • Cleaning principle remains the same; user does not have to switch to aqueous cleaning • Significant recovery of solvent • Reduced operating costs	• High initial capital cost • Slower processing time • Relatively higher energy requirement
Automated aqueous cleaning	• Eliminates solvent use by using water-based cleaners	• Cleaning of small parts	• Eliminates solvent hazards • Reduces water consumption • Cleaning chemicals are reused • Easy to install and operate	• May not be able to replace vapor degreasing for some delicate parts, and requires more space than vapor degreasing • Wastewater treatment required • Relatively higher energy requirement
Aqueous power washing	• Eliminates solvent use by using water-based cleaners	• Cleaning of large and small parts	• Eliminates solvent hazards • Reduces cleaning time	• Pressure and temperature may be too great for some parts • Wastewater treatment required

TABLE 12–11
(continued)

Cleaning/ Degreasing Technology	Pollution Prevention Benefits	Reported Application	Operational Benefits	Limitations
Ultrasonic cleaning	• Eliminates solvent use by making aqueous cleaners more effective	• Cleaning of ceramic, aluminum, plastic, and metal parts; electronics; glassware; wire; cable; and rods	• Eliminates solvent hazards • Can clean in small crevices • Cost effective • Faster than conventional methods • Inorganics are removed • Neutral or biodegradable detergents can often be employed	• Part must be immersible • Testing must be done to obtain optimum solution and cavitation levels for each operation • Thick oils and grease may absorb ultrasonic energy • Energy required usually limits parts sizes • Wastewater treatment required if aqueous cleaners are used
Low-solids fluxes	• Eliminates need for cleaning and therefore eliminates solvent use	• Soldering in the electronics industry	• Eliminates solvent hazards • Little or no residue remains after soldering • Closed system prevents alcohol evaporation and water absorption	• Conventional fluxes are more tolerant of minor variations in process parameters • Possible startup or conversion difficulties • Even minimal residues are unacceptable in many military specifications
Inert atmosphere soldering	• Eliminates need for flux and therefore eliminates solvent cleaning	• Soldering in the electronics industry	• Eliminates solvent hazards • Economic and pollution prevention benefits from elimination of flux	• Requires greater control of operating parameters • Temperature profile for reflow expected to play more important role in final results

Source: *EPA Guide to Cleaner Technologies: Cleaning and Degreasing Process Changes.* EPA/625/R-93/017, p. 5.

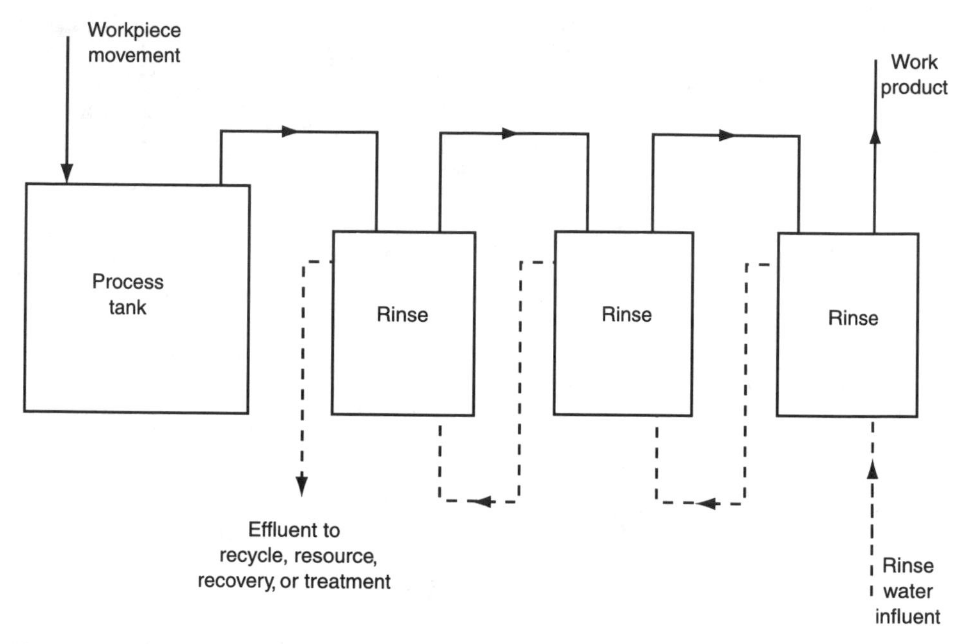

FIGURE 12–3

Source: *EPA Guides to Pollution Prevention: The Metal Finishing Industry.* EPA/625/R-92/011. October 1992. P. 5.

3. Reduce the volume of rinse water. Rinse water volume can be reduced by as much as 90 percent by increasing the efficiency of the rinsing process.[24] Parallel rinses should be converted to a multistage counter-current system (see Figure 12–3). This system allows more contact of the workplace with the rinse water.
4. Extend the life of the electroplating baths by: good housekeeping (this refers to the maintenance and inspection of the bath tanks and equipment); initiating electrolytic recovery or chemical treatment and filtration (metal contaminants and solids that might interfere with the effectiveness of the bath should be removed); reducing **dragin** (liquids adhering to the workplace from the last bath can contaminate subsequent baths, so rinsing between baths should be done to reduce dragin); using distilled, deionized, or reverse osmosis water (contaminants in the water supply can concentrate in the rinse water, while dragout is usually recovered from the rinse water and can affect bath chemistry, so as a result, bath and rinse water may need to be treated to prevent plating problems);[25] properly maintaining racks; and using purer anodes and bags.
5. Reduce dragout loss. Dragout volume should be reduced to prevent transfer of bath chemicals to rinse water.

Recycling and Resources Recovery Processes

1. Reuse the rinse water effluent. Waste streams should be analyzed to determine if rinse water can be reused in the rinsing process. Figure 12–4 shows a typical multiple-use rinse system. If the original rinse systems for the alkaline and acid

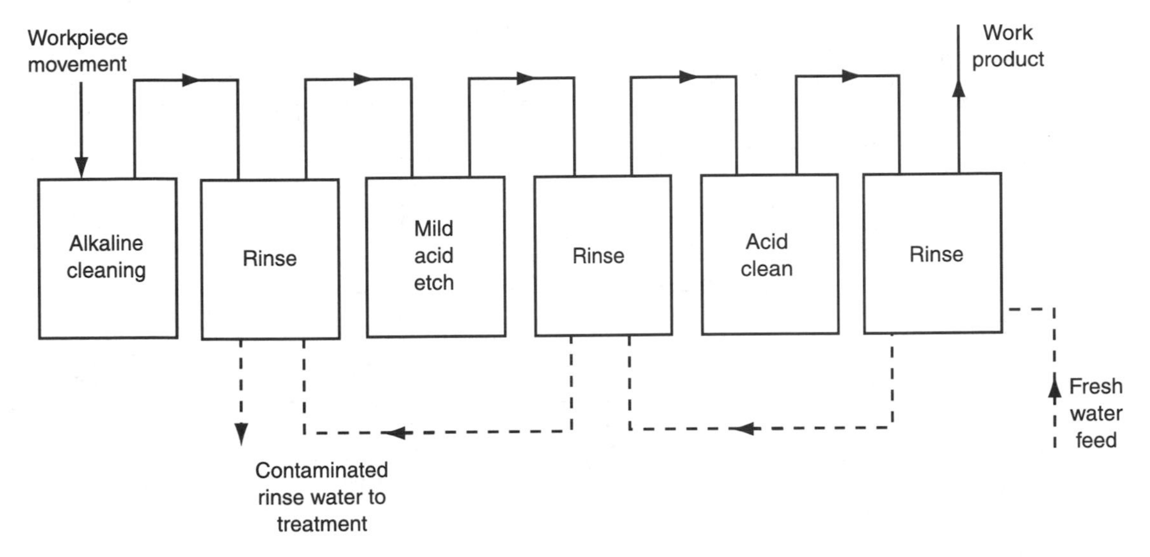

FIGURE 12–4

Source: *EPA Guides to Pollution Prevention: The Metal Finishing Industry.* EPA/625/R-92/011. October 1992. P. 5.

rinses used the same volume of water, 50 percent less rinse water would be used in the new system.

2. Implementing material reuse techniques.
3. Regenerating spent-process bath solutions. Baths that have been reduced in effectiveness by use often can be replenished with new chemicals and water. This procedure can be used until contaminants accumulate, making it costly to reuse the bath. Automated monitoring systems are available to manage bath chemistry.
4. Recycling process bath chemicals and rinse water solutions through use of chemical recovery technologies including:
   a. **Recycling rinse water**—Rinse water can be recycled in systems that use clean water for the last rinse (open loop) or use rinse water for the last rinse (closed loop)[26] (see Figure 12–5). The closed loop system is more efficient since the treated effluent is returned to the waste system.
   b. **Evaporation**—This is one of the simplest and cheapest recovery techniques. Rinse water is evaporated, thereby concentrating the dragout chemicals, which are then returned to the process bath.
   c. **Reverse osmosis**—Rinse water is concentrated by using this technique. The rinse water is pressurized and forced through a semipermeable membrane,[27] which will not allow larger molecular weight substances to pass through. The application of reverse osmosis is limited by the ability of the membrane to withstand pH extremes and sustained pressure.
   d. **Ion exchange**—This technique uses ion exchange resins to extract cations and anions. Cationic resins bind positively charged ions (nickel, copper, sodium, cadmium) and release hydrogen ions in exchange. Anionic resins bind negatively charged ions (chromate, sulfate, chloride) and release hydroxyl ions in exchange. The metals are removed from the resin with acid or alkali solutions.[28]

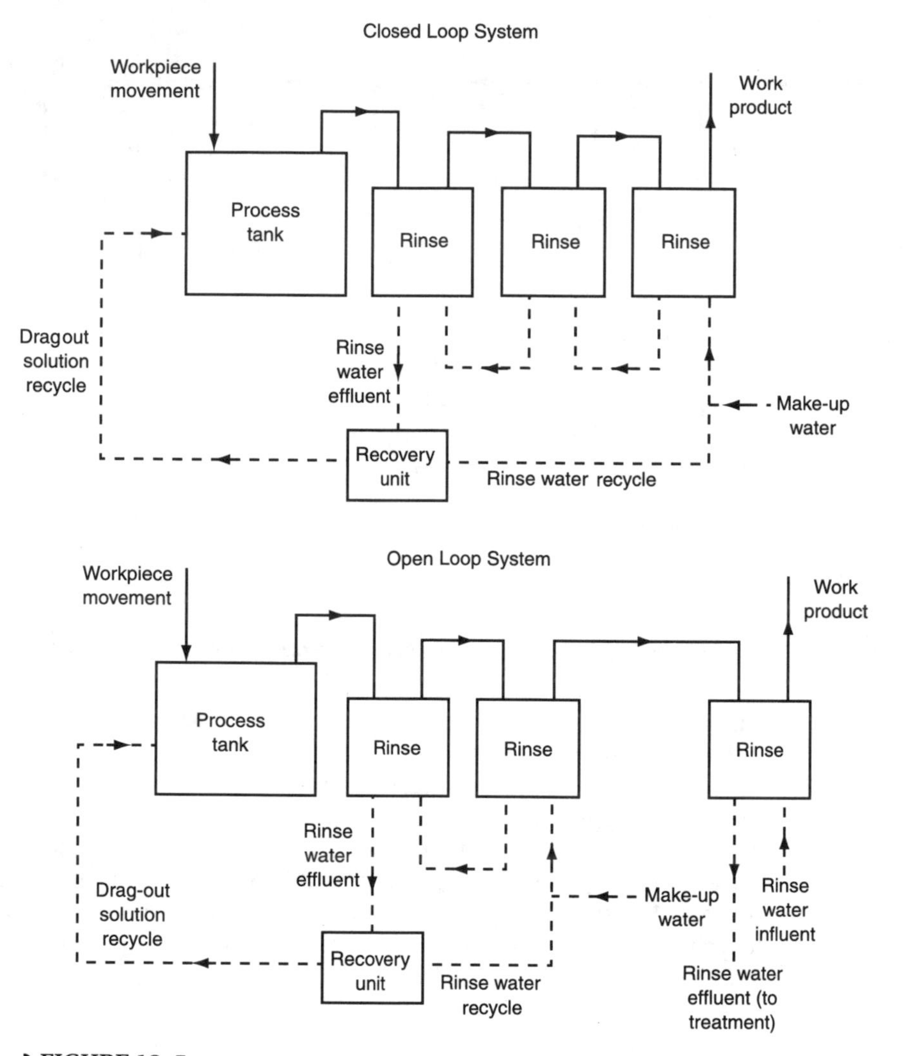

FIGURE 12–5

Source: *EPA Guides to Pollution Prevention: The Metal Finishing Industry.* EPA/625/R-92/011. October 1992. P. 5.

e. **Electrolysis/electrowinning**—Electrowinning is really an electroplating process. Electrodes are placed into the waste stream, direct current is applied to them, and the metals deposit on the cathode. When the cathode is saturated with metal, it can be stripped to recover the metal. Electrowinning can be used in combination with ion exchange to increase the recovery efficiency (see Figure 12–6).
f. **Electrodialysis**—The rinse water is passed through a series of alternately spaced cation and anion permeable membranes. An electrical current is applied across the membranes. The electrical potential across the membranes cause ions to move through the membranes. The ions are concentrated and can be removed and returned to plating baths (see Figure 12–7).

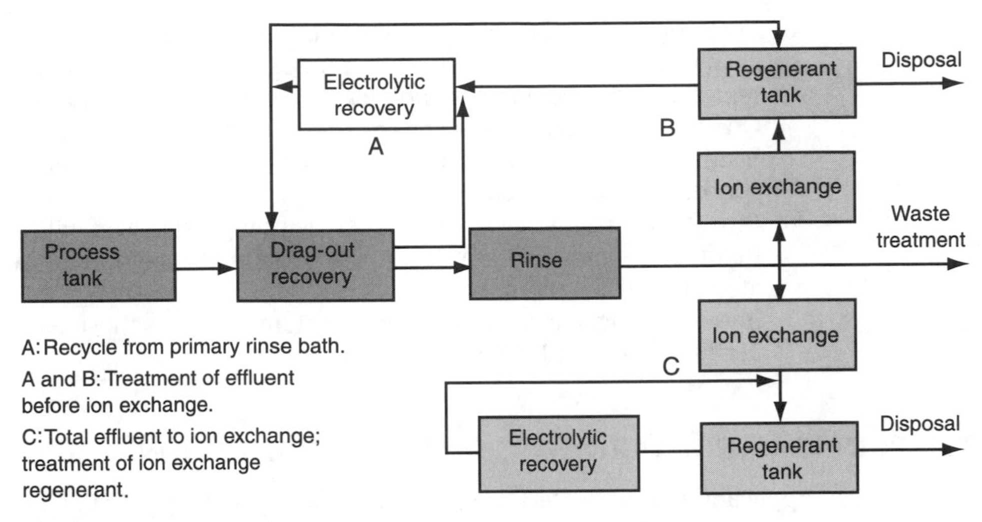

FIGURE 12–6

Source: Schulte, S. 1994. "Recovery/Recycling Methods for Platers." *Products Finishing*: 59 (2A). P. 296.

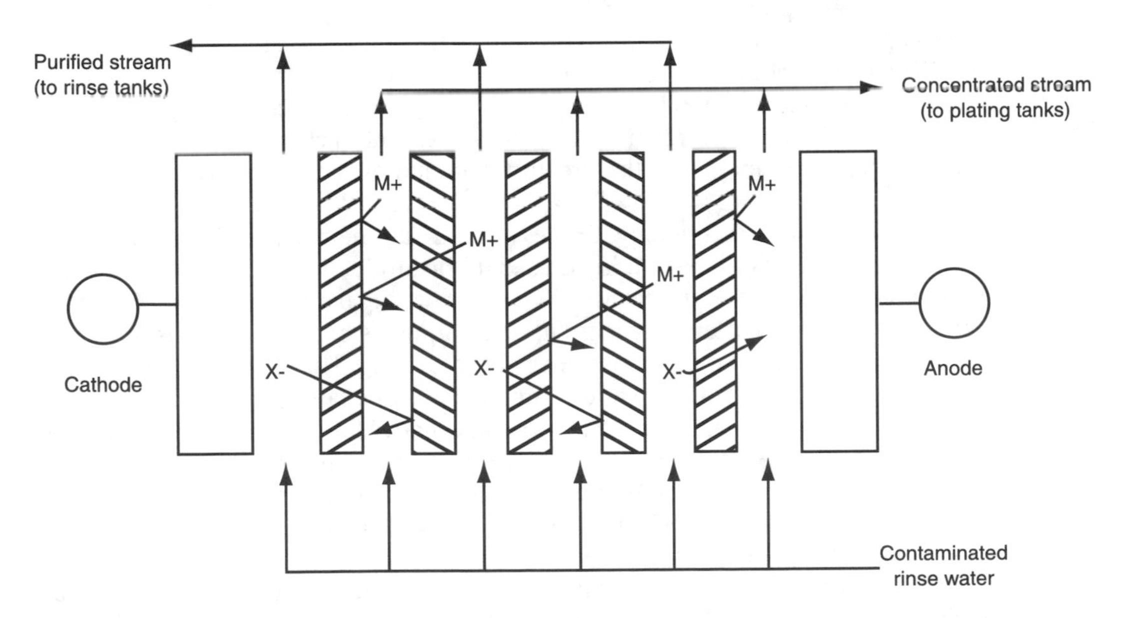

FIGURE 12–7

Source: Schulte, S. 1994. "Recovery/Recycling Methods for Platers." *Products Finishing*: 59 (2A). P. 296.

5. Recycling spent solvents by **distillation.** Many companies have already adopted cleaner technologies for cleaning and degreasing processes. Some companies rent degreasing equipment and pick up the used solvents for off-site recycling. On-site reclamation of the solvent from other liquid contaminants is done by heating the solution above the solvent's boiling point. The solvent can then be condensed and recycled.
6. Separating various waste streams for recycling, selective treatment, and batch treatment. Separating the waste streams can reduce the volume of sludge generated during treatment.
7. Implementing alternative treatment systems (such as those used for recycling, discussed previously).

EMERGING TECHNOLOGIES FOR POLLUTION PREVENTION

Application of cleaner technologies in the plating industry have the potential of eliminating or significantly reducing the generation of hazardous materials. There are a few areas, however, where new, emerging technologies are needed to replace current ones that produce toxic pollutants. A review of some of these technologies follows:

Catalytic Wet Oxidation Cleaning is an alternative to the use of chlorinated solvents for cleaning and degreasing. In this technology, oxygen-rich air is pumped into the cleaning solution to break down organics adhering to the substrate.[29]

Zinc-alloy electroplating—alloys of zinc can replace cadmium plating in many applications. Cadmium is a heavy metal and is highly toxic. Cadmium plating baths usually contain cyanide, which is also very toxic. Zinc-nickel and zinc-cobalt non-cyanide processes may have the greatest potential as an alternative to cadmium plating.[30]

The **Blackhole Technology Process** is an alternative to electroless copper plating.[31] Electroless copper plating is used in the production of printed wire boards. Formaldehyde is used in the electroless copper plating baths and is suspected to be carcinogenic. The Blackhole process uses an aqueous carbon black suspension to prepare printed wire boards for copper electroplating. This technology eliminates the formaldehyde waste stream and related environmental/health risks.

Physical Vapor Deposition (PVD) is an alternative for replacing chromium electroplating.[32] The toxicity and carcinogenic effects of hexavalent chromium are well known. PVD uses thermal energy and ion bombardment to convert the source material into a vapor that is deposited on the workplace. Titanium nitride is a possible replacement coating for chromium using PVD.

Two other processes that have been developed to replace chromium plating are nickel-tungsten-boron alloy plating and nickel-tungsten-silicon carbide plating.[33, 34]

MANAGING ELECTROPLATING WASTE STREAMS FOR THE FUTURE

Most efforts to minimize electroplating/metal-finishing waste streams are focused on source reduction technologies, including process changes (alteration of existing technologies to increase recovery efficiency), process substitutions (using cleaner technologies, for instance, substituting noncyanide plating in place of the cyanide-based

TABLE 12–13
Aspects of Industry-Environment Interactions.

Activity	Time Focus
Remediation	Past
Treatment, storage, disposal	Present
Industrial ecology	Future

Source: Graedel, T., and B. Allenby. 1995. *Industrial Ecology.* Englewood Cliffs, NJ: Prentice Hall. P. 10.

plating for the same metal), and alternate finishes (replacement of traditional coatings with new, more environmentally friendly technologies); and recycling and resource recovery (better technologies for reusing and treating waste streams). These methods of waste minimization coincide with the three stages that industry can interact with the environment[35] (see Table 12–13). Industry must deal with past mistakes in the form of cleaning up hazardous sites (remediation). The current stage industries are in is waste minimization, reducing or eliminating use of toxic substances, and regulating end-of-pipe emissions. The theory of industrial ecology suggests that the electroplating industry should progress to the stage of industrial ecology. Industrial ecology requires products be designed for the environment as well as for other business considerations. This means products and industrial processes would be designed not to pollute, therefore, cleanup is not an afterthought. Industrial ecology produces waste minimization in the design process and then applies the technologies of waste minimization to the resulting waste streams. Adoption of industrial ecology in the metal-finishing industry could be a challenging task. Some of the questions that might be asked by an industrial ecologist are:

1. Do we really need to electroplate this product?
2. Is another nonmetal-coating process more environmentally friendly?
3. If the electroplating is for decorative purposes, is the plating absolutely necessary considering the potential environmental hazards?
4. Should metal plating be eliminated completely in our product and replaced by new technologies (organics, composites) because mining of metals and related industrial processes are not environmentally sustainable?
5. Should metal-plating processes be reduced or eliminated because consumers view them as environmentally harmful?

Some of these questions are already being asked today. Perhaps in the future more time will be spent designing products to eliminate waste streams and less time managing our wastes.

SUMMARY

- The electroplating process generates wastewater, solid wastes, and air emissions.
- The electroplating industry is one of the most regulated industries in the United States.

- The EPA and OSHA are the primary federal agencies regulating electroplating emissions.
- Electroplating uses electrical current to deposit a coating of metal on a metallic surface.
- Electroless plating uses chemical reducing agents instead of electrical current to produce a metallic coating.
- Identification and characterization of waste streams are the first steps in the waste-minimization process.
- The two methods of waste minimization currently used are source reduction and recycling and resource recovery.
- The electroplating industry should implement industrial ecology to further minimize waste streams.

QUESTIONS FOR REVIEW

1. What industries may use the electroplating process?
2. Outline the government regulations affecting the electroplating industry.
3. Define **solvent** and **cleaning solution.**
4. Why should new cleaning technologies be applied in the electroplating industry?
5. What are the two categories of "clean technologies" used in cleaning/degreasing operations?
6. List the five source-reduction processes used in electroplating waste control.
7. List the recycling and resource recovery (treatment) processes used in electroplating.
8. What is "dragout" and why is this important in electroplating waste streams?
9. What chlorofluorocarbons are used by the electroplating industry?
10. How can industrial ecology minimize electroplating waste streams?
11. Why is the EPA concerned about chromium electroplating processes?
12. What are the MACT standards and how do they apply to the electroplating industry?

ACTIVITIES

1. Contact your state government's environmental regulatory agency and obtain a list of electroplating industries in your state. Write the industries and get information on the type of electroplating process they use and how they achieve regulatory compliance.
2. Plan a field trip to a local industry that uses electroplating. Obtain permission to analyze their waste stream using the worksheets in Appendix G.
3. Obtain an issue of a trade publication specializing in the electroplating industry. Write a report on the content of the publication. What articles focused on environmental compliance? What techniques have you learned about in this text that were discussed in the article(s)?
4. What possessions do you own that required electroplating during their production? (Remember, most electronic devices contain electroplated circuitboards.) Make a list of goods in your home that were electroplated. What kind of electroplating is used in these goods? Is any of this electroplating unnecessary?
5. Do a search on the Internet (use the Internet resources listed on page 264) for information on electroplating and electroplating regulations. Pay particular attention to changes in the current regulatory environment and emerging technologies for pollution prevention. Write a report summarizing your findings.

NOTES

1. Bleeks, T. 1994. "Metal Stripping." *Products Finishing:* 59 (2-A). P. 217.
2. Environmental Protection Agency. 1994. *Guide to Cleaner Technologies: Alternative Metal Finishes.* Washington, D.C.: EPA.
3. Environmental Protection Agency. 1985. *Environmental Regulations and Technology: The Electroplating Industry.* Washington, D.C.: EPA.
4. Environmental Protection Agency. 1992. *Guides to Pollution Prevention: The Metal Finishing Industry.* Washington, D.C.: EPA.
5. Ostler, N., and J.T. Nielsen, Eds., 1996. *Prentice Hall's Environmental Technology Series, Environmental Regulations Overview, Vol. 2.* Englewood Cliffs, NJ: Prentice Hall. P. 70, 81.
6. Environmental Protection Agency. 1994. *Guide to Cleaner Technologies: Cleaning and Degreasing Process Changes.* Washington, D.C.: EPA.
7. Miller, C. 1994. "Preparing metals for plating." *Products Finishing: 59(2A).* Pp. 116–117.
8. Environmental Protection Agency. *Guide to Cleaner Technologies: Alternatives to Chlorinated Solvents for Cleaning and Degreasing.* Washington, D.C.: EPA.
9. Ibid.
10. Environmental Protection Agency. 1990. *Guides to Pollution Prevention: The Fabricated Metal Products Industry.* Washington, D.C.: EPA.
11. Environmental Protection Agency. 1992. *Guides to Pollution Prevention: The Metal Finishing Industry.* Washington, D.C.: EPA.
12. Bayes, M. 1994. "Electroless Nickel Plating." *Products Finishing*: 59(2A). Pp. 118–128.
13. Gallager, L., 1995. "Clean Water Act." In T.F.P. Sullivan, Ed., *Environmental Law Handbook, 13th Ed.* Rockville, MD: Government Institutes, Inc., pp. 135–168.
14. Mabbett, A. "1994. EPA/OSHA Government Regulations Summary for the Finishing Industry." *Products Finishing*: 59(2A). Pp. 273–287.
15. Kushner, J., and A. Kushner. 1994. *Water and Waste Control for the Plating Shop.* Cincinnati, OH: Gardner Publications Inc., pp. 262–272.
16. Ibid.
17. Gallager, L. 1995. "Clean Water Act." In T.F.P. Sullivan, Ed., *Environmental Law Handbook, 13th Ed.* Rockville, MD: Government Institutes, Inc., pp. 135–168.
18. Kushner, J., and A. Kushner. 1994. *Water and Waste Control for the Plating Shop.* Cincinnati, OH: Gardner Publications Inc., pp. 262–272.
19. Ostler, N., and J.T. Nielsen, Eds. *Prentice Hall's Environmental Technology Series, Environmental Regulations Overview, Volume 2.* Englewood Cliffs, NJ: Prentice Hall. P. 106.
20. Lee, R. 1995. "Clean Water Act." In T.F.P. Sullivan, Ed., *Environmental Law Handbook, 13th Ed.* Rockville, MD: Government Institutes, Inc., pp. 225–277.
21. Mabbett, A. 1994. *Products Finishing*: 59(2A). pp. 273–287.
22. Mabbett, A., 1994. "EPA/OSHA Government Regulations Summary for the Finishing Industry." *Products Finishing*: 59(2A). Pp. 273–287.
23. Environmental Protection Agency. *Guide to Cleaner Technologies: Alternatives to Chlorinated Solvents for Cleaning and Degreasing.* Washington, D.C.: EPA.
24. Environmental Protection Agency. 1992. *Guides to Pollution Prevention: The Metal Finishing Industry.* Washington, D.C.: EPA.
25. Schulte, S. 1994. "Recovery/Recycling Methods for Platers." *Products Finishing*: 59 (2A). Pp. 288–300.
26. Environmental Protection Agency. 1992. *Guides to Pollution Prevention: The Metal Finishing Industry.* Washington, D.C.: EPA.
27. Schulte, S. 1994. "Recovery/Recycling Methods for Platers." *Products Finishing*: 59 (2A). Pp. 288–300.
28. Environmental Protection Agency. 1981. "Summary Report: Control and Treatment Technology for the Metal Finishing Industry—Ion Exchange." Washington, D.C.: EPA.
29. Environmental Protection Agency. *Guide to Cleaner Technologies: Alternatives to Chlorinated Solvents for Cleaning and Degreasing.* Washington, D.C.: EPA.
30. Environmental Protection Agency, 1994. *Guide*

to Cleaner Technologies: Alternative Metal Finishes. Washington, D.C.: EPA.
31. Ibid.
32. Ibid.
33. Ibid.
34. Ibid.
35. Graedel, T., and B. Allenby. 1995. *Industrial Ecology.* Englewood Cliffs, NJ: Prentice Hall.

REFERENCES

Durney, L.J. 1984. *Electroplating Engineering Handbook.* New York, NY: Van Nostrand Reinhold Co., Inc.

Graham, K.A. 1971. *Electroplating Engineering Handbook.* New York, NY: Van Nostrand Reinhold Co., Inc.

Hartinger, L. 1994. *Handbook of Effluent Treatment and Recycling for the Metal Finishing Industry.* Materials Park, OH: ASM International.

PRC Environmental Management, Inc. 1989. *Hazardous Waste Reduction in the Metal Finishing Industry.* Parkridge, NJ: Noyes Data Corporation.

PERIODICALS

Products Finishing is published monthly by Gardner Publications Inc., 6600 Clough Pike, Cincinnati, OH 45244-4090. Telephone: (513)231-8020; (800)950-8020; Fax: (513)231-2818.

Metal Finishing is published monthly, (semimonthly in January and May) by Elsevier Science Inc., 655 Avenue of the Americas, New York, NY 10010. Telephone: (212)633-3950; Fax: (212)633-3990

PROFESSIONAL AND TRADE ORGANIZATIONS

The American Electroplaters and Surface Finishers Society, Inc. AESF Headquarters, 12566 Research Parkway, Orlando, FL 32826-3298. Telephone: (407)281-6441; Fax: (407)281-6446. E-mail: aesf@worldnet.att.net

The Electroplaters Society, Inc., 10 South Main St., Pennington, NJ 08534-2896. Telephone: (609)737-1902; Fax: (609)737-2743. E-mail: ecs@electrochem.org

The Institute of Metal Finishing, Exeter House, 48 Holloway Head, Birmingham, B1 1NQ England. Telephone: 0121 622-7387; Fax: 0121 666-6316. E-mail: <ukfinishing@dial.pipex.com>

INTERNET ADDRESSES

Home page of Envirosense: http://es.inel.gov/index.html

Envirosense Metal Finishing Content Guide: http://www.seattle.battelle.org/es guide/metals/ metal.html (This is a very highly recommended site to visit for up-to-date information on the metal-plating industry and regulations.)

Home page of the EPA: http://www.epa.gov

Home page of the EPA Office of Air and Radiation: http://epa.gov/oar

Home page of the EPA Office of Enforcement and Compliance Assurance: http://es.inel.gov/oeca

Rules and Regulations page on the EPA: http://www.epa.gov/epahome/rules.html

Home page of the finishing industry: http://www.finishing.com

Home page of the Institute of Metal Finishing: http://www.ukfinishing.org.uk/IMF/index.html

Home page of Metal Finishing: http://www.metalfinishing.com/

Home page of the National Center for Manufacturing Sciences: http://icon.ncms.org/

Home page of the National Metal Finishing Resource Center: http://www.nmfrc.org

13

Hospital Waste Stream Generation

Erwin Gutzwiller

Upon completion of this chapter, you will be able to meet the following objectives:

- Gain an understanding of the complexities associated with hospital hazardous waste.
- Learn who the principal regulators of hospital hazardous waste are and the relationship of their role in the regulatory process.
- Learn some of the sources of hazardous materials/waste production in the hospital setting and develop an understanding of the intricate nature of the hospital hazardous waste stream.
- Understand the extent of the problem of hazardous materials/waste and their potential effect/threat on the employees, patients, visitors, and the community.
- Learn the requirement for risk assessment/management and health and safety training for hospital employees.

INTRODUCTION

Under the clean and sanitary cover of the hospital setting beats the heart of a serious hazardous waster generator. Visible cleanliness, however, does not assure microbiological cleanliness. A hospital has many potential biological hazards that can pose a serious threat to employees, patients, and visitors. Hospitals are often overlooked as hazardous waste generators because they are not usually included in the list of "typical industrial waste generators." However, all three classes of waste can be found in the hospital, including:

- Chemical waste
- Biological waste
- Radiological waste

Some hazardous wastes are generated by the various hospital processes, while other hazardous material/waste is brought into the hospital setting by the patient. This situation creates a multitude of problems for the hospital staff and requires considerable planning to prepare for.

Like the industrial sector, the hospital community is subject to the usual regulations prescribed by three federal agencies primarily responsible for the regulation of hazardous waste:

- Occupational Safety and Health Administration (OSHA)
- EPA
- Department of Transportation (DOT)

Hospitals are also subject to the Joint Commission on Accreditation of Healthcare Organizations (JCAHO). This rulemaker addresses health, safety, and environmental questions including the generation and disposal of hazardous/toxic waste and has a mission to "improve the quality of care provided to the public."[1] This organization inspects hospitals and grants accreditation to a hospital if it meets the standards prescribed in the *Accreditation Manual for Hospitals,* usually published annually. A hospital's accreditation in not automatically renewed after its initial accreditation, but must undergo a full accreditation survey every three years. The joint commission may also perform an unscheduled or unannounced survey if it "becomes aware of circumstances in an accredited organization that suggest a potentially serious standards compliance problem."[2] JCAHO does not change the regulations of OSHA, the EPA, or DOT in the development of its standards, but adds to or more specifically defines these regulatory agencies' requirements.

Other federal agencies as well as state and local authorities have some input in the regulatory process. These agencies range from the Centers for Disease Control and Prevention (CDC) at the federal level to the municipal sewage and solid waste department in Any Town USA. Like the joint commission, state and local authorities can only further define and strengthen federal regulation; they are generally unable to reduce the requirements set forth by federal agencies or any agency of higher authority.

An example of this is found in OSHA 29 CFR 1910, which "recommends" that a safety officer be designated for a business. The regulation makes no requirement that this be a full-time position or simply an additional duty given to some member of the organization. JCAHO, on the other hand, requires "a safety officer appointed by the chief executive officer or a designer."[3] It also states that other safety activities be conducted on a regulated basis in order to obtain or maintain commission accreditation.

There are many standards addressed in the regulations that may or may not have the force of law. Standards are usually established by professional organizations, such as the American National Standards Institute (ANSI) or the National Fire Protection Agency (NFPA), to address areas of interest that affect the membership of their organizations. At this point, standards are only recommended actions or specifications to help those professionals within the organization. Some standards are picked up by regulatory agencies which "incorporate by reference" a standard into regulation, making it a part of the regulation. It is then necessary for those who are subject to the regulation to learn about the standard and how it fits into the regulatory process they are trying to implement. Many of these situations exist in the field of hazardous materials management and the healthcare setting. It is usually necessary to contact the organization that developed the standard to obtain a copy of the standard to coordinate it with the balance of the regulation.

Another reason that hospitals must be concerned about hazardous materials/ waste management is the obligation of the hospital to protect its patients, visitors, and the environment as well as its employees. Because the public is more aware of environmental issues, there is a greater need to ensure compliance and protect the facility's public reputation. Also, hospitals inspected by OSHA and other regulatory agencies may be subject to substantial fines for inadequate management and handling of chemical, biological, and radiological products and their associated waste.

GENERAL PROCESS DESCRIPTION

Hospitals generate hazardous waste in many different ways. The type and quantity of hazardous waste depends on the hospital. Facilities range in size from small, rural with less then 100 bed primary care sites, to large tertiary care centers associated with university medical schools, private research or government-operated Veterans Hospitals, each having hundreds of patient beds as well as extensive research and laboratory facilities. The larger the facility, the more hazardous waste generated.

Along with an increase in the quantity of waste produced, the type and character of waste can also change. In addition, the following should be considered:

- Hospital laboratories and pharmacies may produce relatively small amounts of hazardous waste individually, but generate a considerable quantity over time collectively.
- Laboratory facilities operating in a hospital setting or a university medical center may produce a number of small quantities of similar chemical wastes. These can be collected in bulk storage facilities until the quantity is sufficient and can be dealt with in a safe and cost-effective manner.
- Hospitals that perform a significant amount of research, for example, can generate many different kinds of unique hazardous wastes that require special handling for both storage on site as well as its final disposal.

Another hazardous waste product of laboratory research is contaminated laboratory animals that range in size from small mice to larger rats or primates. The type of contamination can also present a disposal problem when products such as mercury (Hg) or a mixture of chemical, biological, and radiological agents contaminate the animal carcass. Many chemical hazardous waste sites are not equipped or licensed for contaminated lab animals and are unable to accept these materials for disposal. Use of specialized waste sites adds to the cost through shipping and special handling charges. Many of these items also require special handling and storage at the generator site while awaiting shipping. Radiological materials may be held during the process of half-life reduction until they reach an acceptable level for disposal.

The hazardous waste technician is involved in a range of duties from the management, control, and disposal of hazardous waste to training new employees in these steps. The hazmat technician is usually a member of the in-house response team to aid in the control and cleanup of hazardous spills. Identifying old chemicals found in laboratory drawers, such as Pickerick Acid or shock-sensitive peroxides, can present a real challenge in identification and special handling for disposal.

Hospital hazardous material/waste generation may come from outside sources as a result of community disaster plans. Title III of the SuperFund Amendments and Reconciliation Act (SARA Title III), passed in 1986, also known as the Emergency Planning and Community Right-to-Know Act, requires federal, state, and local governments to provide planning for chemical emergencies. Hazardous materials spills or industrial releases of hazardous chemicals external to the hospital can generate patients who arrive at the hospital covered with chemicals and chemical-soaked clothing. This poses a threat to the patients, the hospital emergency department staff, and transport personnel. If not properly contained, this contamination could threaten other patients and visitors to the facility. Even if containment is confined to the emergency department of the hospital, the emergency room may need to be closed until it can be decontaminated and returned to service. Some community emergency plans require the decontamination of the patient at the scene of the spill, but this has not eliminated the arrival of contaminated patients at the hospital. Therefore, each receiving hospital should assume the patient to be contaminated until it can be determined otherwise.

Decontamination of these patients produces a hazardous waste residue that must be contained, characterized, and ultimately disposed of in accordance with appropriate regulation. Additionally, decontamination of hospital equipment and disposal of protective clothing worn by the hospital staff must also be accomplished. Some of the chemically contaminated protective equipment may be contaminated with biological residue produced during the treatment process. The proper disposal of equipment with multiple types of contaminants can be difficult. Decontamination of such equipment, especially disposable equipment, is out of the question.

Hazmat planning, training, and exercises are mandated by the JCAHO and other regulators. In recent reviews of environmental compliance problems at hospital and medical facilities, a study of the root causes of these issues found that lack of understanding or poor execution of the correct technique were responsible for over 55 percent of the cases. JCAHO mandates two exercises annually, a full-time safety officer, and a safety committee with a requirement to meet bimonthly.

All of the regulators mentioned here have some type of training requirement included in their regulatory requirements. Training should have as a minimum two goals: First, it should emphasize prevention through the proper handling and disposal of hazardous substances. It should be directed toward new employees as a part of their inprocessing, and periodically presented in shortened form to the staff as refresher training. Second, training should prepare the hospital staff to recognize a spill, avoid self-contamination and prevent others from being harmed. If designated, properly trained staff respond to the containment and cleanup of leaks or spills.

The hazardous materials technician could play several different roles in this process. As a responder to the hospital hazardous waste problems, the technician is the trained professional. Cumulative experience can be used to develop and deliver training and to help insure proper execution of the safety policies and procedures in the hospital. Additionally, day-to-day activities within the various departments of the hospital provide the opportunity to observe activities and identify areas or tasks that require more attention. Changes to plans or procedures can result in improved safety.

Hazardous materials markings are also important in the safe management of hazardous materials. Employees who work in central receiving/shipping or any employee who receives packages from outside sources should be given special training in the DOT system of hazardous materials markings. Topics covered should include placards, labels, packaging, and proper methods to handle packages or materials that might show some evidence of a leak or other loss of material.

So far, we have only discussed the problem of hazardous waste in terms of chemicals that exist in the hospital. Biological or infectious waste and radioactive waste products are also common in the hospital setting. It is almost impossible to turn on the radio or television today and not hear something about AIDS, Human Immune Virus (HIV), hepatitis, or other communicable diseases. The hospital stands as the center for the care and treatment of these and other diseases, and must have a program to manage the biological waste generated. The biological/medical waste stream is, for the most part, considered a subset of the solid waste stream and is treated in that manner by many states. There is considerable variation among states on specific regulations, so it is incumbent on the hazardous waste technician to become conversant with local requirements. In general, the medical waste stream is divided as depicted in Table 13–1.

Employee safety is the first and most important reason to develop a good biohazard program to eliminate injuries or illness of the staff. Every year, over 200 health care professionals die of hepatitis they contracted in the health care setting.[3] The CDC reported that a staff member who receives a blood splash or sharp's injury involving

TABLE 13–1
Medical waste stream.

MEDICAL WASTE		
General Waste	**Sharps and <20ml Blood and Body Fluids**	
No special requirements May be handled as general solid waste	Packaging requirement only No treatment required	
REGULATED MEDICAL WASTE		
Pathological Waste	**Microbiological Waste**	**Blood and Body Fluids**
Incineration	Incineration, stream sterilization or chemical treatment	Individual containers in volumes greater than 20ml. Incineration or sanitary sewage

▶ **FIGURE 13–1**
Biohazard symbol.

blood from an HIV or hepatitis patient does face the potential for developing AIDS or hepatitis.[4] On December 6, 1991, OSHA published 29 CFR 1910.1030, its regulation pertaining to blood-borne pathogens. The standard requires that each employer develop a written Exposure Control Plan designed to minimize or eliminate exposures to employees.

The Exposure Control Plan includes a section on the proper handling and processing of biological materials by the hospital staff for later disposal. Specially marked containers are placed in those areas that generate or have the potential to generate biological/biomedical waste. The symbol and label show in Figure 13–1 is printed or placed on the container to identify the proper repository for the these materials.

Use of the biohazard symbol and label varies depending on what stage of handling the biological materials are in. OSHA and DOT provide specific instructions concerning the use and placement of biohazard identification information.

Another reason for concern about how potentially infectious waste is managed is to avoid fines from violating state and federal regulations. Regulatory fines and negative publicity go hand-in-hand. Though it is rare for the size of these fines to seriously jeopardize the financial stability of an institution, the negative publicity can produce more significant financial complications. These range from devaluation of its reputation in a community to the loss of financial support from local contributors. The results of this situation can be far-reaching and more costly than regulatory fines.

Solid waste, technically, is any waste that is not discharged into the air, so the term also includes both liquid and semisolid materials. The American Hospital Association estimates an average hospital produces approximately 20 pounds of solid waste per patient per day. Of that, approximately 10 to 12 percent is considered potentially infectious. For a fully occupied 100-bed hospital, the yield is on the order of 200 pounds of infectious waste per day.

Let us begin by trying to define infectious waste. The terms "infectious waste," "medical waste," and "biohazardous waste" are often used interchangeably. There is some controversy among regulators concerning what items should be included in this category. Table 13–2 provides a look at the various items and which agencies include them in their definition.

TABLE 13–2
Regulatory definitions of infectious waste.

Regulation	Definition
OSHA: 29 CFR 1910.145	Biological hazards, or BIOHAZARD, means those infectious agents presenting risk of death, injury, or illness to employees
29 CFR 1910.120	Any biological agent or other disease-causing agent which after release into the environment and open exposure, ingestion, inhalation, or assimilation into any person, either directly or from the environment or indirectly by ingestion through the food chain, will reasonably be anticipated to cause death, disease, behavioral abnomalities, cancer, genetic mutation, physiological malfuncitions, or physical deformations in such person or their offspring
EPA: 40 CFR 257	Medical waste is any solid waste that is generated in the diagnosis, treatment, or immunization of human beings or animals in research, biological production, or testing; it is divided into seven classes
DOT: 49 CFR 173.4	Class 7—Radioactive Material
49 CFR 173.13	Class 6—An infectious substance means a viable microorganism, or its toxin, that causes or may cause diseases in humans and animals and includes those agents listed in 42 CFR 72.3—Department of Health and Human Services

It is obvious from Table 13–2, that regulators do not agree on one clear definition. Any definition used by a hospital or health care facility must comply with state and federal regulation and must clearly include the OSHA bloodborne pathogens definition. OSHA can and will fine a hospital for deviating from this definition. One of the best sources to help clear up any concerns about the definition of infectious waste is the Environmental Commission in the hospital's state of residence.

Unlike the OSHA requirement to protect employees from the potential effects of bloodborne pathogens, medical/solid waste disposal regulations are usually developed and enforced by the individual state and local governments. The variation from state to state or county to county on dealing with bloodborne pathogens makes it necessary to become familiar with the rules and regulations developed by state and local agencies in this area. As a rule, federal regulations define the packaging, labeling, storage, transportation, and treatment, while local regulations will define disposal of medical waste.

There are some generalizations that can be made about medical waste, assuming that it is free of any other confounder such as chemical or radiological contamination. Much of the infectious solid waste that is generated in the hospital setting can be disposed of as normal waste at the local landfill once it has been properly treated.

The treatment methods accepted by each state may vary, but the following are some of the more universally accepted processes:

- Incineration
- Steam sterilization
- Chemical treatment (sterilization)
- Microwave sterilization

Larger hospitals usually opt to dispose of some of their waste through the process of on-site incineration. This usually requires more staff to accomplish the disposal process but is generally cheaper and eliminates the need to develop large safe-and-secure waste holding areas. As a general rule, the status of animal waste from laboratory facilities is usually treated in the same manner as human waste for purposes of characterization, treatment, and disposal. This can change depending on the characterization report of its content.

It is necessary to clearly define infectious waste to eliminate any confusion by the staff responsible for separating it from regular waste on the patient units and the clinical laboratories. This definition must be incorporated into all employee training programs. The number of materials included by definition will have an impact on the volume of infectious waste a hospital will generate. The cost to manage and dispose of it is based on volume. There is no reason why a hospital should pay a premium price for waste that is really not potentially infectious. It will not create a higher level of safety, but will increase the facility's operating cost. Possible infectious waste items include the following:

1. Biological waste and discarded materials contaminated with blood
2. Excretion or secretion from humans or animals that were isolated to protect others from communicable diseases.
3. Discarded medical equipment and parts that were in contact with blood or body fluids
4. Dialysis wastes that were in contact with blood
5. Laboratory wastes from medical, pharmaceutical or other research, commercial or industrial laboratories that were in contact with infectious agents
6. Surgery or autopsy waste that came in contact with infectious agents
7. Contaminated animal carcasses, body parts, and animal bedding exposed to infectious agents
8. Cultures and stocks of infectious agents and associated biologicals
9. Used sharps

There is another consideration that can be used to establish waste as infectious, and that is designating all waste generated by certain departments as infectious waste. Areas such as the operating room, OB-GYN clinic, emergency department, cardiac care, dialysis, and transfusion services can be designated as infectious because of the quantity of blood-soaked items associated with these areas. The "area specific" approach simplifies decisionmaking by the staff, but usually carries with it an increased cost.

LABORATORY WASTE

Clinical or research laboratories can offer the most challenging waste streams generated by a hospital. This is especially true as it relates to a university medical center or other hospital research units. The typical industrial process for the most part can be determined by the equation:

$$\text{Raw material in} - \text{product} = \text{waste}$$

The laboratory process, however, creates a wide range of waste products that generally require specialized lab packing for disposal by the appropriate hazardous waste method. This differs greatly from the bulk accumulation of waste generally used by industry. The lab packing process is generally more costly per cubic foot than bulk waste. One reason for the increased cost of lab packs is the requirement to place a container of the material designated for disposal in a secondary container for shipment. The second container is necessary when the primary container is susceptible to breakage or leakage, or is not acceptable as a shipping vessel under the DOT shipping container regulations. This adds to the volume and weight of the shipped material and, ultimately, its cost.

Disposal of chemically, biologically, or radiologically contaminated laboratory animals and cage bedding is another aspect of the laboratory waste stream that must be considered. Intermediate storage conditions that provide both odor control as well as prevent further putrefaction of these products is the first concern. Next, the type or types of contaminants and the quantity can have a profound effect on how and where disposal can be accomplished. In many cases, it is necessary to establish this quantity in parts per million (ppm) and even parts per billion (ppb) and will require the assistance of the researcher to establish this information. In some cases, where the exact amount of material cannot be determined, worst case estimates must be used. This can drive up the disposal cost and limit the number of disposal agencies that can take the waste. Disposal sites have license limitations restricting the volume or level of radiation they may accept. The fewer disposal companies available means a less-competitive price for the disposal.

RADIOLOGICAL MATERIALS

Radioactive materials are used in many activities in the hospital setting. These applications vary widely and require special control and disposal methods. Such considerations as the definition of low- and high-level radiation or the quantity in millicuries or microcuries are important to the establishment of the program and the operating procedures that will be put in place. Some of these departments found in the hospital are:

Nuclear medicine
Radiology
Research laboratories (P^{32} and other tracer materials)
Pharmacy

These departments and others use radioactive materials in their day-to-day activities and require considerable controls for the material's storage, use, and disposal. It should be noted that the departments specified here may operate under many different names and over a very different range of patients, but the important issue is the use of radioactive substances.

The first area of consideration, once the proper licenses for the storage and use of radioactive materials are in place, is the proper procedure for receiving a package of radionuclides. Some radioactive materials can be shipped by FedEx, UPS, or the U.S. Postal Service. Controls must be put in place to insure the content of the package as well as determining the condition of the container that stores the radioactive substance during shipment. All packages containing radionuclides must be carefully inspected upon receipt for evidence of damage or leakage. This must be done within three hours of arriving at the licensee's facility if it is received during the licensee's normal working hours. If received after normal working hours, the process must be accomplished no later than the first three hours of the start of the next working day/shift. Rules regarding the handling of shipments vary, depending on the local circumstances. Measures must be taken to ensure that a package received at the licensee's location is always placed in a designated, secure location until processed and opened by the properly trained and appointed personnel. A record of receipts of radionuclides must be maintained.

Once a package containing radionuclides has been logged in, the package should be monitored for any external radiation. Protective plastic gloves should be worn during this period. Next, a wipe-test of the package should be performed for removable contamination. The following steps should be accomplished to monitor for leakage of both beta and gamma emitters:

1. Note radiation units stated on package, then verify and record them in receipt log.
2. Place the package in a vented hood.
3. Open the outer package, remove the packing slip, then open the inner package and verify that the contents agree in name and quantity with the packing slip.
4. Measure radiation field of the unshielded container. If necessary, place the container behind shielding to reduce exposure to allowable limits.
5. Check for possible breakage of seals or containers, loss of liquid, or change in color of surrounding material.
6. Wipe-test inner contents and document any pertinent findings on the packing slip.
7. Record any activity found on the receiving log.
8. If contamination, leakage, or shortages are observed, notify the radiation safety officer.

Packages must not be left on the floor, unsecured, or unattended when they are delivered to a facility. They must be placed in a designated secure location and the responsible parties must be promptly notified if they are not present to receive it.

Radioactive wastes can only be stored in restricted access areas. Liquid waste should be stored in shatterproof containers, such as polyethylene bottles. If circumstances make this impracticable, an outer container of shatterproof materials must be used. Flammable wastes should be kept to a minimum in the laboratory. Waste containers must be metallic and a fire extinguisher must be located in the vicinity with a sign posted giving its location.

During storage there must be no possibility of a chemical reaction that might cause an explosion or the release of chemically toxic or radioactive gases. This is usually accomplished by the following precautions:

1. Liquids must be neutralized (pH 6 to 8) prior to placement into the waste container.
2. Containers of volatile compounds must be sealed to prevent the release of airborne particles.
3. Highly reactive materials (such as metallic sodium or potassium) must be reacted to completion *before* storage.

Waste disposal of radionuclides can be relatively simple or complex (see the section on mixed waste). Small quantities of wastes may be disposed of by release to the atmosphere, inland or tidal waters, sewage, or by burial. The quantity limits are established by federal and state agencies. Short-lived radionuclides are often stored and allowed to decay until they can be disposed of as nonradioactive wastes. This method is especially useful in the case of mixed waste situations. The Nuclear Regulatory Commission (NRC) requires storage for a minimum decay time of ten half-lives before release as ordinary waste. To put this point in perspective, the following common half-life times are considered:

P^{32}, Potassium 32—14.3 days
H^{3}, Tritium—12.3 years
C^{14}, Carbon 14—5,730 years[5]

Based on these half-lives, P^{32} must be stored for 143 days, H^{3} for 123 years, and C^{14} for 57,300 years. It is easy to see that decay disposal storage techniques are only available in some cases. Longer decay periods are necessary if significant levels of radioactivity are detectable after the prescribed time.

When special waste disposal problems occur, use of a commercial company licensed to handle radioactive material often constitutes the best approach. Detailed records must be kept of the disposal of all radioactive wastes as evidence that regulations have been followed.

Laboratories and pharmacies, especially those in the larger hospital operations, can produce a considerable quantity of chemical material to be managed. Many hospitals that support numerous laboratories, large pharmacies, and research activities find it more cost effective to buy chemicals in bulk and distribute them on a request basis in smaller quantities. These facilities need to create a bulk storage area(s), that is properly marked, secured, and protected against fire, acts of nature, and ventilated to insure that leaks do not build up fumes that could explode or reach toxic levels. Next there is the process of reducing the bulk chemicals to bottle-size quantities. Labeling requirements, distribution and collection of unused or contaminated material must be planned. One assumption that is inherent in this process is that any opened end-use container is considered contaminated and not returnable to the bulk container. Other disposal or reuse methods can be used to facilitate the reduction of the ultimate waste inventory. Unused materials may be offered to other users in the hospital to reduce unnecessary waste.

HAZARDOUS WASTE STREAM IDENTIFICATION

Hazardous waste can be produced almost any place in the hospital. This "general waste" category includes the management of waste products such as batteries and fluorescent light tubes. Since a hospital is unable to just close down during a power failure and send its employees home, it must be prepared to render care even though the normal power source is unavailable. The requirement to have auxiliary power in the event of power outage means storing fuel, oil, and other petroleum products. Flashlight batteries and equipment batteries, many of which contain heavy metals, are abundant in the hospital setting. While a single battery might not be considered a problem, the number and type of batteries accumulated here may require special arrangements to accomplish their disposal.

Storage and disposal of various gases and cryogenic materials are subject to several regulatory practices. Safety considerations under OSHA should be carefully detailed in operating instructions and training manuals. Release of certain gases inside the hospital could pose a serious health threat, create a fire, or become an explosion hazard.

Last, and by no means least, is the management question of the cost and liability for empty containers on the hospital campus. These include empty gas cylinders or special shipping containers for which the hospital is paying rent. Monitoring (including a daily or weekly inventory) can save considerable dollars and reduce the potential for safety violations.

REGULATORY IMPACT

The requirements of OSHA's Hazard Communication Standard (HAZCOM) 29 CFR 1910.1200 are usually administered at the federal level unless an individual state is given that authority. In this case, it is referred to as a "state plan state." It is important to know which conditions exist, since a state plan state often modifies the program by increasing the regulation's requirements. The activities of the state are still supervised by federal OSHA, and the state can *only* make changes to the program that strengthen the regulation.

The primary concern of the HAZCOM to hospitals are the labeling, Material Safety Data Sheets (MSDS), and training. All hazardous chemical containers must be properly labeled and MSDS files must be up-to-date and available to employees. The MSDS must provide specific information about the physical and chemical characteristics of substances to which employees may be exposed and include the precautions if an accidental release occurs. This information is of particular interest to the hazmat technician and presents a serious information management problem. A computer data base, for both rapid retrieval and orderly management, should be implemented.

Keeping up with new chemicals can be a full-time job. Central receiving of all chemicals to facilitate management and control is strongly recommended. Grouping products into categories by hazard type can be especially useful in the implementation of various safety plans. Such references as the *DOT Pocket Guide* or the *NIOSH Pocket Guide* can be useful in developing the data and serve as general reference documents.

MULTIHAZARD WASTES

What are **multihazard wastes?** They are defined as any waste contaminated with more than one of the following materials: medical/pathological/infectious, hazardous chemical, and/or radioactive. Some examples of multihazard wastes are:

- Potentially pathogenic materials containing radioactive material
- Potentially pathogenic materials containing hazardous chemicals
- Hazardous chemicals containing radioactive materials
- Potentially pathogenic materials containing radioactive materials *and* hazardous chemicals

It is obvious that the best solution is to avoid generating mixed waste or combining chemical, radioactive, and infectious waste. However, this is not always possible in the hospital setting. If generation of mixed waste is not avoidable, the following steps should be considered:

1. Keep volume to a minimum.
2. Do not add water to organic wastes.
3. Do not combine strong oxidizers with organic compounds.
4. Keep liquid and solid wastes separate.
5. Separate short half-life isotopes from longer half-life isotopes.
6. Identify all constituents.

In cases where chemical and/or radioactive materials have been mixed with infectious waste, the following procedures should be followed:

1. Inactivate the infectious agent according to the guidelines developed for that procedure. These procedures should be developed *in advance* of any activity where a combination process might occur.
2. After inactivation of the infectious component, consider the waste as radioactive or chemically contaminated according to the situation. *Do not autoclave (sterilize) radioactive materials under any circumstances as this will contaminate the autoclave.*

WASTE REDUCTION/WASTE MINIMIZATION TECHNIQUES

Waste reduction and/or waste minimization is extremely important to all generators since the management and disposal of hazardous wastes can be very costly. The methods and terminologies are somewhat different, so it is necessary to separately define both chemical and radiological waste minimization.

Chemical waste minimization is generally defined as the reduction in the volume and/or the toxicity of chemical waste generated at a specific site. The following general recommendations should be used in developing a waste reduction/minimization program:

- Substitution of less toxic material in specific protocol
- Avoid surplus materials by ordering supplies carefully
- Keep chemical mixing to a minimum
- Avoid mixing chemicals with other types of waste

In addition to these guidelines, one of the best techniques to be applied to the waste stream reduction is planning. During the early stage of experimental and procedure design, review the chemicals to be used and the methods of disposal *before* beginning the project. Often, disposal needs are only considered *after* the waste has accumulated.

Radiological waste minimization is the reduction in either the volume and/or the half-life of a radioactive waste generated at a specific location. Half-life considerations are extremely important since they can vary from a time period of hours, days, or thousands of years. Some recommendations concerning the use of radioactive materials are as follows:

1. Substitute nonradioactive methods when possible.
2. Use short half-life material when possible (Those with a half-life less than thirty days are preferable).
3. Procure and use only the amount necessary.
4. Limit use of radioactive material with hazardous chemicals.
5. Minimize infectious material and radioisotope mixing.
6. Use biodegradable or nontoxic liquid scintillation fluids.
7. Use plastic scintillation vials (this minimizes radioactive ash).
8. Concentrate radioactive material by extraction or evaporation as much as possible.
9. Do not use large volumes of water or solvents for cleaning and diluting.

It is important to note that at present there is no available disposal or treatment method for certain types of wastes that contain both hazardous chemicals and radioactive material. It is therefore very important to minimize the mixing of radioactive and hazardous chemicals. Should this happen, it is necessary to hold the contaminated materials in a radioactive, chemically safe storage area until the radioactive portion has been reduced by sufficient half-lives for proper disposal. At that point, the material can be disposed of according to the appropriate chemical protocol.

SAFETY MANAGEMENT

Management can play an important role in health care hazard control management. First, an effective risk management program can provide valuable information to advance safety effectiveness. Safety management programs, including education, training, and motivation, recognize the role of human factors or human behavior in preventing accidents and reducing losses. Senior management must provide the leadership and demonstrate commitment to the safety program. Use of the newly created ISO 14000 standard is one way for safety and environmental programs to go beyond the minimum requirements of law and regulation. Support is demonstrated not only in word, but by the commitment of personnel and financial support for the programs. A list of recommended safety management program elements follows.

Administrative Support

- Budget funds to evaluate hazards, monitor areas using hazardous materials, implement necessary controls, and conduct health examinations that include baseline physicals.
- Appoint a hazard control coordinator (this can be in addition to a safety officer) and other personnel to perform the necessary functions to support the program.
- Allocate the necessary time for committee meetings, training, and education, and other functions such as surveys and focus groups.

Training

- Develop an orientation program for new employees.
- Conduct refresher training on a regular basis for all employees.
- Train personnel who have been assigned to new positions that are exposed to hazardous materials.

Hazard Identification

- Obtain material safety data sheets (MSDS) on all hazardous materials from the manufacturer.
- Maintain hazardous materials inventories, by department or physical area of the hospital, for use in inspections, planning, and evacuation in the event of a spill or other release.
- Conduct regular site inspections.

Hazard Evaluation

- Use industrial hygiene and safety inspections to determine and monitor the need for safety controls.
- Conduct medical surveillance and medical evaluations for exposed workers.
- Monitor the medical health of the general employee population.

Recordkeeping.

- Publish a safety policy.
- Establish program goals and expectations that are both realistic and measurable.
- Communicate the importance of the program to *all* employees.
- Assign responsibility and authority to qualified personnel.
- Maintain records, surveys, employee medical evaluations, and identify corrective actions. Records must be maintained in accordance with federal, state, and local regulations including the Privacy Act.

Program Review

- Periodically review all data including safety inspections, industrial hygiene surveys, employee health and medical data, and any other information sources that could reveal patterns of exposure and would measure effectiveness of the exposure prevention program.
- Review and update programs as necessary to insure that new materials and procedures are covered.

In spite of management's best efforts to insure a good exposure prevention or accidental release of a hazardous substance, the active involvement of the employees at every level of the hospital and administrative staff is required or the program will not be successful. Willing and knowledgeable participants can build a stronger, more reliable team to provide the necessary network for the safe and responsible handling of hazardous materials. All employees should participate in the safety program from the medical director/CEO to the sanitation staff. The weakest link is the employee who doesn't know or understand what it is they are handling, and as a result can make the biggest mistakes.

SUMMARY

- A constant challenge for the hospital hazardous waste management team is to remain current with the numerous changes in the rules and regulations. The management of this task alone can be a full time job in light of the number of agencies and regulating sources at the federal, state, county and municipal levels.
- It is important to remember that once you are aware of a new regulation or requirement, the next and more difficult step is to implement the regulation in the hospital setting. This is more difficult than the average industrial organizational operation since the hospital setting requires the consideration of staff personnel, patients and visitors.
- Protection of all categories of personnel can be very difficult to coordinate and develop a plan that fits all the circumstances. It is also important to recognize the management of nonambulatory patients when it is necessary to evacuate or move these patients in an emergency situation.
- Waste reduction and waste minimization can be accomplished through bulking of waste materials and constant review of all department orders for chemical, biological, and radiological materials. Numerous other techniques can be useful to accomplish a safer environment.
- Planning the control and disposal of medical waste requires constant attention. Hospitals, by their nature, tend to be decentralized and operate their respective areas in a vacuum. The larger the hospital, the more this tends to be ture. Continuous monitoring and training of both managers and handlers is essentiual for a successful program.
- Avoid mixing waste materials whenever possible (i.e., chemical with biological, etc.). It is exceedingly more difficult and expensive to dispose of mixed waste.
- Management of a hospital's hazardous waste stream is complex and challenging. It requires continuous attention, partly because of the hospital's dynamic environment. New diseases, treatment protocols, research, and infection control requirements provide the HAZMAT technician with many opportunities. It also is necessary for continuous training and education to stay current. Your classroom training is just a beginning, not an end.

QUESTIONS FOR REVIEW

1. Only employees who work in heavy industry use hazardous chemicals.
   a. True
   b. False
2. Management can best communicate safety issues to all levels in the workplace through which of the following ways:
   a. Establishing realistic safety goals
   b. Publishing and training staff/employees on an organization safety plan
   c. Giving managers at all levels the appropriate authority and responsibility to manage
   d. All of the above
3. The Joint Commission on Accreditation of Healthcare Organizations (JCAHO) usually changes the regulations issued by OSHA, the EPA, and DOT to fit the hospital setting.
   a. True
   b. False
4. Which of the following documents provides detailed information about a chemical's hazardous properties?
   a. The Hazardous Communications (HAZCOM) Plan
   b. Units performance and procedure manual
   c. Material Safety Data Sheet (MSDS)
   d. The PDR

5. The regulations from different government sources (i.e., OSHA, the EPA, DOT) are in agreement on the definition of infectious waste.
   a. True
   b. False
6. Hospital waste that contains a mixture of waste from more than one category (i.e., chemical, biological, radiological) can be disposed of based on the most serious threat.
   a. True
   b. False
7. OSHA prefers which of the following methods to control hazards:
   a. Assign personal protective equipment (PPE) to each employee for use around hazards
   b. Use alarms and warning devices to notify that a dangerous condition has occurred
   c. Establish safety rules and workplace practice controls
   d. Use engineering and technical methods/ solutions to reduce or eliminate hazards
8. Radioactive waste is not common in the hospital setting and is of little concern in the development of a hospital safety plan.
   a. True
   b. False
9. Which of the following is the best indicatior of and effective safety program?
   a. A decrease in insurance premiums
   b. Use of outside consultants to prepare reports and other data
   c. Auditing accident/incident data, cost, and loss of information
   d. Wait for JCAHO report
10. Which technique is the best for reducing hazardous waste?
   a. Substitution of a less toxic material in a specific protocol
   b. Avoid surplus materials by careful ordering practices
   c. Avoid mixing chemical waste with other types of waste
   d. All of the above

ACTIVITIES

1. Obtain three material safety data sheets (MSDS's) on the same product from three different manufacturers and compare the similarities and differences. HINT: Check with the chemistry department—they might already have the MSDS's on hand.

2. Plan a fieldtrip to a local hospital and have the safety officer conduct a behind-the-scenes tour of the facility. The objective is to develop an awareness of the hazardous waste sources that might not be readily apparent in a hospital setting as compared to an industrial setting.

3. Review a copy of the Joint Commission on Accreditation of Healthcare Organizations (JCAHO) manual and familiarize yourself with its content. Make a copy of the Table of Contents for future reference.

4. Of the three areas (chemical, biological, and radiological), the discipline of radiation safety is the most unfamiliar to the average HAZMAT technician. Review the following references:
   10 CFR 31.11—General License Requirements
   10CFR 35.900—Radiation Safety Officer Training and Education Requirements

NOTES

1. Joint Commission on Accreditation of Health Care Organizations. 1995. *Accreditation Manual for Hospitals.* Oakbrook Terrace, IL. 1994.
2. Ibid.
3. Ibid., Section 4: PL.1.3.
4. Gavin, M. L. "Infectious Waste Management."
5. Vesilind, Peirce, and Weiner. 1988. *Environmental Engineering,* 2nd Edition. Stoneham, MA: Heinemann (Reed) Publishing. p. 323.

14

Source Reduction in the Semiconductor Fabrication Processes

Kelly Erin O'Brien, R.E.A.

Upon completion of this chapter, you will be able to meet the following objectives:

- Understand the semiconductor fabrication process.
- Know the sequence of steps in the semiconductor fabrication process.
- Know common chemicals used in the process.
- Understand how wastes are generated from the process.
- Understand what wastes are generated from the process.
- Understand waste treatment processes.
- Understand source reduction alternatives.

INTRODUCTION

Semiconductors play a major role in the operation of virtually all electronic devices. These solid-state electrical devices are designed to perform numerous functions, many of which we take for granted in the equipment and appliances we use daily. Not only do semiconductors handle information processing and display in computer systems, but they also perform electrical power management, serve as control devices in automotive emissions and anti-lock braking (ABS) systems, and convert light energy to electrical energy.

PART 1—HOW SEMICONDUCTORS ARE MADE

The process of manufacturing semiconductors, or integrated circuits (commonly called ICs, or chips), typically consists of more than a hundred steps, during which hundreds of copies of an integrated circuit are formed on a single wafer. Semiconductor manufacturing is best understood by dividing the process into three stages:

- Stage 1—Substrate and wafer production
- Stage 2—Semiconductor device fabrication
- Stage 3—Semiconductor device interconnection

Semiconductor devices are produced in fabrication units, or "fabs." Dust and contaminants can cause malfunctions in the microscopic electrical circuitry. Cleanliness is imperative, since a speck of dust trapped in one of these layers of circuitry can ruin an entire microchip.

Fabs make use of "clean" rooms that are engineered to minimize dust or other contaminants from landing on the wafers during fabrication. A clean room is a confined area in which the humidity, temperature, and particulate matter are precisely controlled within specified units. The "class" of the clean room defines the maximum number of particles of 0.3 micron size or larger that may exist in one cubic foot of space anywhere in the designated area. For example, in a Class 1 clean room only one particle of any kind may exist in one cubic foot of space. Newer clean rooms are typically Class 1–10, and are needed for manufacturing ICs with feature sizes as small as 1 micron.

Semiconductors are etched, or cut into the surface of, an underlying base material, called a substrate. The substrate in about 95 percent of the semiconductors made today is silicon. Other materials, such as germanium, gallium arsenide, gallium phosphate, and gallium arsenic phosphide can also be used to manufacture semiconductors.

The semiconductor manufacturing process involves the creation of eight to 25 or more patterned layers upon and into the substrate. This layering process creates electrically active regions in and on the semiconductor wafer surface. Finally, layers of circuitry are interconnected based on specifications for the ultimate practical application of the "chip."

Even though the fabrication processes for the production of wafers can be explained in general terms, each production run, and even different sections of the same run, may follow different process flows and feature different types of equipment. What follows is an overall discussion of the major processing steps and typical hazardous waste generation points of the semiconductor manufacturing process. Improvements and process changes are continually evolving which may appear to make this discussion archaic. However, the information contained herein will serve as a foundation for understanding the industry and future process variations and improvements.

STAGE 1—SUBSTRATE AND WAFER PRODUCTION

The first step in semiconductor manufacturing begins with production of a substrate and then, a wafer—a thin, round slice of a semiconductor material.

Making the Substrate

Silicon dioxide (raw quartz or sand) is reduced to elemental silicon by melting it in a carbon-arc furnace at temperatures over 3400° F. Once cooled, the silicon is pulverized and heated once again. High-purity hydrogen chloride gas is passed over and through the silicon powder and reacts into trichlorosilane ($SiHCl_3$). Reacting trichlorosilane with hydrogen at a high temperature produces an "electronic grade" silicon with less than 1 ppb of impurities. Impurities in silicon as small as 1 ppm can completely change the substrate's electrical properties. Since device fabrication relies upon selective, precision "contamination" to develop the desired electrical properties, a highly purified substrate is vital to the operation.

The now-purified polycrystalline silicon is heated to its boiling point, using high-frequency radio waves. This takes place inside a sealed, argon-rich quartz reaction chamber, or "ampule." The melt is "doped," or contaminated, with a small amount of boron powder to create silicon with a specified number of electron-deficient areas called "holes." Phosphorus or arsenic is used to dope the melt if an area with an excess of electrons is desired.

Crystal Pulling

Next, in a process called "crystal pulling," a small piece of solid silicon, known as the "seed," is placed on the end of a rod and dipped into molten silicon liquid. The surface tension between the seed and molten silicon causes a small amount of the liquid to rise with the seed and cool. As the seed is rotated and pulled slowly out of the melt, the liquid cools to form a single crystal, called an "ingot." Crystal ingots are commonly two feet long and weigh about two hundred pounds.

The crystal ingot is then ground to a uniform diameter (presently, between three and nine inches, though some companies are designing fabrication units to use twelve-inch ingots), and a diamond saw blade cuts the ingot into thin wafers.

Machining

The wafer is processed through a series of machines, where it is "lapped" (or ground) smooth, acid etched, and chemically and mechanically polished to a mirror-like luster. Today's device fabrication technologies require wafers with 10 angstrom, or better, variation over the surface of the wafer. Upon completion of machining operations, the wafers are cleaned first in a dilute acid and deionized water solution, and then in an ammonium hydroxide solution. Once cleaned, the wafers are ready to be sent to the device fabrication area.

STAGE 2—SEMICONDUCTOR DEVICE FABRICATION

Semiconductors basically function by selectively creating areas of altered electrical conductivity, within the highly pure silicon substrate obtained from the wafer production process. High-purity silicon is a poor conductor of electricity due to stable bonds in the outer shell, which is formed by the four equally shared electrons by neighboring silicon atoms. Selective "doping," or adding to the silicon small amounts of phosphorus or arsenic—materials which have excess electrons in their outer shell—develops electron-rich, negative (n) areas. In the same way, electron-deficient, positive (p) areas can be developed by adding a small amount of an outer-shell electron-deficient material, such as boron. When these two types of materials, p and n type, are joined, a "p-n junction" is formed. The ability to control electron movement across p-n junctions is the basic premise upon which all semiconductor technology is built.

In the device fabrication area, the next step is the buildup of electronic circuitry onto the wafer. Selective doping is conducted through photolithography. The electronic circuitry is built in a predetermined pattern depending on the ultimate use of the wafer. "Masked," or protected, areas between these patterns prevent circuitry buildup and thus provide cutting areas to separate the wafers into self-contained circuits. When the wafer is cut, the resulting self-contained, or integrated, circuits are called "microchips." Depending on the application of the microchip and the size of the wafer, 100 or more microchips may be produced from one wafer.

Epitaxy

Epitaxy, often the first production step in device fabrication, is a process of developing a controlled growth on a crystalline substrate of a crystalline layer, called an "epilayer."

There are two types of epitaxy procedures: homo-epitaxy and hetero-epitaxy. In "homo-epitaxy," the epilayer exactly duplicates the properties and crystalline structure of the substrate, as when silicon layers are grown on a silicon substrate. In "hetero-epitaxy," the deposited epilayer is a different material with a different crystalline structure than that of the substrate, such as silicon on sapphire. Homo-epitaxy is the most common epitaxial process.

The value of epitaxy is that a lightly doped silicon epilayer can be grown upon a heavily doped silicon substrate. Though the crystal structure of the two materials are the same, the sharply differing electrical properties of the materials create an effective insulating junction between the epilayer and the substrate. Thus, the possibility of short circuits between devices built in the layers is minimized.

Almost all epitaxy processes use chemical vapor deposition (CVD) to grow epilayers. CVD is an unheated, gaseous process that deposits insulating films or metal onto the heated wafer surface. Two types of reaction chambers are used: "cold wall" and "hot wall" chambers.

Cold wall reactors are quartz bell jars that handle up to 50 wafers mounted on a metal holder, or "susceptor." The susceptor and wafers are heated by radio-frequency energy to a temperature of about 2,000°F. The bell jar remains unheated; thus, no material is deposited on it. The susceptor and wafers, on the other hand, acquire deposits from the dopant gas, which is introduced into the reaction chamber. Once the deposition process is complete, the walls of the reaction chamber are cleaned by introducing hydrogen chloride (HCl) gas.

Hot wall reactors are long, electrically heated cylindrical tubes, called "furnaces," into which six- to eight-foot-long quartz "boats," containing vertically standing wafers, are placed. Hot wall reactors are more commonly used for deposition of dielectrics or polysilicon rather than in epitaxial processes. Everything inside the reactor furnace is hot; thus, quartz boats, wafers, and the furnace walls all have material deposited on them. The inside of the chamber is cleaned by introducing hydrogen chloride gas at high temperatures.

Low-pressure chemical vapor deposition (LPCVD) can be done to promote the chemical reaction, using either bell jars or furnaces, though the furnace is the more common reactor used with this method.

Many of the toxic gases essential to the semiconductor industry are used in the CVD process. These include the following:

- Hydrogen chloride, for reaction chamber cleaning
- Silane, dichlorosilane, and trichlorosilane, for silicon deposition
- Arsine, phosphine, and diborane, for mild doping

Hazardous wastes generated from this process include vacuum pump oil contaminated with acid and metals, and solid wastes containing arsenic, antimony, and phosphorus.

Oxidation

Oxidation is the process of forming thin films of chemically resistant silicon dioxide on the surface of the silicon wafer. The most common method of oxidation is thermal, and thermal oxidation is further divided into two processes: wet and dry.

Thermal oxidation is conducted in a high-temperature (approximately 2,200°F.) reaction chamber. The dry process grows thin oxides under about one atmosphere of pressure in an oxygen and hydrogen chloride gaseous atmosphere. The wet process is accomplished by reacting hydrogen and oxygen or ultra pure water in the furnace chamber, with the intent of producing thicker oxides on the wafer surface. Waste generated from the oxidation process consists of silicon dioxide, or other raw material, used for wafer fabrication; hydrofluoric acid used to clean the quartz chambers and boats; and rinsing water used to clean the wafers.

Photolithography

The word "photolithography" find its roots in Greek: *phos*, meaning "light;" *lithos*, meaning "stone;" and *graphicos*, meaning "write." Thus, photolithography, also known as "photomasking" or "masking," is a procedure which uses light to transfer an image onto the surface of the silicon wafer. This results in the formation of extremely accurate patterns on the wafer's silicon oxide surface.

The photolithographic process is the heart of the device fabrication process and is done by cyclical repetition of a series of steps basic steps: photoresist application, photoexposure, developing, etching, stripping, and doping.

Photoresist Application

The first step in the photomasking process is covering the semiconductor wafer's oxide coating with a thin protective layer of "photoresist" containing an organic polymer. Wafers travel on a "track," or cushion of air, to the photoresist application station. The wafer is contained inside a cup and spun, while the entire surface of the wafer is coated by dripping an amount of photoresist in the wafer's center.

The coated wafer is now carried through small ovens or upon hot plates that "softbake" at about 190°F, to semiharden the deposited photoresist and promote bonding. Ultraviolet (UV) light—usually from a high-intensity (1.5 megawatt), mercury-vapor lamp—shines through a mask containing the circuit pattern. Patterns in the mask protect selected areas of the wafer from exposure to ultraviolet light and determine where the circuits will be.

There are two different types of photoresists: positive and negative. The two photoresists differ primarily in the exposure process: Positive photoresist polymers are normally insoluble to an alkaline developer solution but become soluble once exposed to

UV light. Negative photoresists become insoluble after exposure to UV light, but otherwise expose patterns by being soluble in developer solution.

Photoexposure

In this step, the photoresist-coated wafer is exposed to high-intensity UV light through the mask. Masks determine the patterns for circuits in the wafer surface.

Since the photoresist is sensitive to ultraviolet light, the photoresist records the ultraviolet light pattern that the photomask imparts, like a photocopy records a document, without affecting the masked area of the wafer.

If positive photoresist is coating the wafer, then the areas exposed to the light undergo a photochemical reaction that will make the photoresist more soluble to an alkaline, caustic, developer solution.

Developing

Removal of the exposed photoresist is done using a developer solution and a stripper containing acids and caustics. Material remaining after development shields regions of the wafer from subsequent etch or implant operations. The photolithographic process produces much of the wastes generated from the semiconductor manufacturing processes. They include the following:

- Hexamethyldisilazane (HMDS) used as an initial coating on the wafer's oxide surface to increase the adhesion of photoresist
- Glycol ethers
- Waste photoresist solvents, such as xylene, freon, and orthodiazoketone
- Non-halogenated hydrocarbons and developers

Etching

Etching is used to remove the silicon dioxide, nitride, glass or other selected layers not covered by the photoresist. Etching is done by either one of two methods: wet etching or dry etching.

Wet etching is a usually a manual process, though in newer clean rooms, the process is automated. Wafers are soaked in acid baths consisting of hydrofluoric, hydrochloric, nitric, or chromic acids.

Dry etching is a more precisely controllable process, where wafers are exposed to an ionized gas in an enclosed, high-vacuum (0.01 atmosphere) chamber. Activated fluorine- or chlorine-based etchants are created by exposing a gas to an electrical discharge. Typical gases include, chlorine, boron trichloride, hydrogen chloride, carbon trifluoride, carbon tetrafluoride, carbon tetrachloride, and boron trifluoride.

These processes expose the silicon surface in preparation for doping with impurities.

Wastes from the etching process include the following:

- Sulfuric acid
- Hydrofluoric acid
- Hydrochloric acid
- Phosphoric acid

- Nitric acid
- Chromic acid
- Buffered oxide etch (BOE), a hydrofluoric acid and ammonium fluoride mixture
- Vacuum pump lubricating oil from pumps used to evacuate reaction chambers in plasma etchers

Stripping

After etching the unprotected silicon dioxide, the photoresist must be removed from the protected silicon dioxide. There are two primary methods of stripping away the resist: dry stripping and wet stripping.

During dry stripping, or "ashing," wafers are placed in a high-vacuum chamber. Oxygen is introduced and subjected to radio frequency power, creating oxygen radicals. The resist oxidizes to water, carbon monoxide, and carbon dioxide, and these wastes are then pumped off.

In wet stripping, very corrosive chemicals, similar to those used in the metal-finishing industry, are used in "wet station" baths, into which cassettes of wafers are immersed. The solvents are heated to about 200°F and the wafers are exposed to ultrasound in order to increase penetration.

Doping

The introduction of a dopant, which contains impurity atoms with specific behavior patterns, is known as doping. In doping, diffusion or ion implantation impurities are introduced into select regions of the wafer substrate to form a boundary between conduction regions. Using the diffusion method, dopants are deposited onto the wafer by stacking the wafers in a long, heated quartz tube and exposing them to gases containing impurities that diffuse into areas of the wafer not masked with photoresist.

Ion implantation is another method used for introducing dopants onto the wafers. This is conducted by bombarding the wafer with ionized impurities. Waste generated from this step consists of solid wastes containing the following elements:

- Arsenic
- Antimony
- Phosphorus
- Arsine
- Diborane
- Vacuum pump oil

STAGE 3—SEMICONDUCTOR DEVICE INTERCONNECTION

Metallization

Metallization is last step in the process and consists of depositing complex patterns of conductive material onto the wafer to interconnect the integrated circuits. Wastes generated from the metallization step differ significantly, depending on the specific plating or coating process used. The waste streams from metallization include solutions of precious metals, heavy metals, and acids.

The most commonly used metal for connecting devices is aluminum; however, other metals are commonly used, including chromium, nickel, gold, silver, titanium, tungsten, platinum, and silicon, as well as copper-chrome-gold and lead-tin combinations.

Included at various stages of the process are rinsing steps that utilize electronic grade, distilled, deionized water and that generate diluted acid/water waste streams. These streams constitute the largest volume of hazardous wastewater generated.

Two common ways of depositing metals on a substrate are physical vapor deposition (PVD) and chemical vapor deposition (CVD).

Physical Vapor Deposition of Metals

Metal deposition historically has been done by either one of two physical vapor deposition methods: evaporation and sputtering.

Evaporation

Evaporation uses a twin-chambered vacuum device: wafers are loaded onto a dome and placed into the upper chamber, and a solid "slug" of the metal that is to be evaporated and deposited is placed in a lower chamber. The metal is evaporated by radiofrequency, thermal, or electron beam heating, and the vapors are deposited on the wafers in the chamber above.

Sputtering

Sputtering is done in high vacuum, where radiofrequency or DC power sources are used to create ionized argon. The argon ions bombard a target metal, causing the metal to release metal ions from its surface. These metal atoms condense on the substrate to form a thin film.

Chemical Vapor Deposition of Metals

Because of the tendency to continually reduce the dimensions of semiconductor devices and their interconnections, current PVD methods are fast becoming outdated. CVD systems allow better control than PVD systems; however, CVD systems are inherently more dangerous, due to the hazardous composition of the gases used in the CVD process.

Trends in the Semiconductor Industry

The semiconductor industry is—by virtue of the tremendous cost of material, facilities, and equipment—continuously seeking more effective and efficient production methods simply to stay competitive in the marketplace. Generally, source reduction and efficient processes go hand in hand, one being a yardstick for the other. Production and disposal of waste is money lost; thus, waste disposal is seen for what it is: a net *payout*. Source reduction, on the other hand, is always net *payback*.

New challenges face the industry as fabrication labs are built to produce larger wafers at increasingly lower unit cost. Serious source reduction techniques must be innately designed into systems while they are still on the engineer's drawing board.

Design for environment (DFE) is a growing trend in many industries, as "loss control" and process efficiency become increasingly important, and source reduction is seen as a profit center rather than red ink on the bottom line.

PART 2—SOURCE REDUCTION MEASURES FOR SELECTED WASTE STREAMS

Volume Reduction and Conservation Measures

The State of California's Source Reduction and Management Review Act of 1989, also known as "SB-14" after its state senate bill number, requires generators of an annual threshold quantity of hazardous waste to develop a source reduction plan and a periodic source reduction progress report.

The California EPA's Department of Toxic Substances Control (DTSC), the organization overseeing the SB-14 program, reviewed source reduction plan and waste-specific questionnaires distributed to generators in the semiconductor industry, and as a result, a number of viable source reduction measures were identified.

The following section describes source reduction methods applied by several semiconductor facilities, addressing typical major wastestreams. Listed below are the methods that will be discussed:

- Aqueous wastes (containing low concentrations of hydrofluoric, phosphoric, hydrochloric, and sulfuric acids)
- Solvents
- Stripper wastes
- Photoresist waste
- Contaminated solid wastes

Aqueous Wastes

Contaminated process wastewater traditionally has been the largest waste stream generated by the semiconductor industry; thus, less water generated means fewer chemicals used. Reduction in water consumption has been achieved by a number of methods, such as fabrication operations switching from continuous flow to "on-demand" rinsing, redesign of water reclamation systems, improved maintenance methods for leaks, and use of reclaimed water in scrubbers and cooling towers.

The manufacturing of highly purified, deionized (DI) rinse water consumes large quantities of sulfuric, hydrochloric, and other acids that are used to balance pH and revitalize purification media. Water conservation practices within the semiconductor industry have led to a substantial decrease in the use of sulfuric and hydrochloric acids.

Deionized water is produced by first treating municipal supply water, typically groundwater, with a reverse osmosis (RO) process prior to a final purification treatment, called "ion exchange." RO removes many of the water impurities, thus greatly extending the recharge period, or service life, of the ion exchange beds used in this process.

Substitution of thin film composite (TFC) membranes for cellulose acetate (CA) membranes in the reverse osmosis process has resulted in better water quality and reduced quantities of sulfuric acid used to adjust the pH of the RO feedwater.

Since acid is also used in the regeneration of ion exchange resin, membrane substitution has improved the quality of the RO-generated DI water, reduced the amount of acid used for pH adjustment and doubled the service life of the resin. One semiconductor manufacturer has reported an approximate 80 percent reduction in water and chemical usage as a result of a switch to high-efficiency TFC membranes.

Reducing acid bath waste by extending the service life of rinse baths is another alternative in use in the semiconductor industry. Wafers are rinsed in acid baths consecutively in several cleaning cycles. Typically, baths had been disposed of and replaced on a set time schedule, rather than testing the bath to meet an engineering quality standard. Many semiconductor companies have rewritten criteria for acid bath replacement applying statistical process controls. It is now more common to replace baths after a specific number of wafers have been processed, or after physical or chemical changes in bath composition have been observed, rather than after a set time period.

Quality control procedure revisions can often prove rewarding. One company reported a reduction of as much as 80 percent in the use of phosphoric acid, after internal studies indicated that higher bath water particle counts could be accomplished without affecting product quality. As a result, phosphoric acid bath life was extended from eight hours to one week.

Additionally, employee training in process methods to reduce cross contamination and revised specifications for bath rotations can substantially reduce quantities of waste generated.

Sulfuric Acid

Among the generators surveyed, a net reduction in the overall discharge of sulfuric acid has been realized by using an acid reprocessor to purify spent sulfuric acid used in clean sinks and strip sinks. Since the reprocessed sulfuric acid is recycled and reused, a significant reduction in the quantity of purchased process material can be realized.

As discussed in Part 1, the manufacture of integrated circuits involves an operation in which silicon wafers are processed through a series of steps that build layers of material onto the silicon substrate. The layers ultimately form the devices, or circuits, which will conduct electricity and perform the desired functions. It is extremely critical that the wafers be cleaned before each process step to avoid contamination. Unwanted impurities, films, and particulate matter have a devastating effect on electrical characteristics of semiconductor devices.

Wafer cleaning is accomplished using a mixture of concentrated sulfuric acid and hydrogen peroxide. The mixing of these two chemicals produces an exothermic reaction, which heats the mixture and accelerates the cleaning process. As the bath temperature rises, the hydrogen peroxide decomposes to form water and oxygen. The water product dilutes the mixture demanding additional hydrogen peroxide to maintain the desired cleaning rate and mixture temperature. A mixture is considered spent and ready for discharge to a wastewater treatment unit when: (1) the addition of hydrogen peroxide can no longer achieve and maintain the desired bath temperature; or (2) the number of particles present in the bath exceeds process specification requirements.

A sulfuric acid and hydrogen peroxide cleaning mixture is used in clean sinks and strip sinks. In a clean sink, as the name implies, the wafers are washed before critical process steps are performed. In a strip sink, the mixture is used to remove photoresist, used to mask certain portions of the wafer prior to etching.

Rather than discharging the spent sulfuric acid and hydrogen peroxide mixture to a wastewater treatment unit, a sulfuric acid reprocessor can be used to reclaim the mixture for reuse in the sinks. Automated computer-controlled vacuum distillation is used to purify and concentrate the sulfuric acid. As part of the computer control, an on-line monitoring system analyzes and compares the concentration of the sulfuric acid produced and the particulate count. Reprocessors are user programmable to precise, process-specific requirements. Off-specification product is returned to the vacuum distillation chamber for reprocessing.

As sulfuric acid is purified, peroxydisulfuric acid (PDSA) can be produced in situ or physically added to replace hydrogen peroxide. Integrated circuit manufacturers have found PDSA to be an acceptable replacement for hydrogen peroxide in both sulfuric acid cleaning and stripping sinks.

PDSA produces a smaller quantity of water during decomposition when mixed with sulfuric acid; therefore, substitution of PDSA for hydrogen peroxide essentially eliminates bath dilution effects. Additionally, on a volume-to-volume basis, less PDSA is required to obtain the same cleaning characteristics achieved by using hydrogen peroxide.

Along with eliminating the need for hydrogen peroxide, lower grade sulfuric acid and PDSA can be purchased and purified to the desired specifications in the acid reprocessor, resulting in process material cost savings. Once the system is in operation, the monthly sulfuric acid consumption by the sulfuric acid sinks can be reduced by as much as 90 percent. The remaining ten percent is the sum of mixture evaporation, wafer carryover, and a waste stream used to remove particulate matter buildup from the reprocessor.

In a case study, a well-known semiconductor manufacturer, in its baseline year of 1990, used approximately 743,000 pounds of sulfuric acid in production processes. This constituted 82 percent of the corrosive liquid waste stream generated by the facility in the baseline year. In early 1993, the company began to reprocess sulfuric acid using an alternative system. The alternative system processed about 6,400 gallons of sulfuric acid per week, resulting in an annual savings of $2.9 million in sulfuric acid purchases. Additional savings were also obtained by reducing the load on the wastewater neutralization system. The company reported that the alternative system resulted in 28 percent reduction in sulfuric acid waste generated in 1993, despite an increase in production in that same year of at least 25 percent.

Another measure for reducing the overall use of sulfuric acid is the replacement of standard diffusion sinks with filtered, recirculating units. Recirculation and filtration allow considerable extension of the acid bath service life, significantly reducing the quantity of sulfuric acid discharged for wastewater treatment.

In general, unfiltered sulfuric sinks in the diffusion section of the fabrication area use three parts sulfuric acid to two parts hydrogen peroxide. As noted previously, the exothermic reaction that takes place between these two chemicals heats the bath and accelerates the cleaning process. The bath mixture must be changed frequently, because it becomes diluted as the hydrogen peroxide decomposes to water.

The acid life can be greatly extended by using sinks that are able to heat, filter, and recirculate the sulfuric acid. If the acid is heated, much less hydrogen peroxide is required to achieve the desired cleaning temperature. Filtration further extends service life. With a full production load, a sink that would normally be changed every 30 minutes might last a week. The net result is a decrease in the quantities of sulfuric acid and hydrogen peroxide needed to clean the wafers. Hydrogen peroxide is eliminated, and lower grade sulfuric acid and PDSA can be purchased and purified in the acid reprocessor, thus lowering materials costs.

A facility in Santa Clara reduced sulfuric acid usage by 60 percent using a three-pronged approach: reduce the frequency of acid bath replacement from every four hours to every eight hours, consolidate two wet stations for sulfuric acid operations to one station, and decrease bath size. These measures resulted in a reduction of sulfuric acid usage from 300 gallons a week to 50 gallons a week.

Hydrofluoric Acid

Hydrofluoric acid (HF) is used extensively by the semiconductor industry during the etching process. This is mainly because of its ability to remove silicon dioxide and metals from the wafer's surface. Due to the lack of a compatible substitute at this time, hydrofluoric acid cannot be eliminated in the semiconductor manufacturing industry. However, by evaluating source reduction technologies, both the concentration and volume of HF used in semiconductor operations can be greatly reduced. HF at a concentration of about 49 percent is typically used for complete and fast etching of wafers. Studies have shown, however, that lower concentrations such as 10:1 or 100:1 HF can be used effectively, but would require a longer process time. Etching operations that require a less aggressive oxide removal rate can use buffered oxide etch (BOE), a mixture of ammonium fluoride and hydrofluoric acid. BOE mixtures generally have a much lower total fluoride concentration.

A major source reduction measure currently being applied by many semiconductor facilities to reduce HF waste involves using dry, or plasma, etching, which uses etchant gases rather than acid baths. After the wafers are placed in a quartz vacuum chamber, a gas containing an etchant gas, typically fluorine, is introduced into the vacuum chamber. In addition to reducing the liabilities associated with managing hydrofluoric acid, this source reduction measure has improved the masking process by producing more precise, finer lines in the photolithographic process. Some of the major drawbacks to this measure are the high cost of the equipment required for the dry etching process and the potential for gas emissions.

Another alternative to the wet chemical etching bath is the use of anhydrous HF vapor etching technology. This technique combines anhydrous HF, nitrogen, and water vapor into a controlled etching system.

Automated etching and cleaning can also reduce the volume of HF used. In batch processing, once the chemicals are mixed, the mixture has a specific lifetime, after which the cleaning ability of the solution is substantially reduced. HF acid spray processors can be used for cleaning and etching wafers instead of standard batch processing in an acid bath. Acid spray processing can reduce chemical consumption by using the cleaning chemicals in an "on-demand" mode.

There are, however, some minor drawbacks in using spray processors. For example, non-uniformity during the silicon dioxide etch process is slightly worse than that from batch processing. This problem could be rectified by using new tool configurations and chemical mixtures to improve spray uniformity.

To extend the useful life of HF solutions, recirculating filter systems can be installed in each cleaning and etching bath. The filters would remove contaminants, enabling the acid solution to be reused. A microelectronics firm in Rancho Bernardo, California, has installed new wet bench stations using high-flow filtered recirculating etch baths. The new filtered and recirculating etch baths successfully maintained the uniformity of etch rates, while decreasing ionic and particulate contamination and increasing product yield. This source reduction measure resulted in a 1,500-gallon per-bench annual reduction in the use of hydrofluoric acid, leading to annual cost savings of $21,600 per bench.

The presence of ionic and particulate impurities in the manufacturing process can cause serious reliability and yield problems. To avoid ionic and particulate contamination, high-purity semiconductor grade chemicals are often used once and then disposed of. Recently, HF systems that integrate ion exchange, particle filtration, and process control have been introduced. For purity, economic, and safety reasons, ion exchange is the preferred technology for HF reprocessing systems. These systems appear to be ideal since they are able to satisfy both the chemical quality and processing concerns of production engineering and environmental source reduction goals. HF reprocessing systems, however, can only be used in limited applications and can not handle rinsewaters or chemical mixtures such as buffered oxide etch (BOE).

Solvent Wastes

Solvents are the second largest waste stream generated by the semiconductor industry. Waste reduction methods to minimize the generation of solvent wastes in semiconductor processes include waste segregation, tape substitution, and the use of nonhazardous cleaners rather than hazardous products. Segregating hazardous wastes from nonhazardous wastes is a common source reduction technique that can lead to significant reductions in the total amounts of hazardous wastes generated. Waste segregation can also facilitate increased recovery and recycling of wastes. Separating chlorinated from nonchlorinated solvents and aliphatic from aromatic solvents enables companies to ship these wastes to recycling facilities.

Tape substitution eliminates the use of hazardous solvents in the backlapping stage of fabrication. Backlapping is the process of thinning out the back of the wafers to facilitate the packaging of the semiconductor devices. For many years, electroplating tape was attached by adhesives to the back of the wafer during the backlapping process. Solvents such as isopropyl alcohol and acetone were then used to remove the adhesive and its residue. By substituting a new type of tape that does not require the use of solvent-sensitive adhesives, semiconductor companies found that they were able to eliminate the solvent waste generated during tape and residual adhesive removal. The new tape is completely removable, following exposure of the tape's adhesives to ultraviolet light.

As noted above, solvents are commonly used during semiconductor fabrication to clean equipment and parts. In recent years, companies have been using a nonhazardous cleaner instead of acetone and isopropyl alcohol. The application of the nonhazardous substitute, however, is limited to cleaning parts and equipment; the product cannot be used to clean wafers because of its potential for introducing contamination.

A more recent development, and an offshoot of DI water production, is the use of ultrapure water (UPW) for parts, equipment, and wafers. Ultrapure water is made in a process similar to DI water—at least as far as the use of reverse osmosis is concerned—but the process is taken a step further. DI water is put through a microfiltration process, which removes even more of the impurities carried in the water from its source. The result is highly polarized water—a "magnet" of sorts—that will agressively pick up impurities, including metals, and carry them off the work surface.

Acetone has also been successfully eliminated in the edge bead removal (EBR) process. This has been achieved by switching to a less hazardous solvent, such as a propylene glycol monomethyl ether acetate. By switching solvents, the quantity of hazardous constituents discharged to the air, water, and land has been reduced.

Part of the integrated circuit fabrication process requires that patterns be imaged onto silicon wafers. To accomplish the imaging, the wafers are coated with a light-sensitive material called photoresist. The images are created in the photoresist by an exposure to ultraviolet light followed by development.

The photoresist is dispensed onto the surface of the wafers by automated equipment called coat tracks or spin tracks. After dispensing several milliliters of photoresist onto a silicon wafer, the wafer is spun at high speed to produce a uniform, thin layer of photoresist. In this process, most photoresist is spun off the wafer. As the photoresist is spun off, a thickened bead of resist forms around the circumference of the wafer. The bead often extends over the lip of the wafer and onto the back. If the bead is not removed it will cause the wafer to adhere to the equipment during the softbake step which is required to prepare the photoresist for photolithography.

To remove the edge bead, a process step is incorporated into the coat track operation called "edge bead removal" (EBR). A controlled spray of solvent is applied to the back side of the wafers to remove the unwanted photoresist bead prior to softbake.

Acetone has been successfully replaced by EBR-10, making it a major source reduction measure for acetone. Microposit EBR-10, a propylene glycol monomethyl ether acetate, is commerical formulation compatible with the current photoresist systems and existing spin track equipment. Because EBR-10 has a lower vapor pressure than acetone, and almost no solubility in water, less EBR-10 is emitted into the atmosphere than acetone, and very little EBR-10 is discharged to water.

Sending empty solvent containers back to the suppliers to be refilled is another measure implemented by a number of semiconductor companies. In order to dispose of containers as nonhazardous waste, California generators must triple-rinse containers. Generators are not required, however, to rinse containers prior to sending them back to the supplier for refilling. This measure has resulted in a reduction in the volume of solvent cleaning waste generated at facilities adopting this practice.

Another source reduction measure currently under evaluation by Los Alamos National Laboratory (LANL) is the use of a new "dry" cleaning technology. Dry cleaning technology is a substitute for "wet" chemical cleaning methods, which are currently used for the removal of particulates and impurities during the manufacturing operations of high density integrated circuits and other electronic devices. The use of a low pressure, gaseous plasma has been proposed to lift and remove fine particles and chemically oxidize surface organic contaminants. This new approach can result in significant reduction in solvents used for cleaning within the semiconductor industry.

Improved cleaning effectiveness can also result by incorporating plasma clean processes directly into the tooling required for processing and by avoiding the gradual buildup of particles and waste products that are typical of wet chemical batch processing. In addition, the use of a nontoxic feed gas, such as oxygen, has the benefit of producing a nonhazardous effluent. Ultimately, particles are captured by standard gas filters, thereby producing minimal solid waste volume without the need for specialized monitors or handling techniques.

Stripper Waste

Waste stripper solution is generated as a result of the photolithographic process. Stripper solution is used to dissolve undeveloped photoresist and remove it from the wafer. Until a few years ago, semiconductor facilities used phenolic strippers containing phenol (a very toxic material) in the photolithographic process. A nonphenolic stripper has been recently introduced by several semiconductor facilities. The use of less toxic stripper solution as a substitute for phenolic strippers has contributed to reduction in disposal costs and liabilities.

Another source reduction method for stripper wastes involves the substitution of the nonphenolic stripper with a biodegradable stripper material. A Santa Cruz-based semiconductor manufacturer was able to extend the bath life of the stripper solution and increase the number of wafers processed per bath by using a new product with extended capabilities. The company anticipates that the new stripper solution will reduce stripper usage by approximately 50 percent. The new rinse will replace isopropyl alcohol (IPA) in the rinse process, reducing IPA usage by 75 percent.

Photoresist Waste

There is a strong incentive in the semiconductor industry to search for waste reduction techniques for photoresist waste. Not only is the virgin material expensive to purchase, but it is also costly to dispose of.

Several waste reduction methods for dispensing photoresist waste have been introduced by the semiconductor industry. These methods include the following:

- Automated dispensing
- Dispensing by means of a positive displacement pump
- Installation of gravity feed delivery systems

A major drawback to manual dispensing is that operators have to inspect photoresist bottles to decide when a new one is needed. In order not to interrupt the photoresist dispensing process, the operators replace photoresist bottles before they are completely empty, resulting in large residual photoresist remaining in the containers. Automated dispensing was introduced as an alternative to manually dispensing photoresist onto silicon wafers, allowing the bottles to completely drain before they are replaced.

Several semiconductor facilities are currently testing a range of duration times to determine whether smaller quantities of photoresist per wafer can be used without sacrificing product quality. One manufacturer estimated a 50 percent reduction in photoresist being applied during the photolithographic process. Another manufacturer achieved approximately 25 percent reduction of photoresist per wafer by increasing equipment calibration frequency from once a week to daily.

Increasing the equipment calibration frequency resulted in a 2 ml reduction of photoresist per wafer—the manufacturer applies 4 ml of photoresist to each wafer instead of 6 ml—and this measure alone resulted in over $100,000 savings per year for the company.

Significant savings in the photolithography stage have been realized by some companies. One Quality Improvement Team (QIT) used a formalized structure and a systematic approach to identify and eliminate defects. The use of more process chemical than necessary was identified as a defect. As a result, the team managed to cut costs of photolithography chemicals by an estimated $400,000 per year.

This cost savings was accomplished through the following measures:

- Installation of meters in the photomasking area that regulated—and reduced—the amount of developing material used for each wafer
- Implementation of process improvements that resulted in a 15 percent reduction in the amount of developing chemicals; the team also set up a new process to cut resist use by 25 percent
- Negotiation with chemical suppliers to reduce prices on purchased chemicals

The QIT, part of the company's quality improvement system, conducted comparison tests to ensure that defects didn't arise as a result of cutting chemical usage. The payback from their efforts was a cut in photomasking costs from $2 to $1.25 per wafer.

Contaminated Solid Wastes

Large quantities of solid contaminated wastes are generated during the wafer fabrication processes. This waste stream is mainly comprised mainly of used gloves, acid contaminated plastics (tubing, wafer cassettes, and containers), and acid rags from equipment wipe-downs, small spills, and drips.

Several source reduction measures for solid contaminated wastes have been introduced by the semiconductor industry. These measures include training employees on hazardous waste handling; rinsing contaminated solid wastes; and adopting improved policies and procedures for hazardous waste handling.

By increasing employee awareness through training, the amount of contaminated solid waste can be reduced. Training offered to employees and managers on proper waste segregation (segregating nonhazardous solid waste, which includes glove wrappers, wafers, razor blades, and plastic boxes, from hazardous solid wastes such as contaminated acid gloves and tubes) and proper equipment operation and maintenance can lead to significant reductions in solid contaminated waste.

Solid contaminated hazardous wastes such as wafer cassettes, tubing, and lightly contaminated gloves can also be made nonhazardous by rinsing them with water. Although rinsing solid wastes will result in the generation of hazardous rinseate, the net hazardous waste reduction will be worthwhile since, at most facilities, the rinseate will be neutralized in the wastewater treatment facility.

Better operating practices and procedures is another major source reduction measure for handling solid hazardous wastes within the semiconductor industry. The following is a list of such practices and procedures:

- Using signs to distinguish hazardous and nonhazardous trash cans
- Installing pH paper stations for use in testing spills to ensure that a spill is truly hazardous prior to disposing of it as such
- Using a vacuum for larger liquid spills to minimize the use of wipes for large spill cleanups
- Implementing bulk chemical distribution, and
- Improving waste handling procedures and reporting requirements to document and correct equipment problems and leaks

A Santa Clara company used a seven-step Total Quality Management (TQM) process to isolate problems and identify weaknesses. A number of action items were implemented, such as additional employee operator training, pH paper stations, glove-washing procedures, and the purchase of an acid spill vacuum. These changes resulted in significant reduction in the generation of acid contaminated solid waste at the facility.

PART 3—GENERAL SOURCE REDUCTION MEASURES

In addition to source reduction measures for specific waste streams, several administrative source reduction measures with potential applicability to a wide range of semiconductor operations are useful to review. Steps taken by employers to raise the level of awareness among employees and managers regarding hazardous waste source reduction are the measures commonly reported by California-based semiconductor companies. Such administrative steps include improved inventory control, improved training and supervision programs, formation of source reduction task forces, and hazardous waste management costs chargeback.

Improved Inventory Control

Improved inventory control is a commonly adopted source reduction measure. When oversupplies of limited-shelf-life materials are reduced and virgin chemical products are distributed on a first-in, first-out basis, the quantity of expired-shelf-life chemicals disposed of as hazardous waste is significantly reduced.

Improved Training and Supervision Programs

Offering periodic training classes for employees and managers on inventory management, housekeeping, waste segregation, and equipment operation and maintenance has proven to reduce hazardous waste generation resulting from leaks, accidental spills, and poor management habits.

Formation of Source Reduction Task Forces

In general, the role of a source reduction task force is to research, prioritize, and present new source reduction measures; oversee the implementation of feasible measures; and report the success of completed measures. The task force also reviews potential problems associated with source reduction, sets up source reduction pilot studies, and participates in trade organizations to obtain ideas and keep up with state-of-the-art practices. Some companies have instituted an incentive system, whereby awards are granted for successful program implementation.

Hazardous Waste Management Costs Chargeback

In-house accounting methods that charge costs associated with waste management to departments that generate the waste have been proven to speed up efforts of each department to consider and implement source reduction measures.

SUMMARY

- Semiconductor fabrication is a multistep, precision process
- A number of highly hazardous materials are used in the process
- The process can produce large quantities of hazardous and solid wastes
- Standardized, rather than calendar-based, process controls produce easy to accomplish improvements
- Replacing wet processes with dry ones are sometimes possible and fruitful
- Many solvent-cleaning alternatives are currently under review and in use

QUESTIONS FOR REVIEW

1. How does dust, or other contaminants in the air, degrade the quality of semiconductor devices?
2. Why is pure silicon a poor conductor of electricity? What materials can be introduced into silicon to enhance electroconductivity?
3. What is the basic premise of semiconductor operation?
4. Name two hazardous wastes generated from the developing (photoresist removal) step. What hazards do the wastes pose?
5. How are patterns laid down in the surface of a silicon wafer for circuits? How is excess pattern material removed?
6. Which of the semiconductor manufacturing processes generate the greatest variety of hazardous wastes.
7. What's the inefficiency in calendar-based rinse bath changes? What alternative method would you suggest?
8. Discuss the merits of wet hydrofluoric acid etching versus plasma etching systems.
9. What roles does ultrapure water (UPW) serve as a source-reduction measure in the semiconductor process?
10. Why does photoresist waste provide an incentive for waste or source reduction techniques?

ACTIVITIES

1. Using the world wide web, search for and compile data on at least two pollution prevention (P2) opportunities currently introduced into the semiconductor or printed circuitboard industries.
2. Ultrapure water (UPW) and Nanopure water (NPW) are promising alternatives to traditional semiconductor cleaning solvents. Contact a semiconductor manufacturer using one or both of these solvents. What are the characteristics of these materials? What are the pros and cons of the use of these materials? In what processes have these materials been introduced?
3. Block-diagram the semiconductor fabrication process outlined in this chapter. Show where hazardous materials are consumed by the process and where hazardous wastes are generated. Indicate each step in the process where a substantial dollar benefit may be realized, if a pollution prevention alternative is introduced.

REFERENCES

Sullivan, J.B., G.R. Krieger, and M. Harrison. 1992. "Semiconductor Manufacturing Hazards," *Hazardous Material Toxicology*. Williams & Wilkins.

Cal-EPA, "Assessment of the Semiconductor Industry Facility Planning Efforts," Document No. 530. Scaramento, CA: Department of Toxic Substances Control, Office of Pollution Prevention and Technology Development, Technology Clearinghouse.

Cal-EPA, "Waste Minimization for Printed Circuit Board Manufacturers," Cal-EPA Access Document No. 209-WMP&. Sacramento, CA: Department of Toxic Substances Control, Office of Pollution Prevention and Technology Development, Technology Clearinghouse.

Appendix A

ISO 14000 Environmental Management System (EMS) and Overcoming Certain Objections

Neal K. Ostler

ABSTRACT

Now that the International Standards Organization (ISO) has released its 14000 series of standards, there is a reasonable amount of disillusionment and discontent on the part of some within the environmentally conscious community. Two of the most obvious objections are (1) that it doesn't require a company to be in compliance but merely to be making a measurable effort, and (2) that it doesn't provide a global standard or a level playing field for companies that become "Registered."

This brief paper is an attempt to respond to those objections and, hopefully, to provide the reader with some insight into the potential positive outcomes from the implementation of ISO 14000.

Environmental protection means different things to a lot of different groups, each having different though closely related concerns. Some groups focus on the depletion of resources and the pollution that arises from the use of organic fuels, and they encourage conservation and the development of renewable sources of energy. Others focus on the protection of our forests, our wilderness areas, our waterways, our endangered species, etcetera. Still others are concerned about the depletion of the earth's natural resources of water and nutrients, and their conservation efforts focus on an attempt to change the behavior of humankind's seeming need to consume and the belief that more and bigger is better. It is my observation that there are common objectives of all these different groups, and I believe the two primary ones are that of (1) the prevention of pollution and (2) sustainability.

BACKGROUND

In the United States, the compliance system that has been in place for the past 25 years is "command and control" and, for numerous reasons, private industry is no longer willing to abide by this type of system. In the Fall of 1995 and Spring of 1996, the U.S. regulatory agencies took a serious beating by Congress, and the consequence is the current overhaul taking place throughout the federal government. The human and corporate reaction to the coercive approach is eventually, backlash, and even then the tendency is to do only as much as each is forced to do and no more.

There is further evidence to demonstrate that what is covered by law is not nearly adequate to protect the environment. While there may be extensive laws to protect the environment from dangerous emissions, from releases of effluents to surface water, and from irresponsible handling of hazardous waste, there are no regulations on water usage, energy consumption, environmental design of products, carbon dioxide emissions, material and other resource solid waste, employee, transportation, and the innumerable things that each of us does that could be done more efficiently with fewer resources.

ISO 14001 offers a change in the approach away from coercion and is designed to achieve progress through an environmental management system (EMS). In order for any EMS to work, it must be complemented by performance standards and must direct itself toward the achievement of established performance levels. ISO 14001 establishes those standards and performance levels and requires a commitment to comply with the relevant laws and processes that bring and maintain that compliance. The primary purpose of ISO 14001 is for organizations to implement management disciplines that will improve their performance in all ways. While it is true that being registered to ISO 14001 does not mean the organization has reached environmental perfection, it does mean the organization has implemented a process that will bring it, over time, to some proximity of perfection. It also presents strong evidence that this is a "Do or Die" type of concept: organizations that fail to implement an EMS will voluntarily be cutting themselves out of their share of the marketplace.

PREVENTION OF POLLUTION

The EPA's definition of "Pollution Prevention" is equated with source reduction. Within ISO 14000, however, the objectives of Prevention of Pollution do not necessarily require source reduction but also may include such issues as recycling, treatment to reduce the intrinsic hazards, energy conservation, life cycle assessments, and product design. ISO 14000 promotes the idea that proactive pollution prevention as the design level is the most effective and desirable goal to instill within an organization; therefore, emphasis is placed on the proactive concept rather than on any one specific approach such as source reduction.

On August 3, 1993, President Clinton issued Executive Order 12856 ("Federal Compliance With Right-To-Know Laws and Pollution Prevention Requirements"). The 1993 Government Performance & Results Act requires performance measures, focusing on the outcome of programs. And on March 7, 1996, President Clinton signed the Technology Transfer Advancement Act (P.L. 104-113 Section 12), allowing government agencies to replace or adopt government standards with private sector standards. It is easy then for this author, and most anyone involved with ISO 14000, to conclude that it would be completely senseless for the federal government to reinvent the wheel, and that it would be only sensible and right to adopt the standards developed, in part, by the American National Standards Institute (ANSI) and the American Society for Testing of Materials (ASTM): the ISO 14000 EMS.

An objection to ISO 14000 is that up front it will cost the company more money to implement. ISO 14001 promises to weed out waste and inefficiencies so that the added cost could be easily made up, possibly even with just one or two cycles of production. In addition, any organization will experience fewer chemical incidents—hence fewer needs for an emergency response, fewer cleanups, and fewer citations and fines. It can be reasonably speculated that, if regulators make progress in observing compliance, there may be the advantage of regulatory flexibility for the organization. If customers (and we are not talking retail sales but corporate and government) demand conformance, the organization will have a competitive advantage. There is no end to the possible benefits (see Benefits) of ISO 14001, but the most important one is greater operating efficiencies and savings for the organization.

SUSTAINABILITY

It can be argued that the primary objective of most every environmentally conscious human being is sustainability. It is not only important to prevent pollution of our water resources but also to protect those resources through conservation efforts. The forests need to be protected to assure a habit for its occupants as to assure its viable role in the cycle of nutrients. The earth's elements and fossil fuels are also a finite resource that must be used wisely and not allowed to be "hazardous wasted" or otherwise squandered.

The New Bottom Line, A Bi-Weekly Column on Business and the Environment is published by Gil Friend and Associates and is distributed internationally by The Los Angeles Times Syndicate (See http://www.eco-ops.com). Gil Friend prefers to call his environmental management system "Strategic Sustainability." *The New Bottom Line* proposes that every organization can change its own bottom line and increase its net profits by making a shift in its emphasis on the environment from that of nuisance to that of an operational function. *The New Bottom Line* further proposes that each organization make this shift a core strategic driver of everything from initial business strategy to product design and marketing plans.

The ISO 14000 standard provides a voluntary EMS that explains how adhering to such standards can help consumers, governments, and companies monitor ways to reduce their environmental impact and increase their long-run sustainability, thus creating "Strategic Sustainability," to borrow the phrase.

In mid-January 1997, the International Institute for Sustainable Development, based in Winnipeg, Canada, published a report that tracks the development of ISO 14000 and provides interpretation of the standard. The report, *Global Green Standards,* intends to facilitate the understanding of what EMS standards are and how they can be effectively applied in most any organization. The report explains why people and companies should pay attention to such voluntary standards as ISO 14000: it will make any company more profitable in its operations, more attractive in the marketplace, and thus afford the company a significant advantage, i.e., it will affect the company's bottom line.

Many persons who developed ISO 14001, a group comprised of interested professionals from 50+ nations, believe that EMS takes the concept of sustainability and operationalizes it for any organization through continuous improvement efforts that become the real fruit of sustainability, doing it better than anything else available. This certainly is "Strategic Sustainability."

It is expected that ISO 14000 will foster an environmental ethic not only voluntarily within the implementing organization itself but also within every individual employee who will be helping their employer become more profitable through greater efficiency and more environmentally responsible.

POSSIBLE BENEFITS

The following list was compiled from queries and answers fielded by Joe Cascio, head of the U.S. ANSI delegation to ISO, during a Satellite Conference on October 24, 1996. The conference was sponsored by the American National Standards Institute (ANSI), along with the University of Missouri-Columbia, the American Society for Testing and Materials (ASTM), and the Global Environmental Technology Foundation (GETF).

It is expected that ISO 14000 will:

- Lead to the harmonization of national rules, labels, and methods;
- Minimize trade barriers and related complications;
- Promote predictability and consistency;
- Lead to improvement of environmental performance;
- Lead to effective maintenance of regulatory compliance;
- Establish a framework to move beyond compliance;
- Assist companies/organizations to demonstrate commitment;
- Assist companies/organizations to enhance public posture;
- Lead to credibility of performance reporting;
- Support a worldwide focus on environmental management;
- Sensitize the internal culture in organizations to environmental matters; and
- Promote a voluntary consensus standards approach to environmental improvement.

When asked, "What are some of the possible ways that the ISO 14000 series of standards will affect or be used by companies and others in the U.S. and around the World," Joe Cascio responded with the following:

- They may be used by companies and organizations to better manage their environmental affairs and to show a commitment to environmental protection.
- Implementation of ISO 14000 could become a condition of business loans to companies that aren't even involved in international trade.
- Insurance companies may lower premiums for those who have implemented the standard. ISO 14000 may become a condition of some customer/supplier transactions, especially in Europe and with the U.S. government.
- Evidence of conformance to ISO 14000 may factor into alternative regulatory programs, the exercise of prosecutorial and sentencing discretion, and into government consent decrees and other legal instruments.

- In the courts, ISO 14000 may become a standard of due care in assessing whether a company was in good faith making consistent and diligent efforts to manage its environmental impact. In multilateral trade agreements, there is a high probability that the ISO 14000 standards will become a factor in establishing whether governments are actually making an attempt to improve the environmental situation within their countries.
- In terms of international aid and loans, the World Bank and other financial institutions may qualify their loans to less-developed countries and begin to use the ISO 14000 standards as an indicator of commitment to environmental protection.

SUMMARY AND CONCLUSION

ISO 14000 is a new proactive approach to environmental protection. Rather than rely on the old system of command and control, the ISO 14000 requires a paradigm shift that will bring all employees and managers into a system of shared, enlightened awareness and personal responsibility for the organization's environmental performance. This new system will rely on positive motivation (above list of benefits) and on the desire to do the right thing (environmental ethic), rather than on the punishment of errors and coercion to comply.

It is not only expected that regulatory compliance will become the natural result of management strategy, but organizations will go well beyond mere compliance as they discover, and prove to themselves, the outcomes and effects on their bottom line. Regulations themselves cannot force a change in mental attitude; however, the effects of voluntary compliance will champion such a paradigm shift in the mental attitude of all employees.

It is acknowledged that conformance to ISO 14000 is not likely to result in overnight or immediate compliance with regulations, but as Joe Cascio states, "The result, over time, is a shift in culture to one that is as sensitive to the environment as to production schedules and product design."

AUTHOR'S OPINION

Having written a number of articles on compliance with environmental regulations, I believe that, at the worst, individuals and companies will react to such efforts with avoidance and manipulation of the regulatory system. At the best, such efforts will be met with passive resistance and with an approach of doing just the bare minimum that is necessary to comply. It may even involve the stretching or embellishment of data to conceal real outcomes or evidence of misbehavior.

Even if all companies and organizations were in total compliance with U.S. regulatory standards, recent industrial accidents have proven that it would still not be enough to ensure against environmental degradation (prevention of pollution) or even come close to attaining the goal of sustainability.

It is this author's personal belief that ISO 14000 holds the final hope for the rescue of our planet. Such a rescue is not going to happen with persons handcuffing themselves to railroad tracks, chaining themselves to trees, singing to the choir in "Green" conferences and seminars in an effort to change the human urge to acquire, publishing doomsday papers announcing the eventual demise of the planet, or blocking the operation of a state-of-the-art hazardous waste facility. While each of these activities is meant to at least catch attention and point to problems, they are not activities that offer solutions; they are merely pundits. The earth's rescue is also not going to happen by empowering a bunch of environmental cops to shut down or interfere with companies behaving the worst environmentally. Self-elimination will occur when those bad actors put themselves out of business by no longer having any competitive advantage or attractiveness in the marketplace.

The rescue of our planet is going to require a paradigm shift in the mental attitude of its inhabitants. ISO 14000 can bring that paradigm shift, and even requires it, and aims to accomplish it directly through the same place where each of us obtains our livelihood and much of the description of our own self-worth: our employer: our work!

RECOMMENDED READING

Cascio, Joseph, Gayle Woodside, Phillip Mitchell. *ISO 14000 GUIDE: The New International Environmental Management Standards.* McGraw Hill Companies, Inc., 1996

Gil Friend and Associates. "The New Bottom Line A Bi-Weekly Column on Business and the Environment." Ecological Operating Systems Homepage, http://www.eco-ops.com/eco-ops

ANSI/GETF ISO14000 Integrated Solutions. "Your Complete ISO 14000 Resource." IIS On-line, http://www.iso14000.org/

The Green Disk. "Guide to a Sustainable Future." *The Green Disk Journal,* http://www.igc.apc.org/greendisk/gsus.html

IISD Announcements. "Global Green Standards: ISO 14000 and Sustainable Development." The International Institute for Sustainable Development (IISD), http://iisdl.iisd.ca/about/announce

Appendix B
The Flow of Materials

INTRODUCTION

The flow of materials analysis is a fundamental tool of the Hazardous Materials Technician and is basic to any industrial process regardless of the type of industry or level of operation.

Essentially this involves compiling data of every input or resource, balancing it against the output or product, and analyzing what has happened to all of the materials in between.

THE EXTERNAL MATERIAL BALANCE

STEP #1: Input Materials (resources)

Since all of the materials used in the process are bought and paid for by the company, they are also accurately measured and inventoried. This information can be obtained from purchasing or accounting. Operations may also be a source of data. Serious consideration must be given to the following:

1. Raw materials delivered by truck or pipeline. Examples are mineral ores, crude oil, and chemicals.
2. Treatment chemicals. Examples are acids, caustics, and solvents.
3. Utilities. Examples are gas, oil, coal, and electricity.
4. Natural resources.

- Water—often carrying effluents and blowdowns in proportion to the amount of water used.
- Air losses through emissions can be substantial.

STEP #2: Output (product) Inventories

Essentially the same sources of information can be utilized for the gathering of data about the product going out the gate, which is also accurately measured, weighed, packaged with careful records kept by the sales/accounting department.

STEP #3: Loss Analysis (balance)

Now that we have a gross measurement of what has been consumed in the process and what has been produced, we can begin to investigate and ferret the differences.

Classic physics tells you "matter cannot be either created or destroyed." With this as a guide, the Hazardous Materials Technician goes on in his search for the following:

I. Transformation of Energy/Emissions.

A. Burning of fuel—such as in furnaces to run distillation equipment, boilers to make steam, heat for buildings, or to run compressors, etcetera—appears as air emissions. They are found in the boiler stacks as CO_2, CO, H_2O, and SO_2.

NOTE: The Clean Air Act tightly monitors these air emissions, therefore, it is important for the Hazardous Materials Technician to accurately define them.

B. Water, either evaporated or blown down from the cooling towers, may be clear to the human eye and appear quite clean yet constitute a significant portion of the process material balance. Water molecules are large, very solvent, and often link with molecules of other hazardous substances for transport into the environment by runoff and by evaporation.

II. Losses.

Precious metals and other valuable substances can become very small molecules by attrition (knocking about with other molecules in the process) and can be lost from a leak or go up the stack as fine particles of dust. The more valuable the material the more closely it is monitored; however, the disappearance of any material is cause for concern of the Hazardous Materials Technician.

III. By-products.

The production of by-products appears in processing and is often easy to measure. These materials frequently have value and can be (and are) sold. As an article of commerce, purchasing, sales, or accounting will have accurate records of the quantities, etcetera.

IV. Wastes.

Much attention is given to waste minimization because it is a very expensive commodity to dispose of, creating a liability to the company. Consequently today, wastes are weighed, measured, analyzed, shipped, and disposed of very carefully. Wastes divide into three (3) categories:

1. RCRA Wastes—those wastes listed as hazardous or are hazardous as determined by criteria that is tested to be (1) ignitable, (2) corrosive, (3) reactive, or are specifically (4) EP toxic.
2. Solid Wastes—constitute garbage, refuse, sludge, and other discarded materials not subject to the Clean Water Act and not considered hazardous. Examples are bricks, boards, and plant trash.
3. Other Wastes—include "spoiled chemicals" or "junk liquids" that have become accidentally contaminated and are thus of no use to the process. These wastes can be sold to others for reprocessing or other such uses. It has value to someone and thus is still considered to have commercial value. Because such wastes prove useful to someone, they are still considered to have commerce value.

STEP #4: Flow Analysis

We've now compiled the materials balance, and for it to be useful, a flow analysis of hazardous materials must be developed. The purpose of this step is to permit the analyst to make many important conclusions if the balance is accurate. These conclusions are listed below:

I. Inputs and outputs from the plant can be evaluated in terms of hazard potential using appropriate MSDS forms.
II. Inputs and outputs of all separate (1) wastes, (2) losses, (3) by-products, (4) effluents, and (5) emissions.
III. Suggestions can be made for a solution or identified problems.

- Suggestions for substituting chemicals in the process to get an effluent or waste out of regulation altogether.
- Find control devices to get an emission or effluent back into compliance.

The following diagram illustrates a flow analysis diagram for chromated copper arsenate used in the treatment of wood.

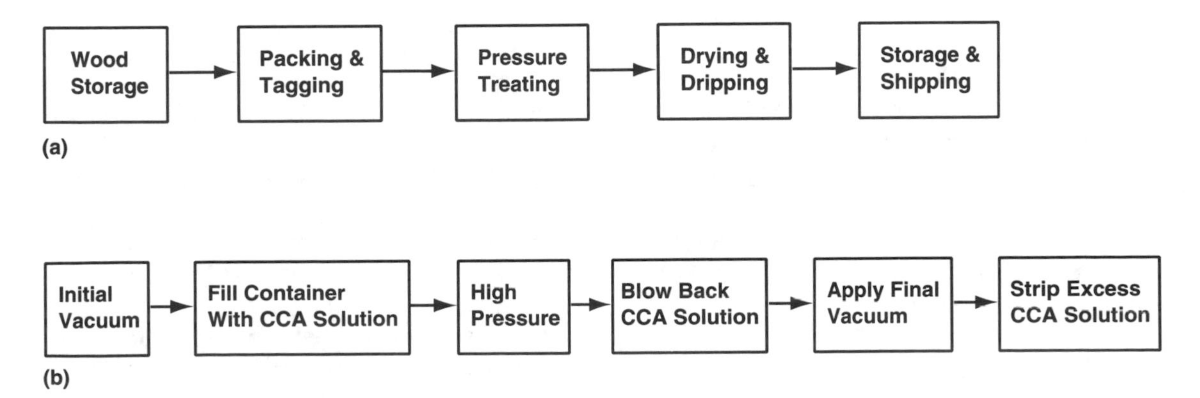

FIGURE 1
(a) Overall Material Flow Diagram for Chromated Copper Arsenate (CCA) Wood Treating
(b) Pressure Treating Cycle

THE INTERNAL UNIT BALANCE

All processing plants, no matter how small or how large and complex, consist of separate phases or units of operation that can be separately identified for analysis by their contribution to the operation as a whole.

The Hazardous Materials Technician can select particularly hazardous streams from within each unit based upon the following criteria:

I. The potential to substitute another material to reduce costs or to remove it from a classification as hazardous either as a resource material, by-product, or waste.
II. The identification of a problem area where significant waste reduction is an optimal solution.

The Technician should prioritize his efforts to define the above two areas, which is why he is present in the process environment.

Complete Unit Material Balances may never be available, but select streams can be targeted for study and characterization.

The Hazardous Materials Technicians must think for themselves, recognize limited resources and time, minimize the generation of unnecessary or unuseful information, yet provide adequate data to define the problem and suggest its solution.

EXCEPTIONS

In many operations, it is impossible to obtain any meaningful material balances due to the nature of business. Often the next process, job, or batch to be run is determined by the next job order or contract brought in by the sales force.

Examples of such operations are "job shop" electroplating, metal finishing, chemicals, pharmaceuticals, and the dyestuff production industries.

It is important that the Hazardous Materials Technician recognizes these types of operations immediately and not attempt a materials balance where none can be accurately obtained. Instead, a selection of particular hazardous streams and previous recipes or formulations should be the target of analysis.

SUMMARY

If a material balance has been prepared properly and is truly accurate, the inputs will equal the outputs and a balance can be obtained.

The data should be recorded in a spreadsheet, diagram, or drawing in an organized fashion.

All units of measurement should be the same to allow for easy interpretation of the data. An example is to place everything in pounds per day, or tons per week, etcetera.

Be accurate. Do not guess; if you must guess, establish tight tolerances for error. Be precise. Ask questions, questions, and more questions. Investigate. Be sure of your data.

Appendix C

Regulatory Agenda Overview

INTRODUCTION

The complex, overlapping, and often confusing environmental regulations are designed to protect human health and the environment from the adverse affects of hazardous materials and their toxic wastes. These regulations meet one or more of the following goals:

1. To serve and protect the general public from factory waste streams, discharges, effluents, and emissions.
2. To protect workers or employees from health and human safety hazards in the workplace.
3. To protect community health and safety from the adverse affects of unauthorized releases (spills) of toxic substances and wastes.

Public awareness and outcry over the presence of hazardous materials in the community and concern for the affects from pollution of the environment has prompted the U.S. Congress to pass numerous acts or laws in response. These laws have provided a framework of responsibility and either created or burdened an existing agency with their implementation and enforcement. These agencies then develop a set of proposed rules and regulations from the provisions of the Law and publish them in the *Federal Register.* Following public comment, which may change or throw out some of the rules, the final regulations are codified and published in the Code of Federal Regulations.

State and local governments follow similar steps, but with few exceptions, they pretty well adopt the federal standards and codes.

OVERVIEW OF ACTS & REGULATIONS

Essentially the regulations can be divided into three major categories: Hazardous Materials, Environmental Protection, and Hazardous Wastes.

I. Hazardous Materials

The manufacturing and use of hazardous materials in production of goods for the consumer necessitates the presence of hazardous materials in the workplace, in transportation, and in the everyday environment of home and work. Several essential acts and their provisions are as follows:

- Consumer Product Safety Act—this act was passed in 1972 in response to heightened consumer and public concern for the safety of products that could be purchased anywhere at the check-out counter. This act requires extensive testing before a product can be marketed and also provides standards for the product labeling by manufacturers, importers, distributors, and retailers. The act provides a mechanism that consumers and others can use to evaluate the safety of products and protect themselves from such risks.
- Hazardous Materials Transportation Act—passed in 1975, this act gives the DOT authority to regulate shippers, carriers, and transporters of hazardous materials, as well as anyone involved in the packaging of such materials for transportation. Found in 49CFR 171–177, the regulations discuss in detail the identification, communication, packaging, and shipping of hazardous materials in commerce. These regulations pertain to materials having a commercial value.
- Occupational Safety & Health Act—found in 29CFR1910, this act provides for standards in the workplace to ensure employee safety from unreasonable risks, covering the entire spectrum of risks from working surfaces and illumination/ventilation to hazardous materials. These standards apply to virtually every employer. Probably the most important rule made through passage of this act is The Hazardous Communication Standard or employee right to know, which requires the employer to inform and train all employees in the hazards present.
- Toxic Substances Control Act—this act empowers the EPA to regulate the manufacture, processing, distribution, and use of existing chemicals. Passed in 1976 and found in 40 CFR 700–799, this act essentially requires that the EPA be notified before marketing and be given time to evaluate the hazards of any new chemical or the new use of an existing chemical before they are placed in commerce.
- Federal Food, Drug & Cosmetic Act—this act established the Food & Drug Administration (FDA) under the U.S. Department of Health and Human Services in 1938. The FDA regulates the manufacture and sale of drugs and cosmetics to ensure they are effective in their intended use and not harmful to the consumer. This is accomplished through research, establishment of standards, oversight of food additives, and by promotion of sanitary practices in public eating establishments. The FDA is also the primary federal support agency for regional poison control centers.

- Sara Title III-Emergency Planning & Community Right-To-Know. A part of SARA, this title is entirely separate and provides the public and government with the information and resources necessary to plan for all contingencies involved from the risk inherent of hazardous materials in the community.

II. Environmental Protection

For many decades government permitted industry to engage in commerce without regard for the effects on the life-sustaining, economic, and recreational value of the air, land, and water. To restore and protect these very essential natural resources, Congress has empowered the EPA with laws to accomplish these goals. Those of most current significance include the following:

- The Clean Air Act (CAA)—passed in 1970 and amended in 1983, this act is the basic federal law for controlling air pollution. It requires the EPA to keep an up-to-date list of serious industrial air pollutants and set an acceptable (with ample margin for safety) level of emissions for those considered hazardous to health. The act establishes permissible programs and provisions for oversight of state programs for enforcement.
- The Clean Water Act (CWA)—passed in 1972 and amended in 1977 and 1987, this act has as its objective the restoration of the integrity of the nations waters through the prohibition of hazardous waste and toxicants being discharged into our water ways. The act provided for permitted discharges governed by the National Pollution Discharge Elimination System (NPDES), standards for Public Owned Treatment Works (POTWs), and goals for dealing with non-point sources such as rain water runoff. The act also stipulates powerful penalties for violators.
- Safe Drinking Water Act—this act protects the quality of all sources of drinking water—surface water, ground water and aquifers—and establishes maximum levels of drinking water contaminants. States are directed to comply with minimum standards and to ban lead content in plumbing materials.
- Federal Insecticide, Fungicide, and Rodenticide Act (FIFRA)—passed in 1947 and last amended in 1978, this act regulates the manufacture, production, and marketing of pesticides in the United States and provides oversight of state programs for certification or applicators of restricted-use products.
- Noise Control Act—passed in 1972 and referenced in 33 USC 1415, this act establishes noise emission standards for commercial products, transportation vehicles, and mechanical equipment, as well as directs the EPA to research the effects of noise. Federal moneys have expired and noise is deemed a state and local problem.
- Asbestos School Hazard Abatement Act of 1984, and Asbestos Hazard Emergency Response Act (AHERA)—the latter, passed in 1986 and effective October 1987, this act requires public schools to have certified persons inspect for asbestos, conduct analysis in accredited laboratories, re-inspect, and keep records. Guidelines for cleanup are also provided. It should be noted that the EPA enforces regulations to prevent contamination of the environment and OSHA provides protection for asbestos workers.
- Rivers and Harbors Appropriations Act—passed in 1989, this act provides for the unauthorized obstruction or pollution of the navigable waters in the United States. Ref: 33 USC 403–411.

- National Environmental Policy Act (NEPA)—this act establishes the Council of Environmental Safety (CES), which requires the submission of extensive research papers for projects of most any type that may have an impact on the environment. Categorical exclusions are enumerated. Once submitted, the CEQ examines the documents and then releases their own Environmental Impact Statement.
- Atomic Energy Act—passed originally in 1954, this act formed the Nuclear Regulatory Commission (NRC) and the Department of Energy (DOE). The regulations for these agencies are found in 10 CFR. The NRC regulates the use of atomic energy, while the DOE provides the framework for a comprehensive national energy plan. Radioactive wastes are under the jurisdiction of DOE, which also trains individuals in both public and private sectors in the handling, removal, transportation and disposal of radioactive contaminants.

III. Hazardous Wastes

Although the government has been concerned with the protection of wildlife and other natural resources, it was not until the 1970s that Congress started giving much attention to pollution of the land, which by then demonstrated that contamination of the land not only threatens future uses of the land itself but also affects the quality of surrounding air, surface water, and the aquifers. Without question, the future of our biosphere and the earth itself critically depends upon the establishment of a number one priority in cleaning up existing uncontrolled waste sites and in managing our current hazardous waste problems to prevent similar disorder in the future.

- Uranium Mill Tailings Radiation Act—this act authorizes the Department of Energy, with concurrence of the NRC and in accordance with EPA standards, to investigate remediation of designated uranium mill tailings sites and gives the NRC jurisdiction over active sites. States are co-participants.
- Marine Protection Research & Sanctuaries Act—requires the EPA to protect oceans from waste by authorizing and permitting safe sites for dumping hazardous wastes.
- Resource Conservation & Recovery Act (RCRA)—this act was passed in 1976 and amended in 1984 as part of an overall control mechanism for the management of hazardous materials from “cradle to grave.” The goals of RCRA are to protect human health and the environment, promote waste minimization, and encourage the conservation of natural resources. The program is administered at the state level through permitting and provides an intricate network for inspection of record-keeping requirements of any activity involving hazardous waste.
- Comprehensive Environmental Response, Compensation, and Liability Act (CERCLA)—passed in 1980 to deal with the unlawful dumping of hazardous waste, this act authorizes the EPA to respond to such sites, prepare a priority list of those deemed hazardous to health or the environment, and to remediate or clean up the listed sites. The law also provides a powerful backbone for its enforcement by providing civil and criminal recourse.
- Suspended Amendments and Reauthorization Act (SARA)—this act re-authorized CERCLA and laid down the future framework for continuing the cleanup of waste sites and provided for $8.5 billion to accomplish its priorities.

Appendix D

The EPA Common Sense Initiative

References:	EPA Office of Public Affairs: Speech by Carol M. Browner, Administrator, July 20, 1994
Substance(s) Covered:	Multimedia protection: land, air, and water
Regulatory Agency(s):	EPA and states
Significant Date:	July 20, 1994
Related Legislation:	NA

INTRODUCTION

The Common Sense Initiative (CSI) is a fundamentally different approach to the complex issue of environmental protection. Instead of the pollutant-by-pollutant approach of the past, the CSI is an industry-by-industry approach for the future. The plan would provide each industry with a one-step shop for all of the environmental guidance, regulations, and reporting requirements.

LEGISLATIVE HISTORY

In the 25 years since the creation of the EPA, there is much that has been accomplished. Many of the more obvious problems have been solved: rivers no longer catch on fire, the skies are much cleaner, hazardous waste is being managed, abandoned land fills are being remediated, and U.S. environmental expertise and technology are in demand throughout the world.

Since the 1970s, the regulatory mechanism has evolved from an effort to address issues. By necessity the current system grew up on an emergency basis as this nation attempted to resolve issue by issue, crisis by crisis, and pollutant by pollutant as needed. Too often, however, our activities have become compartmentalized, and too often, instead of pollution prevention, the problems have merely been shuffled from one place to another: from land to air, from air to water, and from water to land.

The result of this fragmentation and conflict has been too little protection at too high a cost, and the resulting frustration has fueled a backlash in some business circles, calling for rollback of public health protection. This is not legitimate, and to suggest we don't need both environmental and occupational health and safety protection is irresponsible.

Drawing upon the lessons of the past 25 years, the need to integrate the management of air, water, and land is now recognized by the EPA: nature is a system and so should be our environmental regulatory mechanism. While changing the means, there is still an ongoing need to preserve and strengthen the principles of environmental protection. The EPA must move beyond the adversarial process and start to inform and involve the communities, industries, and people of this country in their regulatory process of these same principles.

OBJECTIVE: The Common Sense Initiative grows out of the EPA's recognition of the lessons learned and yet also builds on the accomplishments of the past to shape a new generation of true environmental protection for the future.

PROVISIONS

Government officials at all levels will come together with industry leaders to create cleaner, cheaper, and smarter strategies to protect the health of people in the U.S. and to protect our natural resources.

- Cleaner: working together with all industries should provide more realistic and measurable results.
- Cheaper: tailoring environmental protection on an industry-by-industry basis will save billions of dollars.
- Smarter: working together should provide more creativity and ingenuity in the application of pollution prevention.

For the first time, every stakeholder will be at the table together: industry, regulator, environmentalists, and the communities. Every major topic will be on the agenda.

The administrator of the EPA, Carol M. Browner, has pledged to implement any changes and to work together on having any revisions made in the law if necessary.

Six Model Industries

1. Automobile Assembly: This industry accounts for a huge percentage of the national consumption of natural resources, with VOCs and toxic air emissions

being the leading concerns. The industry faces development of new standards for effluents and emissions and is regulated by the Clean Air Act, Clean Water Act, RCRA, Community Right-To-Know, CERCLA, and the Toxic Substance Control Act.

2. Computers and Electronics: Characterized by a rapid turnover in technology and a large number of different hazardous materials (over 1/2 of 700) as basic ingredients, the EPA is currently developing a Maximum Achievable Control Technology (MACT) for the semiconductor industry under the Clean Air Act.
3. Iron and Steel: Water pollutants include toxic organics, metals, cyanides, fluoride, sulfide, ammonia, and oil and grease. Coke ovens and electric arc furnaces are air toxic sources, and the industry is a major source of particulate emissions. This industry is looking for consistent interpretation and enforcement of the regulations.
4. Metal Plating and Finishing: This industry is characterized by waste streams, which include effluents, solids, and air emissions from a wide variety of acids, solvents, metals, and cyanides that used in metal finishing and processing. Facilities are subject to more than 20 separate training and reporting regimes.
5. Petroleum Refining: Refineries are characterized by emissions that include VOCs, CO_2, NO_x, SO_2, as well as particulates and water pollutants that include benzene, toluene, ammonia, oil and grease. This industry is subject to regulations over air, water, land, and emergency response. Pollution abatement is a major cost to this industry.
6. Printing: Printers use a large variety of chemicals that can lead to significant emissions of air toxics and volatile organic compounds. Solid wastes include spent film, solvent laden rags, waste ink, used plates, and photoprocessing chemicals containing silver. Some printers face more than 30 different reporting and recordkeeping requirements every year.

At the time of this writing, there have been positive statements of support released by representatives of each of the industries discussed.

The Basic Blueprint

Administrator Browner will convene a high-level team from each of the six model industries. The Common Sense Teams will work together to find ways to achieve cleaner, cheaper, and smarter performance in the following areas:

Regulations: Improve *new* rules through increased coordination.

Pollution Prevention: Actively promote P_2 as a standard business practice and as a central ethic.

Reporting: Make it easier to provide, use, and publicly disseminate relevant pollution and environmental information.

Compliance: Assist those companies who seek to obey or exceed legal requirements and consistently enforce the laws against those that do not.

Permitting: Make the process more efficient, encourage innovation and public participation, and develop one stop for air, water, and land.

Environmental Technology: Provide incentives and flexibility for each industry to develop new technologies that meet and exceed environmental standards, while also cutting operating costs.

COMPLIANCE & ENFORCEMENT

It is not expected that fundamental change will happen overnight; however, within a year (from July 20, 1994) each Common Sense Team is expected to have developed tangible recommendations to achieve better and more efficient environmental protection within their Industry. Consensus solutions will be implemented within the EPA immediately, and if legislative change is agreed upon, Congress will be approached for amendments and changes to the current legislation.

Appendix E
Environmental Leadership Program (ELP)

INTRODUCTION

The EPA wants to recognize such facilities for their leadership and encourage others to join them. The Environmental Leadership Program (ELP) is designed to demonstrate innovative approaches to establishing and assuring compliance with environmental requirements. The projects test a range of activities, including the following:

- Implementation of an environmental management system;
- Routine audits through independent third-party verification or self-certification;
- Support for mentoring projects to help small businesses achieve compliance;
- Involvement of employees and the sharing of environmental performance information with the public; and
- Pollution prevention practices.

Goals: Listed below are the goals of ELP:

- Better protection of the environment and human health by promoting a systematic approach to managing environmental issues and by encouraging environmental enhancement activities (e.g., biodiversity, energy conservation);
- Increased identification and timely resolution of environmental compliance issues by ELP participants;
- Multiplying the compliance assistance efforts by including industry leaders as mentors; and by fostering constructive and open relationships between agencies, the regulated community, and the public.

ELIGIBILITY TO PARTICIPATE

Any public, private, or federal facility that meets the following ELP criteria for environmental leadership can apply to be an ELP participant:

- A facility must have a mature environmental management system (EMS) that conforms to the ELP EMS. The criteria for an ELP EMS are outlined in the ELP EMS fact sheet (EPA 305-F-96-011).
- A facility must have a compliance and EMS auditing program.
- As part of its EMS, a facility must participate in community outreach/employee involvement programs. Such programs foster the development of relationships between facilities and two of their major stakeholders—local communities and employees.
- A facility must submit facilitywide compliance audit results and EMS information (data or results documentation) obtained within the previous two years. In addition, the application should include the dates and a summary of the findings from any agency regulatory inspection(s) conducted in the previous two years.
- Federal facilities are required to verify that their parent agency endorses the Code of Environmental Management Principles (CEMP) and briefly describe how the applying facility is implementing the CEMP. ELP has been adopted as the Model Installation Program for federal facilities under EO 12856.

BENEFITS OF AN ELP

Benefits to the environment are anticipated from the program's focus on encouraging environmental enhancement activities, such as environmental restoration projects and product stewardship. The program will facilitate an exchange of information and encourage the implementation of best practices related to environmental management systems and pollution prevention activities.

The ELP provides an opportunity to foster constructive relationships between the ELP members, regulators, and the public. Building productive working relationships among environmental stakeholders may lead to tangible benefits for the environment and public health, especially if regulatory resources can be effectively redirected to focus on environmental "bad actors" and expanded compliance assistance efforts. The Formal Recognition includes the following:

- **Public Recognition.** EPA will issue certificates of membership in the ELP and develop programs and activities designed to publicly recognize ELP facilities at federal, regional, state, and local levels.
- **Logo Usage.** Member facilities can use the EPA-issued ELP logo in facility (but not product) advertising, on facility equipment and structures, and internally on stationery, coffee mugs, T-shirts, jackets, etcetera.
- **Inspection Discretion.** Through the EPA's enforcement discretion, participating regulatory entities will reduce and/or modify discretionary inspections.

Due to the leadership and exemplary environmental performance of ELP members, it is anticipated that the following Reduced Regulatory Burden benefits are currently being considered but are not yet available: The ELP EMS is envisioned as one potential agency "gateway" for additional benefits for implementing an EMS, such as allowing ELP facilities an opportunity to expeditiously participate in the EPA's regulatory reinvention projects, which include expedited permits, longer permit cycles, and streamlined permit modifications.

ENVIRONMENTAL MANAGEMENT SYSTEM (EMS)

The ELP EMS requirements are based on an EMS with the characteristics of ISO 14001, an international environmental standard that was published in September 1996, but with explicit inclusion of compliance assurance and other required elements. A copy of ISO 14001 can be obtained from the American National Standards Institute (ANSI). (Please see the "For More Smart Quotes Information" section for ANSI's address.) The purpose of ISO 14001 is to provide organizations with the elements of an effective EMS that can be integrated with other management requirements to help organizations achieve environmental and economic goals. It was developed to raise consistency in managing environmental responsibilities. The overall goal of ISO 14001 is to support environmental protection and pollution prevention in balance with socioeconomic needs.

The ELP EMS will have guidelines and requirements for the following:

- Requirements of the Audit itself,
- Regular internal and external audits,
- Credentialing requirements for auditors, and
- The ELP Annual Environmental Performance Report.

COMPLIANCE

If an ELP member is unable to maintain the required level of environmental performance as defined by ELP, the facility may be terminated or suspended from ELP participation. Below are examples of activities that may result in termination or suspension from the ELP:

- ***Termination:*** (1) Material nonconformance to ELP requirements, (2) any federal or state criminal conviction of the corporation for an environmental offense, (3) appearance of facility or corporation on the EPA's delisting/debarment list, (4) delinquent penalties resulting from previous enforcement action against the facility, and (5) federal or state criminal action pending against facility or corporation for an environmental offense.
- ***Immediate suspension:*** (1) Violations that present an imminent and substantial endangerment or a serious actual harm, (2) pending federal or state environmental judicial action, and (3) pending federal or state environmental administrative action.
- ***Possible suspension:*** (1) Repeat violations as defined under the EPA's Incentives for Self-Policing Policy, (2) pending citizen suits, (3) environmental criminal action pending against an individual officer or employee, (4) criminal conviction for an environmental offense against an individual officer or employee, and (5) investigation of falsification or misrepresentation of the performance of the facility/entity's EMS.

- Misuse or misrepresenting environmental performance in advertising or marketing claims
- Failure to live up to tracking and reporting requirements
- Failure to live up to the ELP EMS principles.

The ELP will also have provisions for reinstatement and for voluntary withdrawals.

CONCLUSION

For the ELP to be truly effective, facilities should be recognized as environmental leaders by the EPA and other participating regulatory entities, as appropriate. The program has been designed with the expectation that the EPA and the state, at a minimum (but also other applicable levels of government) will work in partnership to review applications, participate in on-site reviews, select facilities, and implement the program. It is anticipated there will be a signed agreement between the EPA and other participating regulatory entities detailing respective roles and responsibilities. A model EPA-State agreement is currently being developed.

PILOT PROGRAMS

As part of the EPA's ongoing efforts to improve environmental performance, encourage voluntary compliance, and build working relationships with stakeholders, the EPA developed the ELP. Initiated in April 1995, the one-year pilot program has been completed, and the EPA is making plans to launch its full-scale Leadership Program in 1997.

Progress reports and other information on these pilots is made available on the Internet (http://www.envirosense.com/elp).

Next Steps:

Write and make available a conclusions report, including proposed criteria and incentives for participation in a "full-scale" ELP.

Work with full range of stakeholders, including states, environmental groups, and others to develop and implement "full-scale" program. Publish notice in *Federal Register* and open comment period for a "full-scale" ELP.

SOURCE

EPA's Design for the Environment Homepage: http://www.envirosense.com/elp

Appendix F

Design for the Environment (DfE)

INTRODUCTION

Under the authority of the Pollution Prevention Act of 1990, the EPA created the Design for the Environment (DfE) Program to build on the "design for the environment" concept pioneered by industry. Under this program, the EPA encourages businesses to incorporate environmental considerations into the design and redesign of products, processes, and technical and management systems. The EPA DfE Program forges voluntary partnerships with a variety of stakeholders in an effort to reach two goals:

- Incorporate environmental concerns into the traditional decision-making parameters of the business world: cost and performance.
- Build incentives for behavioral change to encourage continuous environmental improvement.

To accomplish these goals, the DfE Program uses the EPA's expertise and leadership to evaluate the environmental and human health risks, performance, and cost of both traditional and alternative technologies, processes, and materials. DfE disseminates information on its work to all interested parties and also assists businesses in implementing new technologies and processes identified through the program. The program has formed cooperative partnerships with the following groups:

- Industry
- Professional institutions/trade associations

- Academia/research institutions
- Environmental/public interest groups
- Other government agencies

WORKING WITH INDUSTRY

The DfE Program is working with several industry sectors to identify cost-effective alternatives to existing products and processes that reduce risks to workers and the environment while maintaining or improving performance and product quality. A typical DfE industry project includes developing a Cleaner Technologies Substitutes Assessment (CTSA) and a communication and implementation strategy. CTSAs provide detailed environmental, economic, and performance information on traditional and alternative manufacturing methods and technologies. To help industry implement some of the new technologies identified during CTSA development, DfE provides a variety of outreach tools, including fact sheets, bulletins, pollution prevention case studies, software, videos, and training materials.

Printed Wiring Boards. The printed wiring board (PWB) is the building block of the electronics industry. It is the underlying link between semiconductors, computer chips, and other electronic components. The traditional electroless copper process for manufacturing PWBs uses toxic chemicals that pose potential health and environmental risks, generates large volumes of hazardous waste, and uses substantial amounts of water and energy.

The DfE PWB Project has worked with PWB manufacturers, trade associations, research and academic institutions, and public interest groups to examine alternative technologies that reduce or eliminate these impacts. The project evaluated the health and environmental risks, performance, and cost of the electroless copper process and six promising alternative technologies. The project partners are now conducting a similar evaluation of alternative surface finishes that can replace the hot air solder leveling process.

Lithography. The DfE Lithography Project is a voluntary effort between the lithographic printing industry and the EPA. The goal of the project is to provide lithographers with information that can help them design a safer, more cost-effective, and more environmentally sound operation.

There are approximately 54,000 lithographic printing shops in the United States, many of which use petroleum solvents to clean their presses. These solvents, called blanket washes, contain volatile organic compounds and can pose risks to human health and the environment. The partners of the DfE Lithography Project evaluated 37 different commercially available products, focusing on blanket washes. Information was gathered on the performance, cost, and health and environmental risks of each substitute blanket wash. The project partners provide this information to help printers make more informed decisions about the products they use in their shops.

Screen Printing. The DfE Screen Printing Project is a voluntary effort between representatives of the screen printing industry and the EPA. The goal of the project is to provide screen printers with information that can help them design an operation that is more environmentally sound, safer for workers, and more cost-effective.

The DfE Screen Printing Project encourages the nation's 20,000 graphic art screen printers to consider environmental and worker safety concerns, along with cost and performance, when purchasing materials and designing systems. The project focused primarily on the process of screen reclamation and evaluated several different reclamation systems. Information was gathered on the performance, cost, and health and environmental risk trade-offs of the components of each screen reclamation system (ink removers, emulsion removers, and haze removers). The project partners provide this information to help printers make more informed decisions about the products they use in their shops.

Flexography. The DfE Flexography Project is a voluntary effort between the flexographic printing industry and the EPA. Like the lithography and screen printing industries, the goal of the project is to provide flexographers with the information they need to design an operation that is more environmentally sound, safer for workers, and more cost-effective.

Flexography is a process used primarily for printing on paper, corrugated paperboard, or plastic consumer packages and labels. Conventional flexographic inks contain solvents made of volatile organic compounds that can pose risks to human health and the environment. DfE is working in partnership with seven trade associations, representing over 1,600 flexographic printers and the ink manufacturers, to evaluate alternative solvent, waterborne, and ultraviolet-cured flexographic ink technologies. The project partners provide this information to help printers make more informed decisions about the products they use in their shops.

Garment and Textiles. The DfE Garment and Textile Care Program is committed to promoting environmentally benign alternative technologies for garment and textile care through a systems approach to the development, fabrication, manufacture, distribution, and care of garments and textile products.

The expanding program will be developed by involving representatives from the EPA's DfE Garment Care Program, and stakeholders from industry, labor, community action, environmental, trade association, and research organizations. The EPA is in the process of developing specific goals, operations, and measures for achieving the program's vision and mission. These statements will be presented in a DfE Garment and Textile Care Program Strategy currently in progress and are due to be released in late 1997.

Some key elements envisioned for the strategy include the following:

- Reinventing a more environmentally friendly garment and textile care industry.
- Promoting risk reduction throughout the industrial cycle from fiber manufacture to fabric care.
- Directing market forces towards the most cost-effective, energy-efficient, and environmentally benign technologies.
- Emphasizing science and strengthening the EPA's core technological base for this garment care sector.
- Using the EPA's science and technology base to support all industries involved with garment care.
- Leveraging government funds with industry and academic resources to catalyze rapid developments in the garment care industry.

Despite the many scientific, technological, and political uncertainties of the next decade, the EPA believes this approach will allow the DfE Garment and Textile Care Program to retain a leadership role in effecting fundamental change through its voluntary partnership program.

Metal Finishing. Metal surface finishing involves a variety of processes to coat a metallic base material with one or more layers of another metal, paint, or plastic to enhance, alter, or protect the metal's surface. Typical metal-finishing processes produce air emissions, wastewater effluent, and solid waste. The Metal Finishing Project has produced a variety of pollution prevention materials for the nation's 12,500 metal finishers, including an industry profile, a regulatory guide, and waste assessment tools. The project has also initiated a series of demonstration projects to examine emerging pollution prevention alternatives, including chrome electroplating projects at four sites in Michigan and Ohio.

ENVIRONMENTAL PURCHASING

Environmentally Preferable Purchasing (EPP) is an EPA program that promotes federal government use of products and services that pose reduced impacts to human health and the environment. Such purchases are required by Executive Order 12873, Federal Acquisition, Recycling, and Waste Prevention. The Executive Order also directed the EPA to develop guidance to help federal agencies incorporate environmental preferability into their purchasing procedures.

EPP is currently documenting the results of government and private sector pilot projects to demonstrate the effectiveness of incorporating environmental considerations into the purchasing process. Environmental purchasing can reduce environmental and health impacts and save money.

In addition to working with the private and public sector projects, the EPA is developing programs to help homeowners make environmental purchases. The EPA has also begun a pilot program in the use of cleaning products.

ASSESSMENT TOOLS

In addition to working closely with specific industry sectors to help them incorporate environmental considerations into the design and redesign of products, processes, and systems, the DfE Program is developing tools that can be used to assess substitute technologies and other pollution prevention options. These tools include the following:

- Cleaner Technologies Substitutes Assessment (CTSA): A Methodology and Resource Guide—this guide describes the methods and resources used by the DfE program to compose the risk, cost, and performance of alternative products, processes, and technologies.
- Use Clusters Scoring System (UCSS)—UCSS helps identify and analyze groups of chemicals that are used to perform various tasks and prioritize these groups based on environmental, human health, and safety risks.
- Financial Tools—The EPA is working with the accounting, insurance, and finance industries to identify and quantify the economic and environmental savings that can be achieved by implementing innovative pollution prevention options.
- TRI Indicators—The DfE is developing a computer program that uses Toxic Release Inventory (TRI) data to measure the relative impacts of multimedia chemical emissions.

CONCLUSION

Building on the "design for the environment" concept pioneered by industry, the EPA's DfE Program helps businesses incorporate environmental considerations into the design and redesign of products, processes, and technical and management systems. Initiated by the EPA's Office of Pollution Prevention and Toxics (OPPT) in 1992, the DfE forms voluntary partnerships with industry, universities, research institutions, public interest groups, and other government agencies.

Project partners' activities include broad institutional efforts aimed at changing general business practices, as well as cooperative projects with trade associations and businesses in specific industries. The DfE Program ensures that the information developed through these voluntary efforts reaches the people who make decisions—from managers and industrial design engineers to materials specifiers and buyers. This information dissemination promotes the incorporation of environmental considerations into the traditional business decision-making process.

SOURCE

EPA's Design for the Environment Homepage: http://www.epa.gov/opptintr/dfe/epp/index.html

Appendix G

Plant ____________	Waste Minimization Assessment	Prepared By ____________
		Checked By ____________
Date ____________	Proj. No. ____________	Sheet 1 of 2 Page __ of __

WORKSHEET
10a

WASTE STREAM IDENTIFICATION

1. Check the waste streams below that are generated in the plant.

PROCESS	WASTE STREAM	
Machining Operations	Coolant/Cutting Fluid	☐
	Other: ____________	☐
Metal Parts Cleaning and Stripping	Solvents	☐
	Alkaline Wastes	☐
	Acid Wastes	☐
	Abrasives	☐
	Waste Water	☐
	Air Emissions	☐
	Other: ____________	☐
Surface Treatment and Plating	Spent Bath Solutions	☐
	Filter Waste	☐
	Rinse Water	☐
	Spills and Leaks	☐
	Solid Waste	☐
	Air Emissions	☐
	Other: ____________	☐
Paint Application	Leftover Paint in Containers	☐
	Overspray	☐
	Drippings	☐
	Air Emissions	☐
	Other: ____________	☐
Other Processes	Leftover Raw Materials	☐
	Other Process Wastes	☐
	Types of Wastes: ____________	
	____________	
	Pollution Control Residues	☐
	Waste Management Residues	☐
	Other Wastes:	
	____________	☐
	____________	☐
	____________	☐

Plant ____	Waste Minimization Assessment	Prepared By ____
Date ____	Proj. No. ____	Checked By ____
		Sheet 2 of 2 Page __ of __

WORKSHEET 10b

WASTE STREAM IDENTIFICATION

(continued)

2. Which waste streams, if any, contain compounds of:		Identify the Compounds:
Cyanide		
Chlorine		
Bromine		
Sulfur		
Cadmium		
Chromium		
Copper		
Iron		
Lead		
Nickel		
Silver		
Tin		
Zinc		
Other Hazardous Components		

Plant ____________ Date ____________	Waste Minimization Assessment Proj. No. ____________	Prepared By ____________ Checked By ____________ Sheet 1 of 4 Page __ of __

WORKSHEET **11a**

INDIVIDUAL WASTE STREAM GENERATION AND CHARACTERIZATION

1. **Waste Stream Name/ID:** ____________ **Stream Number** ____________
   Process Unit/Operation ____________
2. **Waste Characteristics (attach additional sheets with composition data, as necessary.)**
   ☐ gas ☐ liquid ☐ solid ☐ mixed phase
   Density, lb/cuft ____________ High Heating Value, Btu/lb ____________
   Viscosity/Consistency ____________
   pH ____________ Flash Point ____________ ; % Water ____________
3. **Waste Leaves Process as:**
   ☐ air emission ☐ waste water ☐ solid waste ☐ hazardous waste
4. **Waste Generation is:**
   ☐ continuous ______ ____________
   ☐ periodic ____________ length of period: ____________
   ☐ sporadic (irregular occurrence)
   ☐ non-recurrent
5. **What Determines It To Be A Waste?**
   ☐ Chemical Analysis
   ☐ Process Type
   ☐ Industry Type
   ☐ EPA Chemicals List
6. **What Could Eliminate This Waste Generation?**
   ☐ Improved Operations
   ☐ Material Substitutions
   ☐ Other: ____________
7. **Generation Rate**
   Annual ____________ lbs per year
   Maximum ____________ lbs per ____________
   Average ____________ lbs per ____________
   Frequency ____________ batches per ____________
   Batch Size ____________ average ____________ range

Plant ______	Waste Minimization Assessment	Prepared By ______
		Checked By ______
Date ______	Proj. No. ______	Sheet 2 of 4 Page __ of __

WORKSHEET
11b

INDIVIDUAL WASTE STREAM GENERATION AND CHARACTERIZATION

(continued)

Waste Stream ______

6. Waste Origins/Sources

Fill out this worksheet to identify the origin of the waste. If the waste is a mixture of waste streams, fill out a sheet for each of the individual waste streams.

Is the waste mixed with other wastes? ☐ Yes ☐ No

Is waste segregation possible? ☐ Yes ☐ No

If yes, what can be segregated from it? ______

If no, why not? ______

How is the waste generated?

- ☐ Formation and removal of an undesirable byproduct compound
- ☐ Removal of an unconverted input material
- ☐ Depletion of a key component (e.g., drag-out)
- ☐ Equipment cleaning waste
- ☐ Obsolete input material
- ☐ Spoiled batch and production run
- ☐ Spill or leak cleanup
- ☐ Evaporative loss
- ☐ Breathing or venting losses
- ☐ Other: ______

Plant ____ Date ____	Waste Minimization Assessment Proj. No. ____	Prepared By ____ Checked By ____ Sheet 3 of 4 Page __ of __

WORKSHEET
11c

INDIVIDUAL WASTE STREAM GENERATION AND CHARACTERIZATION
(continued)

Waste Stream ____

7. Management Method

Leaves site in
- ☐ bulk ____
- ☐ roll off bins ____
- ☐ 55 gal drums ____
- ☐ other (describe) ____

Disposal Frequency ____

Applicable Regulations[1] ____

Regulatory Classification[2] ____

Managed
- ☐ onsite ☐ offsite
- ☐ commercial TSDR ____
- ☐ own TSDR ____
- ☐ other (describe) ____

Recycling
- ☐ direct use/re-use ____
- ☐ combusted for energy content ____
- ☐ redistilled ____
- ☐ other (describe) ____

Reclaimed material returned to site?
☐ Yes ☐ No ☐ used by others

residue yield ____

How is the residue managed? ____

Note[1] list federal, state & local regulations, (e.g., RCRA, TSCA, etc.)
Note[2] list pertinent regulatory classification (e.g., RCRA - Listed K011 waste, etc.)
TSDR - Treatment, Storage, Disposal or Recycling Facility

Plant ____	Waste Minimization Assessment	Prepared By ____
		Checked By ____
Date ____	Proj. No. ____	Sheet 4 of 4 Page __ of __

WORKSHEET
11d

INDIVIDUAL WASTE STREAM GENERATION AND CHARACTERIZATION
(continued)

Waste Stream ____

7. Management Method (continued)

Treatment
- ☐ biological ____
- ☐ oxidation/reduction ____
- ☐ incineration ____
- ☐ pH adjustment ____
- ☐ precipitation ____
- ☐ solidification/stabilization ____
- ☐ other (describe) ____

Which final disposal is involved in management of the waste or its residue?
- ☐ landfill ____
- ☐ pond ____
- ☐ lagoon ____
- ☐ deep well ____
- ☐ ocean ____
- ☐ other (describe) ____

What is the projected date for phasing out this disposal practice? ____

Costs as of ____ (quarter and year)

Cost Element:	Unit Price $ per ____	Reference/Source:
Onsite Storage & Handling		
Pretreatment		
Container		
Transportation Fee		
Disposal Fee		
Local Taxes		
State Tax		
Federal Tax		
Total Disposal Cost		

Plant ____________ Date ____________	Waste Minimization Assessment Proj. No. ____________	Prepared By ____________ Checked By ____________ Sheet 1 of 1 Page __ of __

WORKSHEET 12

WASTE STREAM SUMMARY

Attribute	Description[1]		
	Stream No. ____	Stream No. ____	Stream No. ____
Waste ID/Name:			
Source/Origin			
Hazardous Component			
Annual Generation Rate (units ________)			
Overall			
Component(s) of Concern			
Cost of Disposal			
Unit Cost ($ per: ________)			
Overall (per year)			
Method of Management[2]			

Priority Rating Criteria[3]	Relative Wt. (W)[4]	Rating (R)	R x W	Rating (R)	R x W	Rating (R)	R x W
Regulatory Compliance							
Treatment/Disposal Cost							
Potential Liability							
Waste Quantity Generated							
Waste Hazard							
Safety Hazard							
Minimization Potential							
Potential to Remove Bottleneck							
Potential By-product Recovery							
Sum of Priority Rating Scores		Σ(R x W)		Σ(R x W)		Σ(R x W)	
Priority Rank							

Notes:
1. Stream numbers, if applicable, should correspond to those used on process flow diagrams.
2. For example, sanitary landfill, hazardous waste landfill, onsite recycle, incineration, combustion with heat recovery, distillation, dewatering, etc.
3. Rate each stream in each category on a scale from 0 (none) to 10 (high).
4. A very important criteria for your plant would receive a weight of 10; a relatively unimportant criteria might be given a weight of 2 or 3.

Glossary

Active exchange program A system whereby laboratories redistribute their surplus chemicals through a paid intermediary or broker.

Alpha decay Decomposition of an atom's isotope in which a helium nucleus is released with a relatively short range of travel.

Air pollutant An airborne gas, aerosol, or particulate occurring in concentrations that may threaten the health of organisms or disrupt the natural processes upon which they depend.

BARRT Best Available Recovery or Recycling Technology

Beta decay Decomposition of an atom's isotope in which a particle the size of an electron is produced, with either a positive or negative charge and an intermediate range of travel.

Biohazard See "biological hazard."

Biological hazard Any viable infectious agent that presents a risk or a potential risk to the well being of organisms. Also called "biohazard."

Blue baby syndrome An anemic condition that occurs in children between ages six months and one-year-old, resulting from a reaction between high nitrate levels and oxygen-carrying hemoglobin in their blood.

Carbon monoxide A product of incomplete combustion (resulting from such things as emissions from utility plants, vehicles, and incinerators) that, once inhaled, affects the body's ability to absorb oxygen and causes drowsiness, headache, and even death.

Chemical treatment One that involves the use of chemicals to alter waste properties by destroying the hazardous component or preparing the still-hazardous material for further processing.

Chelating agent A substance that removes metal ions from a solution.

Chlorofluorocarbons (CFCs) Ozone-depleting substances regulated by the Clean Air Act due to their potential for harming the stratospheric ozone layer above the earth's surface.

Cleaner A water-based or water-soluble substance that is used to displace, dissolve, or chemically alter a contaminant.

Closed-loop recycling A form of recycling in which the waste is never allowed to escape the original process vessel or other container where it was generated.

Coal-tar creosote A residual product from the distillation and processing of coal tar that is a blend of naphtahlene, drain oil, wash oil, anthracene drain oil, and heavy distillate oil.

Composite mesh pad mist elimination A common scrubbing device in the metal-finishing industry used to control and neutralize gaseous emissions through the removal of chromic acid from the air stream by slowing the air's velocity, causing the entrained chromic acid droplets to impinge on the fiber pads.

Conditionally exempt small-quantity generator An EPA identification status determined by how much waste a facility generates each month. Generators of less than 100 kg. per month of hazardous waste (or up to 1 kg. of acutely hazardous waste per month) are in this category, which exempts them from regulation of their on-site hazardous waste handling and disposal.

Design for the Environment (DfE) A program that sets initiatives for promoting good environmental design of products and processes, especially through encouraging the deployment of risk reduction and P2 concepts at the earliest design stages.

Direct dischargers Facilities that discharge pollutants directly into the waters of the United States, and are regulated under the Clean Water Act through the National Pollutant Discharge Elimination Systems (NPDES).

Distillation A process used to separate and/or concentrate chemical mixtures in the liquid state.

Dragin The amount of liquid present on a piece before entering a plating bath.

Dragout The amount of liquid remaining on a piece removed from a plating bath.

Electrodialysis A chemical separation technique using a semipermeable membrane. Electrical current is used to increase the speed of the process.

Electrolytic recovery An electrochemical process used to recover metals from process baths and solutions whereby metal ions are removed from the waste stream by passing the stream through an electrolytic cell involving closely spaced anodes and cathodes.

Electrowinning A process used to recover metal from solutions. Electrical current is used to deposit the metal on an electrode, from which the metal can be removed.

Federal Register A publication updated daily by the government that lists all federal regulations, proposed regulations, and notices. The Register serves as a valuable source of information regarding current activities of all federal agencies.

Gamma decay Decomposition of an atom's isotope in which an X-ray like particle of high to very high energy is produced, with a potentially great range of travel.

Gaseous waste stream A waste stream made up of gases, vapors, or airborne particulates resulting from such operations as fiber spinning, spraying, and processing.

Green Lights A program that offers encouragement and incentives for companies to study their lighting needs to see where substantial energy savings may be realized.

Ground level ozone A gaseous air pollutant, strong pulmonary irritant, and a principal component of smog that is formed by the emission of VOCs and nitrogen oxides into the ambient air.

Half-life The time it takes for half of a radioactive isotope to decay.

Incineration The thermal treatment of hazardous waste using high temperatures to change its chemical, physical, or biological characteristics or composition.

Indirect dischargers Waste generators who send their effluent out for pretreatment before it can be released back to the environment.

Information exchange A clearing house for information on supply and demand.

In-process ion exchange A reversible chemical reaction that exchanges ions in a feed stream for ions of a like charge on the surface of an ion-exchange resin or solid substance. This process is used for for water recycling or metals recovery.

Large-quantity generator An EPA identification status determined by the amount of waste a facility generates each month. Large-quantity generators produce more than 1,000 kg. per month, or greater than 1 kg. of acutely hazardous waste per month.

Liquid waste stream A waste stream made up of nonaqueous waste, which can result from solvent recovery systems, vessel washings, condensate, and other such activities.

Listed waste One that appears in the Code of Federal Regulations (Part D, Subpart 261, Title 40).

Major source facilities As defined by the Clean Air Act, those industrial facilities that emit or have the potential to emit 10 tons per year or more of any hazardous pollutant, 25 tons per year or more of any combination of hazardous air pollutants, or 100 tons per year of any air pollutant.

Manifests Hazardous waste shipping documents on which generators are required to certify that they have a waste minimization program in place at their facility.

Mass balance Used to quantify losses or emissions and to determine a baseline for tracking the progress of a facility's pollution prevention efforts; to gather data used to estimate the size and cost of additional equipment and/or other modifications needed in its P2 program; and to gather data for evaluating economic performance.

Material balance A method for determining whether the amount by weight of all raw materials coming into the production process is equal to the total amount of material leaving it.

Material exchange A facility that takes temporary possession of a waste for transfer to a third party.

Material flow-chart The first step in a waste survey, the material flow-chart maps a complete picture of an operation's production process by outlining all raw materials used, additives, end products, byproducts, and liquid and solid wastes produced.

Mixture rule In simplest terms, a rule that states that any substance mixed with a listed waste may also carry the hazardous waste designation.

Multihazard waste Any waste contaminated with more than one of the following materials: medical/pathological/infectious, hazardous chemical, and/or radioactive.

National Pollutants Discharge Elimination System (NPDES) A system for issuing permits to facilities with point source discharges of contamination that allow them to discharge their contaminants into water.

Nuclear fuel cycle The movement of fuel in the nuclear industry, including the extraction, conversion, separation, purification, use, treatment, storage, and disposal of both nuclear fuel and radioactive waste.

On-site releases Releases made by a facility to either the air, water, land, or underground injection wells.

Passive exchange program A system for redistributing a laboratory's surplus chemicals, whereby a list of available chemicals is distributed to potential users. Interested parties then direct their inquiries to the designated group responsible for the inventory, or pass on the information to the person in possession of the chemical.

Pathogens A type of water pollution which includes viruses, bacteria, protozoa, and parasitic worms that cause such diseases as cholera, dysentery, gastroenteritis, hepatitis, typhoid fever, and other serious illnesses. Pathogens mainly enter water through the feces and urine of infected people and animals.

Photoprocessing pollution Waste, including spent developer, fixer, and rinsewater, created from chemicals used to develop and fix negatives, transparencies, and prints used to prepare copy.

Platemaking pollution Hazardous waste and process wastewater created during the platemaking process of printing, which involves the use of acids, solvents, metal salts, and photopolymer solutions. Such wastewater is caustic and may contain heavy metals.

Pollution prevention (P2) An environmental technology concept that incorporates both the practice of the conservation of natural resources and the practice of sustainable agriculture.

Post-consumer recycling Separating materials that have already served their intended end-use from the solid waste stream, and returning them to their producers or centers for recycling.

Pre-consumer recycling Using or reusing raw materials, byproducts, products, and waste streams within the original production process before the materials ever reach consumers.

Reclamation The processing of recovering waste for its value as a material, for regeneration, or (if permitted) for burning to generate power or recover energy.

Recycle To reuse, use, or reclaim a material by either returning it to the original manufacturing process as an original input material or as a substitute for an input material, or to use it as an ingredient in an entirely different manufacturing process.

Reprocessing The process of recovering desirable uranium or other radioactive elements from nuclear fuel in which the rods are crushed or sheared to expose the uranium in their center, then nitric acid is used to leach out the uranium and plutonium by solvent extraction.

Reverse osmosis A membrane-separation technology used for chemical recovery in which a pressurized solution is forced through a membrane that separates the solution into a permeate (which passes through the membrane) and reject stream (which contains most of the dissolved solids, so is deflected by the membrane).

Sampling station A consistent site used to gather information regarding a facility's waste stream, such as peak-time releases and concentration of the chemicals being released.

Sediments Particles of soil, sand, and minerals that wash from the land into the water. In large amounts, these can cause a problem as they may eventually fill stream channels, harbors, and reservoirs.

Semiannual Regulatory Agenda Published by the EPA in the Federal Register, this is a listing of rules the agency expects to propose and finalize in the future.

Sewer map A chart detailing the location of a facility's water, wastewater, sanitary, storm, and drain lines.

Small-quantity generator An EPA identification status determined by how much waste a facility generates each month. Facilities in this category generate between 100 and 1,000 kg. of hazardous waste per month, and up to 1 kg. of acutely hazardous waste per month.

Smog Ground-level ozone, comprising the most serious air quality challenge for most urban areas.

Solid waste stream A waste stream made up of solid matter, including sludges and slurries, resulting from such things as filter cakes, distillation fractions, spent catalysts, vessel and tank residues, bag house dusts, and sludge from an on-site wastewater facility.

Solidification The process of surrounding a hazardous compound with a nonreactive, nonhazardous media to create a stable, presumably nonleaching compound that is suitable for disposal.

Solvent The liquid portion of a solution in which other substances may be dissolved.

Source reduction The development of operating practices and procedures that can lead to the elimination of a waste at its source.

Sulfur dioxide A gas (released when sulfur-containing fuels such as coal and oil are burned) that can cause respiratory problems and that also may be converted to atmospheric acid, a component of acid rain.

Suspended particulate Pieces of ash, smoke, soot, dust, or liquid droplets released to the air by the burning of fuel, industrial processes, agricultural practices, or a number of natural processes. If inhaled, these can lodge in the lungs and lead to respiratory disease.

Thermal pollution Raising of water temperatures to levels that may harm aquatic organisms.

Toxics Substances that pose a direct threat to organisms.

Toxics Characterization Leachate Procedure (TCLP) A test for measuring the levels of hazardous chemicals contained in an acidified water sample that is passed through the waste material.

Treatment In waste management, "treatment" is any practice other than recycling that is designed to alter the physical, chemical, or biological character or composition of a hazardous substance, pollutant, or contaminant in order to neutralize it.

Validation The scientific study of a process to prove that it is under specified control, and to determine process variables and acceptable limits.

Waste broker A facility that charges a fee to locate buyers or sellers for another facility's waste. A waste broker does not take actual possession of the waste for transfer to a third party, but acts as a go-between only.

Waste ink Ink generated during blending or application to printing hardware and substrates that can contain toxic metal pigments, organic compounds, oils, and solvents.

Waste Shipment Record (WSR) A document detailing the movement and ultimate disposition of a hazardous waste.

Wastewater stream Aqueous waste that results from raw water usage.

Wet-packed scrubber A common scrubbing device used in the metal-finishing industry to control and neutralize gaseous emissions. It consists of a spray chamber filled with packing material in which process water is continuously sprayed onto the packing and the air stream is pulled through the packing by a fan. Contaminants are absorbed by the water spray and the air is released to the atmosphere.

Zero effluent The elimination of liquid process effluents from a pulp or paper mill without transferring an unreasonable load to the air-shed or soil.

Index

Boldface page numbers indicate figures or tables.

ISBN 0-13-238569-4